Statistical and Computational Methods in Brain Image Analysis

CHAPMAN & HALL/CRC MATHEMATICAL AND COMPUTATIONAL IMAGING SCIENCES

Aims and Scope

This series aims to capture new developments and summarize what is known over the whole spectrum of mathematical and computational imaging sciences. It seeks to encourage the integration of mathematical, statistical and computational methods in image acquisition and processing by publishing a broad range of textbooks, reference works and handbooks. The titles included in the series are meant to appeal to students, researchers and professionals in the mathematical, statistical and computational sciences, application areas, as well as interdisciplinary researchers involved in the field. The inclusion of concrete examples and applications, and programming code and examples, is highly encouraged.

Published Titles

Statistical and Computational Methods in Brain Image Analysis
by Moo K. Chung

Theoretical Foundations of Digital Imaging Using MATLAB®
by Leonid P. Yaroslavsky

Rough Fuzzy Image Analysis: Foundations and Methodologies
by Sankar K. Pal and James F. Peters

Proposals for the series should be submitted to the series editors above or directly to:
CRC Press, Taylor & Francis Group
3 Park Square, Milton Park, Abingdon, OX14 4RN, UK

CHAPMAN & HALL/CRC
MATHEMATICAL AND COMPUTATIONAL IMAGING SCIENCES

Statistical and Computational Methods in Brain Image Analysis

Moo K. Chung

Waisman Laboratory for Brain Imaging and Behavior
University of Wisconsin-Madison, USA

CRC Press
Taylor & Francis Group
Boca Raton London New York

CRC Press is an imprint of the
Taylor & Francis Group, an **informa** business
A CHAPMAN & HALL BOOK

First published in paperback 2024

First published 2014 by CRC Press
2385 NW Executive Center Drive, Suite 320, Boca Raton FL 33431

and by CRC Press
4 Park Square, Milton Park, Abingdon, Oxon, OX14 4RN

CRC Press is an imprint of Taylor & Francis Group, LLC

Publisher's Note
The publisher has gone to great lengths to ensure the quality of this reprint but points out that some imperfections in the original copies may be apparent.

ISBN: 978-1-4398-3635-4 (hbk)
ISBN: 978-1-03-291995-9 (pbk)
ISBN: 978-0-429-09432-3 (ebk)

DOI: 10.1201/b15056

Visit the Taylor & Francis Web site at
http://www.taylorandfrancis.com

and the CRC Press Web site at
http://www.crcpress.com

To my beautiful wife
Hyun Mi Park

Contents

Preface

Brain image analysis is an emerging new field that utilizes various non-invasive brain imaging modalities such as MRI, fMRI, PET and DTI in mapping out the 4D spatiotemporal dynamics of the human brain in both normal and clinical populations in macroscopic level. This discipline emerged about twenty years ago and has made substantial progress in the past two decades. A major challenge in the field is caused by the massive amount of nonstandard high dimensional imaging data that is difficult to analyze using available standard techniques. This requires new computational approaches and solutions.

The main goals of this book are to provide an overview of various statistical and computational methodologies used in the field to a wide range of researchers and students, and to articulate important yet technically challenging topics further. The book is mainly focused on methodological issues in analyzing structural brain imaging modalities such as MRI and DTI. Concepts and methods are illustrated with real imaging applications and examples. Most of the brain imaging data set along with `MATLAB`® codes used in the book can be downloaded from `http://brainimaging.waisman.wisc.edu/~chung/BIA`.

Although there are abundant research papers scattered in various journals, there is not a single research paper or book that covers many quantitative techniques used in the field with detailed illustrations of actual imaging data and computer codes. By making the data and codes available, we tried to make the book more accessible to a wide range of researchers and students. We wish to provide methodological understanding in a manner immediately usable to researchers who actually need it. The book has many examples and case studies, and is structured mainly as a graduate level textbook. The key features of this book are as follows.

- The book presents the coherent statistical and mathematical treatment of underlying methods. Only the methods that have been found to be useful in neuroimaging applications are presented.

- Introduction to the real examples and case studies that have been previously published in journals. Most of the brain imaging data and codes illustrated in the textbook can be downloaded from the website.

- We put a significant emphasis on image visualization using `MATLAB`. We have provided many publication quality visualization examples that have been actually used in publishing journal papers.

Although I am indebted to many colleagues and students in writing this book, I would particularly like to thank Richard Davidson, the director of the Waisman Laboratory for Brain and Behavior, at the University of Wisconsin-Madison for providing endless guidance and support over the years. The book is mainly the result of research done in the brain imaging lab since 2001. This book could not have existed without the dedicated collaborators who tirelessly interacted with me over the years. I would like to especially thank Andrew Alexander, Houri Vorperian, Stacey Schaefer, Kim Dalton, Seth Pollak and Richard Davison of the Waisman Center at the University of Wisconsin-Madison for providing most of the imaging data used as illustrations in this book. Lastly I owe a special thanks to my PhD advisor Keith Worsley for a chance to collaborate on `SurfStat` package that has been extensively explained throughout the book.

Although most figures are produced by myself using `MATLAB`, a few figures are generated by students and postdocs. These figures are identified in the figure legend as such. The book also introduces a few `MATLAB` codes written by others. The authors of the codes are identified in relevant places as well.

Moo K. Chung
January 25, 2013
Madison, Wisconsin

1

Introduction to Brain and Medical Images

The most widely used brain imaging modalities are magnetic resonance images (MRI), functional-MRI (fMRI) and diffusion tensor images (DTI). MRI depends on the response of magnetic fields in generating digital images that provide structural information about brain noninvasively. Compared to the computed tomography (CT), MRI has been mainly used for *in vivo* imaging of the brain due to higher image contrast in soft tissues. Unlike CT, MRI does not use X-ray for imaging so there is no risk of radiation exposure. MRI produces images based on the spin-lattice relaxation time (T1), the spin-spin relaxation time (T2) and the proton density (ρ) [39]. The widely used T1- and T2-weighted imaging weight the contribution of one component and minimize the effect of the other two. The T1-weighted MRI is more often used in anatomical studies compared to the T2-weighted MRI. Structural images you are seeing are most likely T1-weighted MRI.

DTI also uses MRI scanners but it specifically enables the measurement of the diffusion of water molecules in the brain tissues. At each voxel in DTI, a rate of diffusion and a preferred direction of diffusion are encoded as 3×3 symmetric and positive definite matrix called diffusion tensor. In each voxel, there are countless number of neural fibers. So the direction of diffusion follows neural fiber tracts in an average sense within each voxel. By tracing the diffusion direction via tractography, it is possible to represent the white matter fiber tracts visually. fMRI also uses MRI scanners in measuring brain activity by detecting changes in blood flow using the blood-oxygen-level-dependent (BOLD) contrast. For last two decades, fMRI has been dominating brain mapping research since it is easy to scan and there is no exposure to radiation. Compared to MRI, fMRI is more frequently contaminated by various types of noise. In order to obtain the underlying signal, various advanced statistical procedures are used. The resulting statistical maps are usually color coded to show the strength of activation in the whole brain.

In this introductory chapter, various brain image data types obtained from different imaging modalities will be introduced with simple `MATLAB` functions for image processing and visualization. The most difficult part for statisticians and new researchers trying to analyze brain image data is getting the data into a computer programming environment. In this chapter, we briefly go over manipulating volume and surface data.

1.1 Image Volume Data

The brain MRI is a 3D array $\mathcal{I}(\mathbf{x})$ containing tissue intensity values at voxel position $\mathbf{x} \in \mathbb{R}^3$ Figure 1.1 shows a usual example of axial cross-section of MRI. We usually do not analyze tissue intensity values directly in quantifying MRI. MRIs go through various image processing steps such as intensity normalization and tissue segmentation before any quantification is attempted. Image intensity normalization is required to minimize the intensity variations within the same tissue that are caused by magnetic field distortion.

Each voxel can be classified into three different tissue types: cerebrospinal fluid (CSF), grey matter and white matter mainly based on image intensity values. A neural network classifier [211] or a Gaussian mixture model [14] have been used for the classification. Figure 1.1 shows the result obtained by a neural network classifier [211]. The collection of identically classified voxels has been mainly used in voxel-based morphometry (VBM), which will be discussed in detail in a later chapter. VBM as implemented in the statistical parametric mapping (SPM) package (`www.fil.ion.ucl.ac.uk/spm`), is a fully automated image analysis technique allowing identification of regional differences in gray matter (GM) and white matter (WM) between populations without a prior region of interest. VBM starts with normalizing each structural MRI to the standard SPM template and segmenting it into white and gray matter and cerebrospinal fluid (CSF) based on a Gaussian mixture model [18, 14]. Based on a prior probability of each voxel being the specific tissue type, an iterative Bayesian approach is used to get an improved estimate of the posterior probability. This probability is usually referred to as the gray or white matter density. Afterwards the density maps are warped into a normalized space and compared across subjects. VBM has been applied in cross-sectional studies on various anatomical studies: normal development [147, 273], autism [69], depression [285], epilepsy [246] and mild cognitive impairment (MCI) [187].

The most widely used 3D MRI volume image format has been the Analyze 7.5 file format developed by the Biomedical Imaging Resource at Mayo Clinic for Analyze package (`www.mayo.edu/bir/Software/Analyze/Analyze.html`). An Analyze format consists of two files: an image binary file (`*.img`) and a header file (`*.hdr`) that contains information about the binary image file. Various brain image analysis software and `MATLAB` can read the Analyze 7.5 file format.

Figure 1.2 shows an example of MRI and the manual segmentation of the left amygdala, which is colored in black. This manual segmentation is stored as a binary image consisting of 0's and 1's. In `MATLAB`, to read the header file, we invoke the command

```
>> header=analyze75info('left_amygdala.hdr');
```

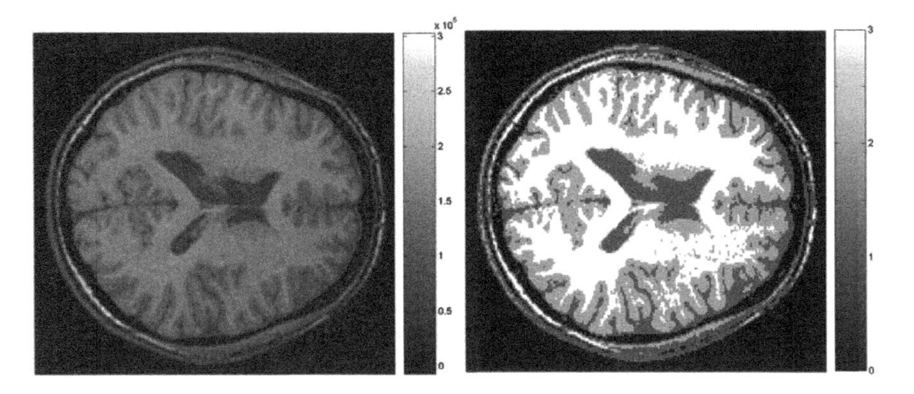

FIGURE 1.1
Left; the axial cross-section of MRI. Right: Tissue segmentation into three different classes (white and gray matter and CSF) using the neural network classifier [211].

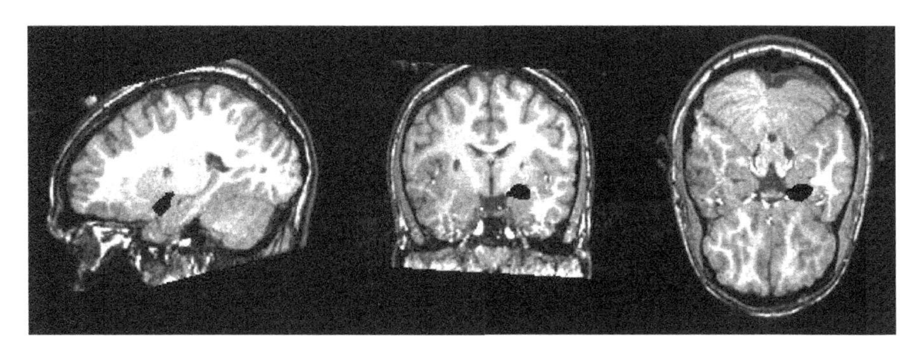

FIGURE 1.2
The manual segmentation of left amygdala at the mid-sagittal, coronal and axial cross-sections. The amygdala was segmented using a prior information of adjacent structures such as anterior commissure and hippocampus [257].

```
header =

        Filename: 'left_amygdala.hdr'
        FileModDate: '21-Sep-2003 15:00:02'
        ColorType: 'grayscale'
        Dimensions: [191 236 171 1]
        VoxelUnits: '  mm'
```

The structure array `header` contains various image information and can be accessed by calling, for instance, `header.Dimensions` which shows the dimension of array `[191 236 171 1]`. The last dimension is reserved for temporally changing 3D volume images such as fMRI but it is not used for this example. To extract the amygdala segmentation and saved into a 3D array `vol` of size $191 \times 236 \times 171$, we run

```
>>vol=analyze75read(header);
```

`vol` is then a 3D binary array of a left amygdala consisting of zeros and ones (Figure 1.2). The array `vol` consists of all zeros except the voxels where amygdala is defined.

1.1.1 Amygdala Volume Data

This is the subset of the 44 subject data set first published in [80] consisting of the both left and right amygdala binary volume of 22 subjects. There are 22 normal control subjects and 24 autistic subjects. The data set is stored as `amygdala.volume.data` containing variables age, total brain volume (`brain`), eye fixation duration (`eye`), face fixation duration (`face`), group variable indicating autistic or normal control (`group`), left amygdala volume (`leftvol`) and right amygdala volume (`rightvol`) data. The data set is fully analyzed in Section 12.7. The main scientific hypothesis of interest is to quantify volume and shape differences between the groups. In this section, we will show how to load the data set and do a simple volume computation. Since the image resolution is $1 \times 1 \times 1$ mm, the amygdala volume is simply computed by counting the number of voxels belonging to the amygdala.

```
load amygdala.volume.mat

left=zeros(22,1);
right=zeros(22,1);
for i=1:22
    left(i)=sum(sum(sum(squeeze(leftvol(i,:,:,:)))));
    right(i)=sum(sum(sum(squeeze(rightvol(i,:,:,:)))));
end;
```

The `group` variable is 1 for autistic and 0 for normal controls. The left amygdala volume is computed and stored as `al` for instance. The amygdala volumes for all subjects are displayed in Figure 1.3 showing no clear separation of the groups.

```
al= left(find(group));
ar=right(find(group));
cl= left(find(~group));
cr=right(find(~group));

figure;
plot(al,ar,'or', 'MarkerEdgeColor','k', ...
```

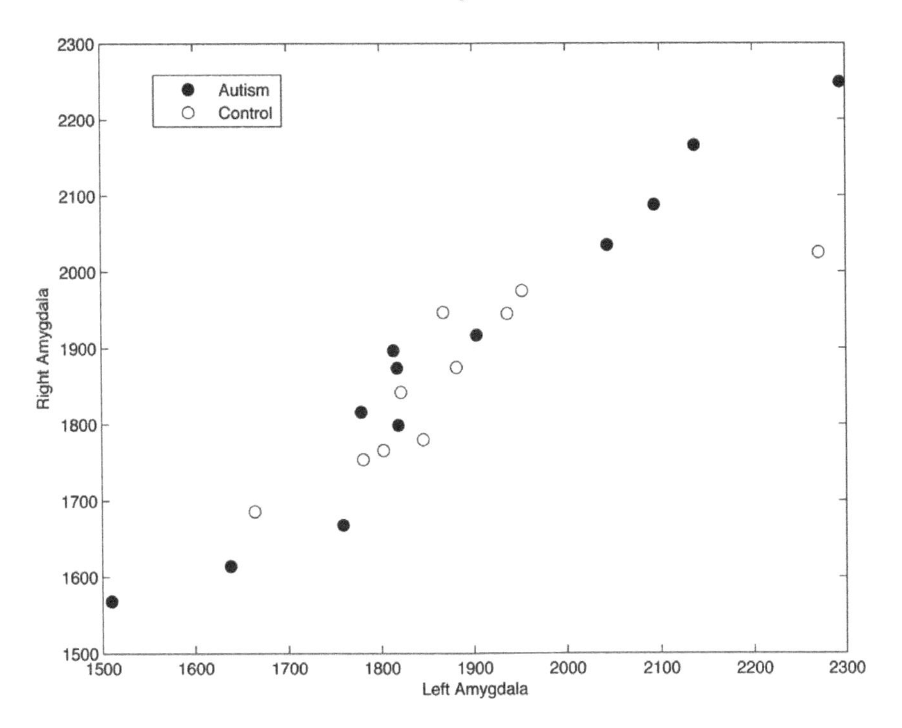

FIGURE 1.3
The plot showing the left and right amygdala volumes of autistic (black) and control (white) subjects. There is no visible group differences in the plot. The volume based method is not effective in discriminating between the groups.

```
'MarkerFaceColor','k', 'MarkerSize',7)
hold on
plot(cl,cr,'ob', 'MarkerEdgeColor','k', ...
'MarkerFaceColor','w', 'MarkerSize',7)
legend('Autism','Control')
xlabel('Left Amygdala')
ylabel('Right Amygdala')
set(gcf,'Color','w');
```

A more advanced technique such as weighted spherical harmonic representation [80] (Chapters 11 and 12) is needed to increase the discrimination power.

1.2 Surface Mesh Data

Simple 3D neuroanatomical objects like hippocampus and amygdala can be represented by their 2D boundary as triangle meshes using the marching cubes algorithm [232], which is implemented in MATLAB.

To extract the surface mesh out of vol, we run

```
>>surf = isosurface(vol)

surf =

    vertices: [1270x3 double]
       faces: [2536x3 double]
```

The MATLAB function **isorsurface** extracts the boundary of the amygdala and represents it as the structure array **surf** consisting of 1270 vertices and 2563 triangles. The visualization can be done with

```
>>figure_wire(surf,'yellow')
```

resulting in a surface mesh like Figure 1.4.

More complex neuroanatomical objects like cortical surfaces are difficult to represent with such a simple algorithm. Substantial research have been done on extracting cortical surfaces from MRI [83, 100, 237]. The human cerebral cortex has the topology of a 2D highly convoluted grey matter shell with an average thickness of 3mm. The outer boundary of the shell is called the *outer cortical surface* while the inner boundary is called the *inner cortical surface.* The outer cortical surface is the boundary between the cerebrospinal fluid (CSF) and the gray matter while the inner cortical surface is the boundary between the gray and the white matters. Cortical surfaces are segmented from MRI using mainly a deformable surface algorithm and represented as a triangle mesh consisting of more than 40,000 vertices and 80,000 triangle elements [83, 237]. We assume cortical surfaces to be smooth 2D Riemannian manifolds topologically equivalent to a unit sphere [100]. The triangle mesh format contains information about vertex indices, the Cartesian coordinates of the vertices and the connectivity that tells which three vertices form a triangle.

1.2.1 Topology of Surface Data

Surface extraction from MRI may cause possible topological defects such as handles and bridges forming on brain surfaces. So it is crucial to able to determine topological defects. Determining topological defects on the surface can be done by checking the topological invariants such as the Betti numbers or Euler characteristics.

Let χ be the Euler characteristic of a surface. The sphere has a Euler

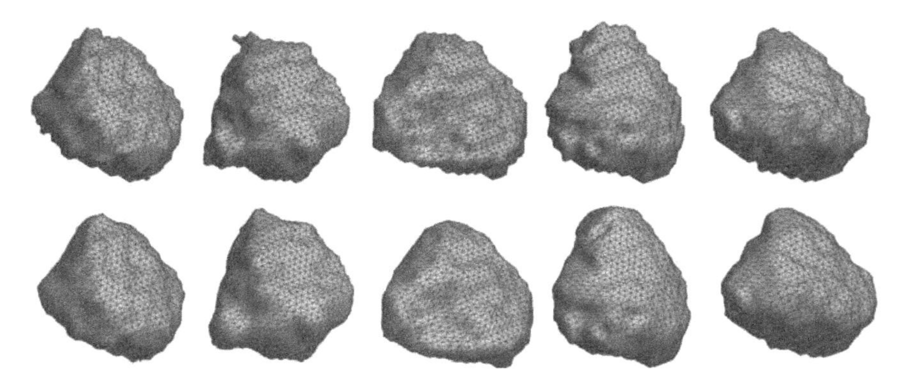

FIGURE 1.4
Top: triangle mesh representation of five different left amygdala surfaces. The mesh noises such as sharp peaks need to be smoothed out using surface smoothing technique. Here we used the weighted-SPHARM representation [80], which will be studied in a later chapter.

characteristic of 2. If the surface has genus g, the number of handles, the Euler characteristic is given by

$$\chi = 2 - 2g.$$

Each handle in the object reduces χ by 2 while the increase of the disconnected components raise χ by 2. On the surface mesh, the Euler characteristic is given in terms of the number of vertices V, the number of edges E and the number of faces F using the polyhedral formula:

$$\chi = V - E + F.$$

Note that for each triangle, there are three edges. For a closed surface topologically equivalent to a sphere, two adjacent triangles share the same edge. Hence, the total number of edges is $E = 3F/2$. The relationship between the number of vertices and the triangles is $F = 2V - 4$. We simply need to compute the Euler characteristic as $\chi = V - F/2$ and check if it is 2 at the end. All binary volumes produced the topologically correct surfaces without an exception. Figure 1.5 shows an example of before and after the topology correction.

For any type of cortical surface mesh, if V is the number of vertices, E is the number of edges, and F is the number of faces or triangles in the mesh, the Euler characteristic χ of the mesh should be constant, i.e. $\chi = V - E + F = 2$. Note that for each triangle, there are three edges. Since two adjacent triangles share the same edge, the total number of edges is $E = 3F/2$. Hence, the

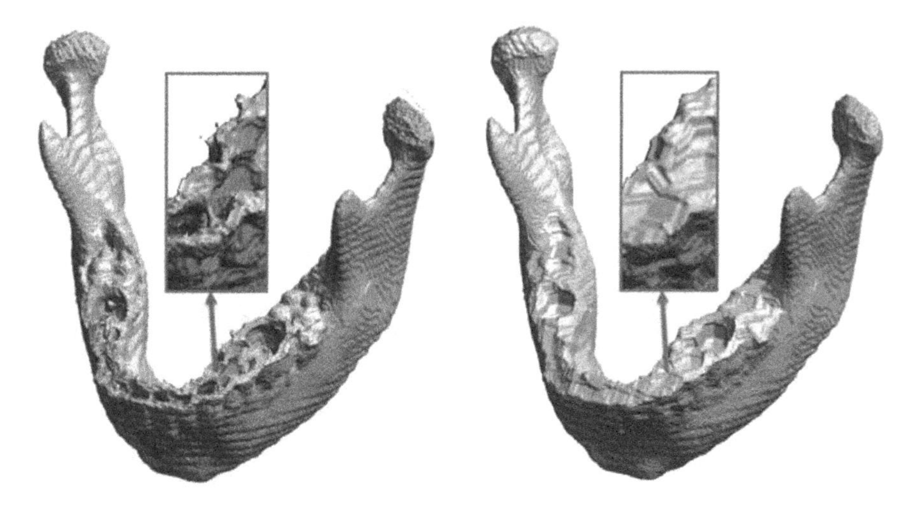

FIGURE 1.5
Topological defects occur in any type of medical image segmentation. Bridges and handles are visible in the teeth regions of the mandible segmentation obtained from CT images. Topological correction is made on mandible binary segmentation and surface. Disjointed tiny speckles of noisy components are removed by labeling the largest connected component, and holes and handles are removed by the morphological closing operation.

relationship between the number of vertices and the triangles is $F = 2V - 4$. In the sample surface, we have 40,962 vertices and 81,920 triangles.

The Euler characteristic computation in brain imaging has been traditionally done in connection with correcting for topological defects in segmented anatomical objects [317, 321]. For instance, the human cerebral cortex has the topology of a 2D highly convoluted grey matter shell with an average thickness of 3mm. The outer and inner boundaries are assumed to be topologically equivalent to a sphere [100, 237]. Image acquisition and processing artifacts, and partial voluming will likely produce topological defects such as holes and handles in cortical segmentation. Since it is also needed to remove the brain stem and other parts of the brain in the triangulated mesh representation of the cortex, the resulting cortical surfaces are not likely to have spherical topology. So it is necessary to automatically determine and correct the topological defects using the topological invariants such as the Euler characteristic.

Topological defects are often encountered in image reconstruction. For instance, since the mandible and teeth have relatively low density in CT, unwanted cavities, holes and handles can be introduced in mandible segmentation [11]. An example is shown in Figure 1.5 where the tooth cavity forms a bridge over the mandible. In mandibles, these topological noises can appear in thin or cancellous bone, such as in the condylar head and posterior palate

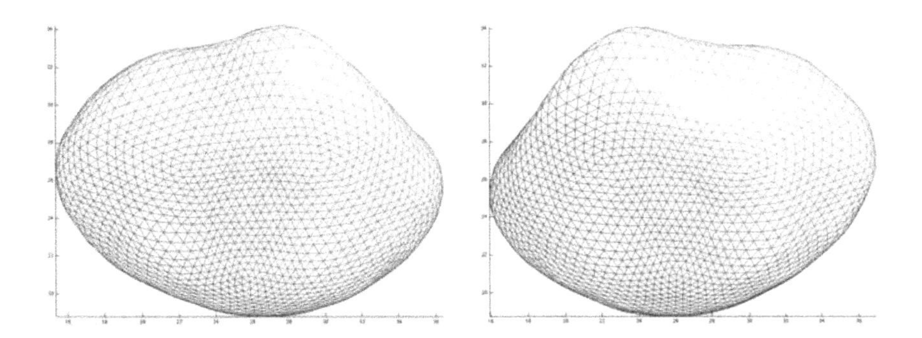

FIGURE 1.6
Left and right amygdala surface template constructed by averaging surface coordinates in the data set `chung.2010.NI.mat`.

[342]. If we apply the isosurface extraction on the topologically defect segmentation results, the resulting surface will have many tiny handles [385, 404]. These handles complicate subsequent mesh operations such as smoothing and parameterization. So it is necessary to correct the topology by filling the holes and removing handles. If we correct such topological defects, it is expected the resulting isosurface is topologically equivalent to a sphere. There have been various topological correction techniques proposed in medial image processing. Rather than attempting to repair the topological defects of the already extracted surfaces [385, 404], we can perform the topological simplification on the volume representation directly using morphological operations [156, 404]. The direct correction on surface meshes can possibly cause surfaces to intersect with each other [385].

1.2.2 Amygdala Surface Data

This is the data set first published in [80] consisting of the both left and right amygdala surfaces of 44 subjects. There are 22 normal control subjects and 24 autistic subjects. The data set is fully analyzed in Section 12.7. All the amygdala surfaces went through image processing and meshes consist of 2562 vertices and 5120 faces. Mesh vertices anatomically match across subjects. So we can simply average the corresponding vertex coordinates to obtain the group average (Figure 1.6).

```
load chung.2010.NI.mat

left_template.vertices = squeeze(mean(left_surf,1));
left_template.faces=sphere.faces;
figure; figure_wire(left_template,'yellow','white');
```

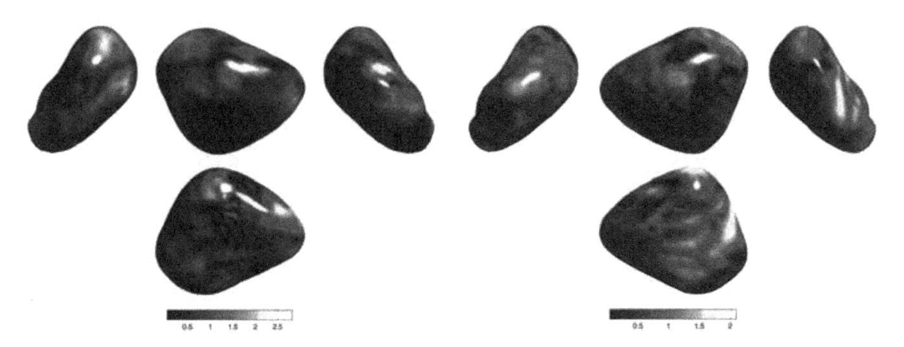

FIGURE 1.7
Displacement that is required to move the template surface to the first subject is displayed in the origami representation [80]. The unit is in mm showing a fairly small deformation scale.

```
right_template.vertices = squeeze(mean(right_surf,1));
right_template.faces=sphere.faces;
figure; figure_wire(right_template,'yellow','white');
```

The displacement vector field of warping the template surface to the 46 subject surfaces is given in $46 \times 2562 \times 3$ matrix. The displacement length of the first subject is shown in the origami representation introduced in [80] (Figure 1.7).

```
temp=reshape(left_template.vertices, 1, 2562,3);
temp=repmat(temp, [46 1 1]);
left_disp= left_surf - temp;

temp=reshape(right_template.vertices, 1, 2562,3);
temp=repmat(temp, [46 1 1]);
right_disp= right_surf - temp;

left_length= sqrt(sum(left_disp.^2,3));
figure_origami(left_template, left_length(2,:))

right_length= sqrt(sum(right_disp.^2,3));
figure_origami(right_template, right_length(2,:))
```

1.3 Landmark Data

Either manual or automatic landmarks that identify the important regions and features of anatomy are often used in landmark-based morphometrics. Landmarks are also used in aligning anatomy using affine registration. The landmark data introduced in this section is manually identified along the mandible surfaces extracted from CT images. The CT images were obtained from GE multi-slice helical CT scanners. CT scans were converted to DICOM format and subsequently Analyze 8.1 software package (AnalyzeDirect, Inc., Overland Park, KS) was used in segmenting binary mandible structure based on histogram thresholding. Then 24 landmarks are manually identified by an expert for two mandible surfaces. The landmarks across different surfaces anatomically correspond.

The mandible surfaces and the landmark data are stored in `mandible-landmarks.mat`. In the `mat` file, `id` contains the subject identifiers. `id(44)` is the subject F155-12-08 while `id(1)` is the subject F203-01-03. The mandible surfaces for these two subjects are shown in Figure 1.8. Our aim is to align them affinely as close as possible.

```
load mandible-landmarks.mat

figure; subplot(1,2,1);
hold on; figure_patch(F155_12_08, 'y', 0.5);
q=landmarks(:,:,44);
C=20*ones(24,1);
hold on; scatter3(q(:,1),q(:,2),q(:,3), C, C, 'filled', 'r')
view([90 60]);  zoom(1.2); camlight;
title(id(44))

subplot(1,2,2);
figure_patch(F203_01_03, [0.74 0.71 0.61], 0.5);
p=landmarks(:,:,1);
C=20*ones(24,1);
hold on; scatter3(p(:,1),p(:,2),p(:,3), C, C, 'filled', 'b')
view([90 60]);  zoom(1.2); camlight;
title(id(1))
```

1.3.1 Affine Transforms

Anatomical objects extracted from 3D medical images are aligned using affine transformations to remove the global size differences. Figure 1.8 illustrates an example where one mandible surface is aligned to another via an affine transform.

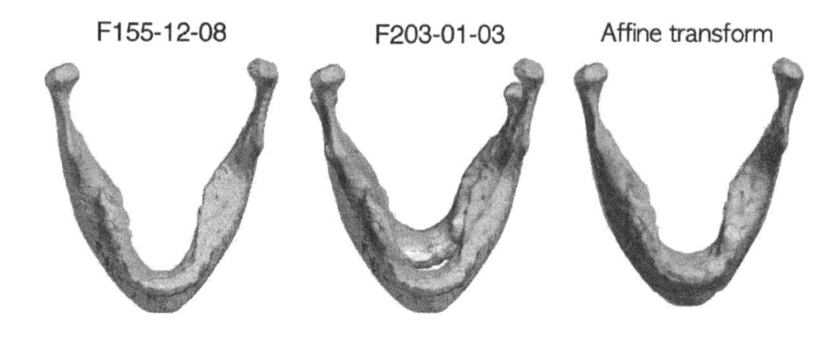

FIGURE 1.8
Mandible F155-12-08 (yellow) is used as a fixed template and other mandibles are affinely aligned to F155-12-08. For example, smaller F203-01-03 (gray) is aligned to larger F203-01-03 by affinely enlarging it. (See color insert.)

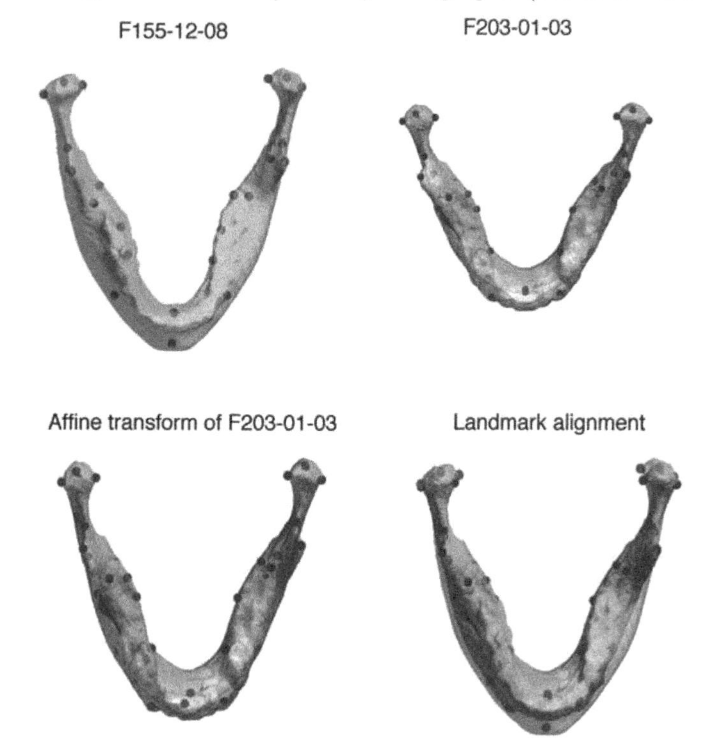

FIGURE 1.9
Mandible F155-12-08 (yellow) is used as a template and mandibles F203-01-03 is affinely aligned to F155-12-08 by matching 24 manually identified landmarks. The affine transform does not exactly match the landmarks perfectly but minimizes the distance between them in a least squares fashion.(See color insert.)

The affine transform T of point $p = (p_1, \cdots, p_d)' \in \mathbb{R}^d$ to $q = (q_1, \cdots, q_d)'$ is given by

$$q = Rp + c,$$

where the matrix R corresponds to rotation, scaling and shear while c corresponds to translation. Note that the affine transform is nonlinear. Note that

$$\begin{aligned} T(ap + bq) &= R(ap + bq) + c \\ &= (aRp + c) + (bRq + c) - c \\ &= aT(p) + bT(q) - c. \end{aligned}$$

Unless $c \neq 0$, the affine transform is not linear due to the translation term.

The affine transform can be easily made into a linear form by augmenting the transform. The affine transform can be rewritten in a matrix form as

$$\begin{pmatrix} q \\ 1 \end{pmatrix} = \begin{pmatrix} R & c \\ 0, \cdots 0 & 1 \end{pmatrix} \begin{pmatrix} p \\ 1 \end{pmatrix} \tag{1.1}$$

Let

$$A = \begin{pmatrix} R & c \\ 0, \cdots 0 & 1 \end{pmatrix}.$$

Then trivially A is linear on $\begin{pmatrix} p \\ 1 \end{pmatrix}$. The matrix A is the most often used form for affine registration in medical imaging.

The inverse of the affine transform is given by

$$p = R^{-1}q - R^{-1}c.$$

This can be written in a matrix form as

$$\begin{pmatrix} p \\ 1 \end{pmatrix} = \begin{pmatrix} R^{-1} & -R^{-1}c \\ 0, \cdots 0 & 1 \end{pmatrix} \begin{pmatrix} q \\ 1 \end{pmatrix}.$$

We denote the matrix form of the inverse as

$$A^{-} = \begin{pmatrix} R^{-1} & -R^{-1}c \\ 0, \cdots 0 & 1 \end{pmatrix}.$$

1.3.2 Least Squares Estimation

For n given points $p_i = (p_{i1}, \cdots, p_{id})'$ and its corresponding affine transformed points $q_i = (q_{i1}, \cdots, q_{id})'$, we can estimate the affine transform matrix A in the least squares fashion. For points p_i and q_i, we have

$$q_i = Ap_i. \tag{1.2}$$

The least squares estimation of A is given by

$$\widehat{A} = \arg\min_{A \in \mathcal{G}} \sum_{i=1}^{n} \|q_i - Ap_i\|^2,$$

where $\mathcal{G}$ is an affine group. We rewrite (1.1) as

$$\underbrace{\begin{pmatrix} q_1 & \cdots & q_n \end{pmatrix}}_{Q} = \begin{pmatrix} R & c \end{pmatrix} \underbrace{\begin{pmatrix} p_1 & \cdots & p_n \\ 1 & \cdots & 1 \end{pmatrix}}_{P}.$$

Then the least squares estimation is trivially given as

$$\begin{pmatrix} \widehat{R} & \widehat{c} \end{pmatrix} = QP'(PP')^{-1}.$$

This can be easily implemented in MATLAB. Then the points p_i are mapped to $\widehat{R}p_i + \widehat{c}$, which may not coincide with q_i in general.

If we try to solve the least squares problem with an additional constraint, the problem can become complicated. For instance, if we restrict R to be rotation only, i.e. $R'R = I$, iterative updates of least squares estimation are needed [336].

Affine transforms from the landmarks p to q is given computed as follows.

```
A =affine_transform(p',q')
F203_01_03a= affine_surface(A, F203_01_03);
figure; subplot(1,2,1);
figure_patch(F203_01_03a, [0.74 0.71 0.61], 0.5);
```

The estimated affine transform is then applied to the landmark coordinates p. The transformed landmarks are named as `paffine` and displayed as follows.

```
paffine= p*A(1:3,1:3)' + repmat(A(1:3,4), 1, 24)';
hold on;
scatter3(paffine(:,1),paffine(:,2),paffine(:,3),...
C, C, 'filled', 'b')
view([90 60]);  zoom(1.2);
title('Affine transform of F203-01-03')
```

Let us superimpose the affine transformed landmarks and the mandible on top of F155-12-08. As we can see, the affine transformed landmarks `paffine` does not exactly match to q. However, the distance between q and `paffine` is the smallest in the least squares fashion.

```
subplot(1,2,2);
figure_patch(F155_12_08, 'y', 0.5);
hold on; scatter3(q(:,1),q(:,2),q(:,3), C, C, 'filled', 'r')
hold on; figure_patch(F203_01_03a, [0.74 0.71 0.61], 0.5);
hold on; scatter3(paffine(:,1),paffine(:,2),paffine(:,3), ...
C, C, 'filled', 'b')
view([90 60]);  zoom(1.2);
title('Landmark alignment')
```

1.4 Vector Data

Vector image data are often obtained in connection with *deformation-based morphometry* (DBM) [14, 16, 81] which does not require tissue segmentation in quantifying the shape of anatomy in MRI [14]. The 3D *displacement vector field* $u(x)$ is a vector map defined at each position x that is required to move the anatomical structure at x in an image $\mathcal{I}_1$ to the anatomically corresponding position in another image $\mathcal{I}_2$. The structure at x in image $\mathcal{I}_1$ should match closely to the structure at $x + u$ in image $\mathcal{I}_2$. The displacement vector field u is usually estimated via volume- or surface-based nonlinear registration techniques. The x-, y- and z-components of displacement vector fields are usually stored in three separate Analyze files. We will study DBM in a later chapter.

Here we briefly introduce the mandible surface deformation vector data set that will be used in the later chapter. `mandible77subject.mat` contains surface deformation vector data for 77 subjects obtained from CT images. The data set contains the vertex coordinates `vertices`, subject id `id`, voxel size `voxelsize`, mesh connectivity information `faces` and affine transform matrices `A` for all 77 subject. Since the surface meshes are constructed from CT images with varying image resolution, it is necessary to keep track of the image resolution information. The surfaces are already affinely registered so there are no global size differences. In order to extract and display the mandible surface of the first subject as in Figure 1.10, we run

```
load mandible77subjects.mat

sub1.vertices=voxelsize(44)*vertices(:,:,1);
sub1.faces=faces;
figure;
subplot(2,2,1); figure_patch(sub1,[0.74 0.71 0.61],0.7);
view([90 60]); camlight
title(id(1))
```

If we are interested in superimposing 20 surfaces (Figure 1.10), we run

```
subplot(2,2,2);
for i=1:20
    i
    surf.faces=faces;
    surf.vertices=voxelsize(44)*vertices(:,:,i);
    hold on; figure_patch(surf,[0.74 0.71 0.61],0.1);
end;
view([90 60]); camlight;
title('20 surfaces')
```

The template, which is the average of all 77 subjects, is given by averaging the surface coordinates in Figure 1.10:

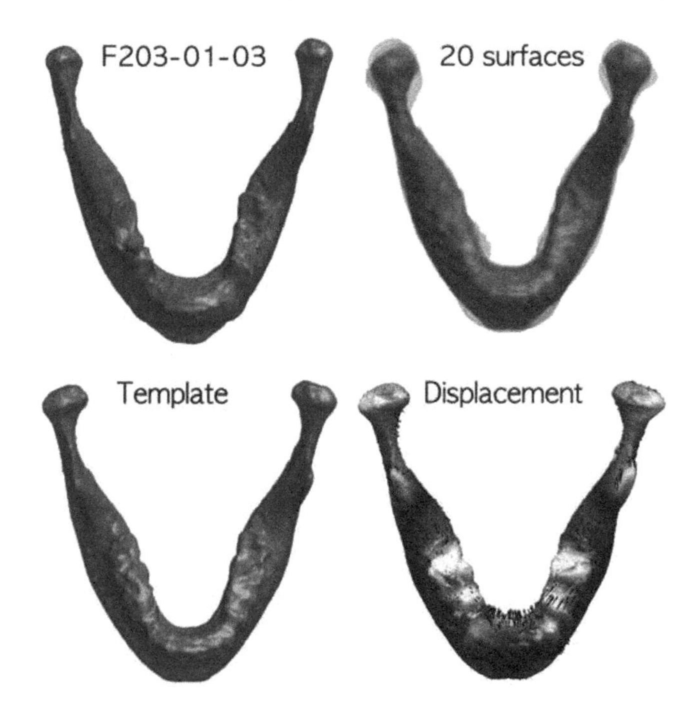

FIGURE 1.10
All 77 mandibles are affinely warped to F155-12-08 space. The affine transformed F203-01-03 surface is shown here. The superimposition of 20 affine registered mandibles shows local misalignment. Diffeomorphic registration is then performed to warp misaligned affine transformed mandibles. The surface meshes are then resampled in such a way that all the mandibles have identical mesh topology so that we can have anatomical correspondence across different mesh vertices. The average of deformation with respect to F155-12-08 provides the final population average template where statistical parametric maps will be constructed. The displacement vector from the template to F203-01-03 surface is shown as arrows.

```
template.faces=faces;
template.vertices=voxelsize(44)*mean(vertices,3);
subplot(2,2,3); figure_patch(template,[0.74 0.71 0.61],0.7);
view([90 60]); camlight;
title('Template')

disp = sub1.vertices - template.vertices;
displength=sqrt(sum(disp.^2,2));
mean(displength)

subplot(2,2,4); figure_surf(template,displength);
```

```
figure_quiver3surface(template, disp, 30);
view([90 60]); camlight
title('Displacement field')
colormap('hot')
```

1.5 Tensor and Curve Data

Tensor and curve data are often obtained as the outcomes of diffusion tensor image (DTI) processing. DTI is a new imaging technique that has been used to characterize the macrostructure of biological tissues using magnitude, anisotropy and anisotropic orientation associated with water diffusion in the brain [29]. DTI provides directional and connectivity information that MRI usually does not provide. The white matter fibers pose a physical constraint on the movement of water molecules along the direction of fibers. It is assumed that the direction of greatest diffusivity is most likely aligned to the local orientation of the white matter fibers [256]. The directional information of water diffusion is usually represented as a symmetric positive definite 3×3 matrix $D = (d_{ij})$ which is usually termed as the *diffusion tensor* or *diffusion coefficients*. The eigenvectors and eigenvalues of D are obtained by solving

$$Dv = \lambda v,$$

which results in 3 eigenvalues $\lambda_1 \geq \lambda_2 \geq \lambda_3$ and the corresponding eigenvectors v_1, v_2, v_3. The principal eigenvector v_1 determines the direction of the water diffusion, and is mainly used in streamline based tractography. Since there are exactly six unique elements in the diffusion tensor D, these six components of DTI are usually stored in six separate Analyze files.

From DTI, we can obtain the fractional anisotropy (FA) map, which is defined as

$$\texttt{FA} = \sqrt{\frac{(\lambda_1 - \lambda_2)^2 + (\lambda_2 - \lambda_3)^2 + (\lambda_3 - \lambda_1)^2}{2(\lambda_1^2 + \lambda_2^2 + \lambda_3^2)}}.$$

FA value is a scalar measure between 0 and 1 and it measures the amount of anisotropicity of diffusion.

We have provided a sample FA-map, which is used as a study template in [66]. The FA-map is stored in NIfTI-1 file format and read using `nii` package written by Jimmy Shen of Rotman Research Institute (`http://research.baycrest.org/~jimmy/NIfTI`). NIfTI-1 format is adapted from the widely used ANALYZE 7.5 file format (`nifti.nimh.nih.gov`).

```
nii = load_nii('mean_diffeomorphic_initial6_fa.nii');
d=nii.img;
surf = isosurface(d);
```

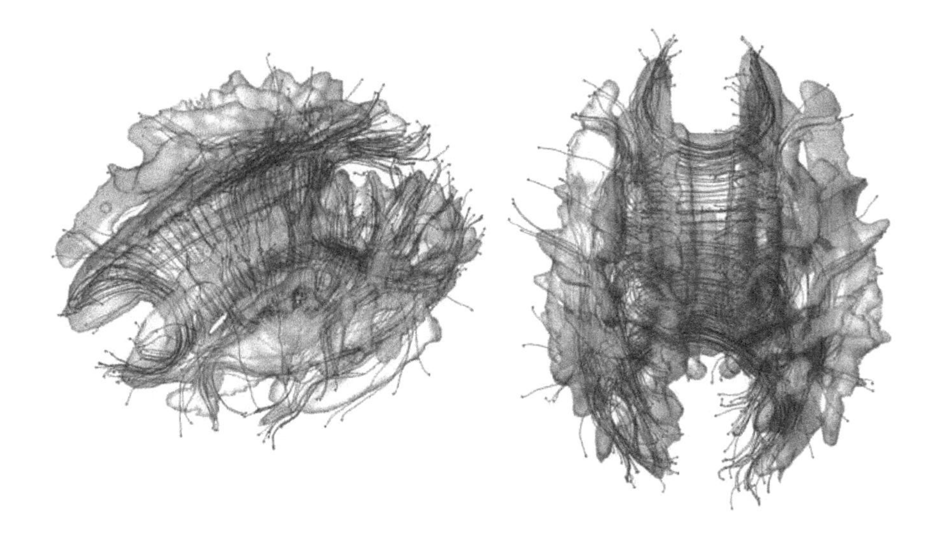

FIGURE 1.11
White matter fiber bundles obtained using TEND algorithm [218, 90]. The surface is the isosurface of the template FA map so most of tracts are expected to end outside of the surface. The ϵ-neighbor method uses the proximity of the end points in constructing the network graph.

```
coord=surf.vertices;
temp=coord;
temp(:,2)=coord(:,1);
temp(:,1)=coord(:,2);
surf.vertices=temp;

figure;
figure_patch(surf,[0.5 0.5 0.5],0.2);
axis off
set(gcf,'Color','w');
```

We have to swap x- and y-coordinates of the image since the `MATLAB` convention is slightly different from the usual brain imaging format.

DTI is usually processed again to obtain the streamline based tractography [218, 90]. Figure 1.11 shows the subsampled tractography result for a single subject. We will superimpose white matter fiber bundles stored in `SL.mat` on top of the FA-map. `SL.mat` contains 10000 white matter fiber tracts stored as the cell data structure. Since there are too many tracts, we subsample for every 30 tracts and plot them.

```
load SL.mat
```

```
hold on;
for i=1:30:10000
    tract=SL{i}';
    if size(tract,2)> 10
        hold on
        plot3(tract(1,:),tract(2,:),tract(3,:),'b')
    end;
end;
```

1.6 Brain Image Analysis Tools

Various neuroimage processing and analysis packages have been developed over the years. The main depository of neuroimaging tools is `www.nitrc.org`. The SPM (`www.fil.ion.ucl.ac.uk/spm`), AFNI (`afni.nimh.nih.gov`) and FSL (`www.fmrib.ox.ac.uk/fsl`) packages have been mainly designed for the whole brain volume based processing and univariate linear model type of analyses. The traditional statistical inference is then used to test hypotheses about the parameters of the model parameters. Although SPM and AFNI are probably the two most widely used analysis tools, their analysis pipelines are mainly based on a univariate general linear model and they do not have a routine for a multivariate or more complex statistical analysis. They also do not have the subsequent routine for correcting multiple comparison corrections for the multivariate linear models yet.

Unlike 3D whole brain volume based tools such as SPM, AFNI and FSL there are few cortical surface based tools such as the surface mapper (SUMA) [310] and FreeSurfer (`surfer.nmr.mgh.harvard.edu`). FreeSurfer is a widely used cortical surface segmentation and analysis tool. It consists of various image processing and segmentation tools. Also the spherical harmonic modeling tool SPHARM-PDM (`www.ia.unc.edu/dev/download/shapeAnalysis`) is available. However, these surface tools mainly do image processing and mesh representation and do not have the support for advanced statistical analyses.

For advanced multivariate linear modeling, for instance, one has to actually use statistical packages such as Splus (`www.insightful.com`), R (`www.r-project.org`) and SAS (`www.sas.com`). These statistical packages do not interface with imaging data easily so the additional processing step is needed to read and write imaging data within the software. Further these tools do not have the random field based multiple comparison correction procedures so the users will likely export statistics maps to SPM or fMRISTAT (`www.math.mcgill.ca/keith/fmristat`) increasing the burden of additional processing steps.

1.6.1 SurfStat

SurfStat package (www.stat.uchicago.edu/~worsley/surfstat) was developed by the late Keith J. Worsley of McGill University utilizing a model formula and avoids the explicit use of design matrices and contrasts, which tend to be a hindrance to most end users not familiar with such concepts. Probably this is the most sophisticated statistical tool for analyzing brain images. SurtStat can import MNI [237], FreeSurfer (surfer.nmr.mgh.harvard.edu) based cortical mesh formats as well as other volumetric image data. The model formula approach is implemented in many statistics packages such as Splus (www.insightful.com), R (www.r-project.org) and SAS (www.sas.com). These statistics packages accept a linear model like

$$\texttt{P} = \texttt{Group} + \texttt{Age} + \texttt{Brain}$$

as the direct input for linear modeling avoiding the need to explicitly state the design matrix. P is a $n \times 3$ matrix of coordinates, Age is the age of subjects, Brain is the total brain volume of subject and Group is the categorical group variable (0=control, 1 = autism). This type of model formula has yet to be implemented in widely used SPM or AFNI packages.

The other major novelty of the SurfStat package is the inclusion of mixed-effects models that can explicitly model the within-subject correlation of image scans of the same subject. SurfStat package is therefore better suited for longitudinally collected study designs than other packages.

1.6.2 Public Image Database

Other than the data sets we provided in the book, a large longitudinal image dataset is also available to the public. Open Access Series of Imaging Studies (OASIS, www.oasis-brains.org) and Alzheimer's Disease Neuroimaging Initiative (ADNI, www.loni.ucla.edu/ADNI) are two widely distributed data sets. The ADNI was initiated in 2003 by the National Institute on Aging (NIA) and the National Institute of Biomedical Imaging and Bioengineering (NIBIB) to determine if longitudinally collected MRI, PET and neuropsychological measurements can predict the progression of mild cognitive impairment (MCI) and Alzheimer's disease (AD) (www.adni-info.org). There is also a Simulated Brain Database (SBD) (brainweb.bic.mni.mcgill.ca/brainweb). The SBD contains a set of realistic MRI data volumes produced by an MRI simulator. These data can be used by the neuroimaging community to evaluate the performance of various image analysis methods in a setting where the truth is known.

2

Bernoulli Models for Binary Images

Here are some basic statistical concepts needed to perform statistical analysis on segmented binary images. Neuroanatomical regions are often segmented manually or automatically using various segmentation tools such as mixture models, Mumford-Shah model or level set. The segmented regions of interest are assigned value 1 while the background is assigned value 0. Then it is natural to model the segmentation error using the Bernoulli distribution. In this chapter, we will introduce Bernoulli models for analyzing such binary images and apply them in detecting the regions of stroke lesions in diffusion-weighted images (DWI).

2.1 Sum of Bernoulli Distributions

Given a sample space S, a random variable X is a rule that assigns a number to each element of S. X is a function that maps element $s \in S$ to a real number, i.e.

$$X : S \to \mathbb{R}.$$

Bernoulli random variable X takes values only 0 or 1. The probability distribution P of a random variable X is a function defined for every number x such that

$$P(X = x) = P(s \in S : X(s) = x).$$

The cumulative distribution function (CDF) of X is is defined as

$$F(x) = P(X \leq x) = \sum_{y \leq x} P(X = y).$$

The expected value (expectation) of X is a linear operator on X defined as

$$\mathbb{E}X = \sum_{x} xP(X = x)$$

The variance of X is defined as

$$\begin{aligned} \mathbb{V}X &= \mathbb{E}(X - \mathbb{E}X)^2 \\ &= \mathbb{E}X^2 - (\mathbb{E}X)^2. \end{aligned}$$

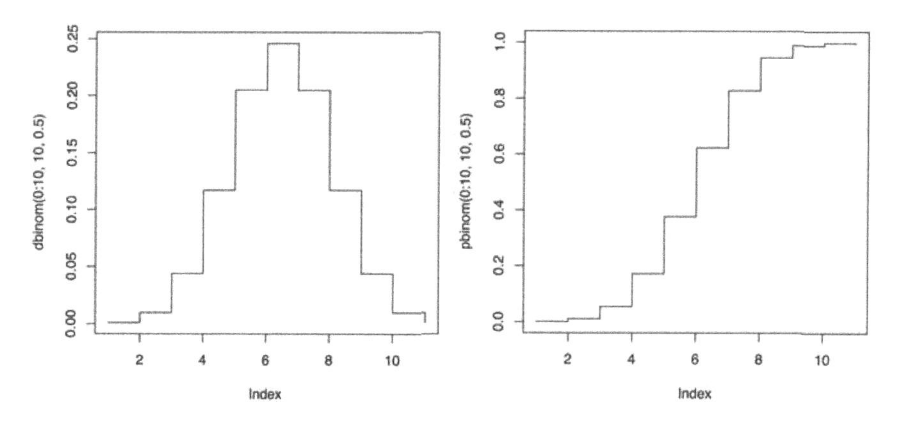

FIGURE 2.1
The distribution of Binomial random variable with $n = 10$ and $p = 0.5$, and its cumulative distribution function. Plots are generated using the statistical package `R`.

X is a Bernoulli random variable with parameter p if

$$P(X = 1) = p, \; P(X = 0) = 1 - p$$

and it will be denoted as $X \sim \mathit{Bernoulli}(p)$. The distribution can be compactly written as

$$P(X = x) = (1 - p)^{1-x} p^x.$$

The expectation and variance of the Bernoulli random variable are trivially

$$\mathbb{E}X = p, \; \mathbb{V}X = p(1 - p).$$

Consider a Bernoulli experiment that give only two outcomes H or T. This can be considered as coin tossing with either obtaining head (H) or tail (T). Let

$$P(H) = p, \; P(T) = 1 - p.$$

Now consider n such independent experiments. This is called a *Binomial experiment*. Let X be the number of total outcomes of H. This is called the Binomial random variable with parameters n and p and denoted as

$$X \sim \mathit{Binomial}(n, p).$$

Then obviously

$$X = X_1 + X_2 + \cdots + X_n,$$

where X_i is the number of outcome H in the i-th experiment. The distribution of X is then given by

$$P(X = x) = \binom{n}{x} p^x (1 - p)^{n-x}.$$

The expectation and the variance of the Binomial distribution can be computed from Bernoulli distributions.

$$\mathbb{E}X = \mathbb{E}X_1 + \cdots + \mathbb{E}X_n = np \tag{2.1}$$
$$\mathbb{V}X = \mathbb{V}X_1 + \cdots + \mathbb{V}X_n = np(1-p). \tag{2.2}$$

The actual distribution and its cumulative distribution function can be displayed using `R` statistical package (Figure 2.1):

```
> plot(dbinom(0:10,10,0.5),type="s")
> plot(pbinom(0:10,10,0.5),type="s")
```

From the central limit theorem, it can be shown that

$$\frac{X - \mathbb{E}}{\sqrt{\mathbb{V}X}} = \frac{X - np}{\sqrt{np(1-p)}} \to N(0,1),$$

the standard normal distribution.

2.2 Inference on Proportion of Activation

We introduce a method for performing statistical tests on binary images using the Binomial random variable approximation to the standard normal distribution. The binary images are obtained from Dong-Eog Kim of Dongguk University Hospital in Korea as a part of an image-based stroke registry. See [204] for details on the images. The diffusion-weighted images (DWI) were obtained for stroke patients and stroke lesions were segmented manually and aligned on a MRI template (Figure 2.2). In DWI, each voxel has an image intensity that reflects a single best measurement of the rate of water diffusion at that location. DWI can be used to better identify severely ischemic brain regions after stroke onset than more traditional T1 or T2 MRI [260].

In Figure 2.2, the first four columns of binary images are lesions of stroke patients with dysphagia who got better ($n_1 = 58$). The next four columns are lesions of stroke patients with dysphagia who did not get better ($n_2 = 23$). We are interested in determining the localized regions of lesions and if there is any pixel showing group differences. This can be done by testing on the significance of proportions.

2.2.1 One Sample Test

For a collection of binary images, each image has regions of activation. The activated and nonactivated regions are encoded as 1 and 0 respectively, which can be modeled as a Bernoulli distribution. At each fixed voxel, we are interested in testing if the activation is statistically significant. Suppose there are

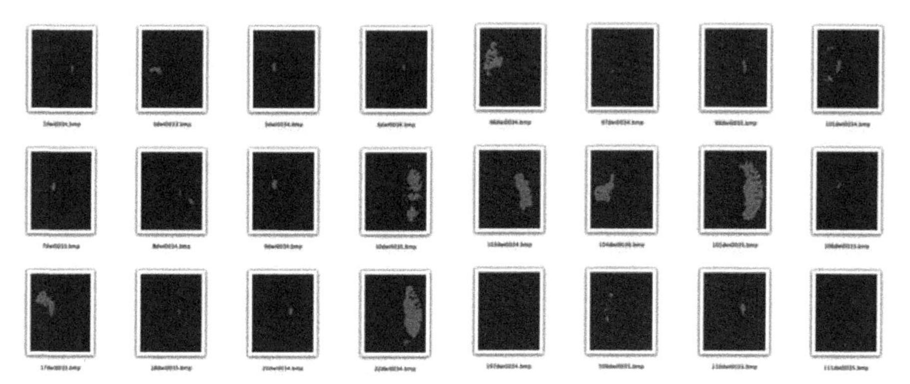

FIGURE 2.2
The lesions of stroke patients with dysphagia who got better (first four columns) and who did not get better (last four columns). We are interested in determining regions of the brain that show group differences.

n binary images and p proportion of of them are activated at the fixed voxel. Unfortunately, p is the population level parameter that is unknown. It can be estimated as

$$\widehat{p} = \frac{\#\text{ of activated images}}{n}.$$

This is referred to as the sample proportion and it is a random variable since the total number of activated images is a Binomial distribution $Binomial(n, p)$. Then trivially

$$\mathbb{E}\widehat{p} = p, \ \mathbb{V}\widehat{p} = \frac{p(1-p)}{n}.$$

We are interested in testing if

$$H_0 : p = p_0 \text{ vs. } H_1 : p > p_0$$

for a specified constant proportion p_0. The often used test statistic is

$$Z_n = \frac{\widehat{p} - p_0 - \mathbb{E}(\widehat{p} - p_0)}{\sqrt{\mathbb{V}(\widehat{p} - p_0)}}.$$

Under H_0, it becomes

$$Z_n = \frac{\widehat{p} - p_0}{\sqrt{p_0(1-p_0)/n}}.$$

This is asymptotically $Z_n \to N(0,1)$ for large n. If the test statistic is large, it is likely that H_0 is not true. So the deviation of Z_n from 0 corresponds to the likelihood of rejecting H_0. This is quantitatively determined using the p-value:

$$p\text{-value} = P\Big(Z > \frac{\widehat{p} - p_0}{\sqrt{p_0(1-p_0)/n}}\Big), \tag{2.3}$$

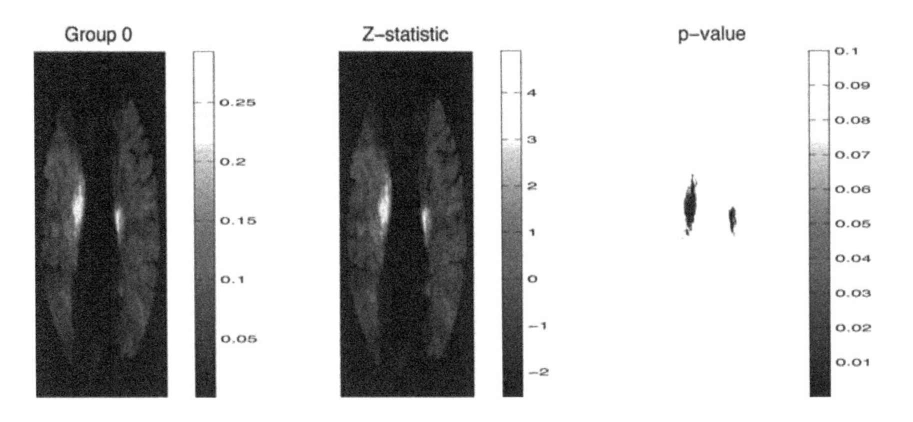

FIGURE 2.3
One sample test result on the first group showing highly significant regions of activation for testing $H_0 : p = 0.1$. The Bernoulli model breaks down testing for $H_0 : p = 0$ so we tested at $p = 0.1$.

where $Z \sim N(0, 1)$ and $\widehat{p}$ is the observed proportion here. For sufficiently large sample size, i.e. $n > 30$, the p-value computation is accurate enough.

Example 1 *Suppose 55 out of 102 images are activated in a particular voxel. Can you conclude that more than half of images are activated? Note that*

$$\widehat{p} = 55/102 = .540 \tag{2.4}$$

$$\frac{\widehat{p} - 0.5}{\sqrt{0.5 \cdot 0.5/102}} = 0.79 \tag{2.5}$$

Note that $P(Z > 0.79)$ *can be computed as*

```
>>1 -  normcdf(0.79)

0.21
```

in `MATLAB`*. This is not small enough so we cannot conclude more than half of the images are activated.*

In order for the test procedure to be valid, p_0 has to be strictly between 0 and 1. If p_0 is 0 or 1, the test statistic breaks down and we are no longer able to compute the p-value using the formula (2.3) (Figure 2.3). Note that Bernoulli models always assume $0 < p < 1$ so testing if the parameter is 0 or 1 is not a valid procedure. The p-value computation for $H_0 : p = 0$ or $H_0 : p = 1$ is trivial. If all pixels are zero, p-value is trivially 0 while at least one pixel is not zero then p-value is one. However, it is not useful to have the test procedure to produce only 0 or 1 p-value. This problem can be remedied if we do not use the Bernoulli model and model the population proportion to follow Gaussian. So we assume the population proportion $\widehat{p}$ to follow

$$\widehat{p} \sim N(\mu, \sigma^2)$$

and test for

$$H_0 : \mu = 0 \text{ vs. } H_1 : \mu > 0.$$

Then the p-value is given by

$$p\text{-value} = P\Big(Z > \frac{\widehat{p}}{\sqrt{p_0(1-p_0)/n}}\Big).$$

This problem can be easily solved if we follow the VBM framework which is developed for handling exactly this type of problem [69]. In VBM, we traditionally smooth binary images with Gaussian kernel and model the smoothed binary images as Gaussian getting rid of the need for modeling it as Bernoulli. It is often suggested to apply the logit transform on the smoothed binary images [14] but once you perform Gaussian kernel smoothing, the images become sufficiently Gaussian so the logit transform is not really needed.

2.2.2 Two Sample Test

The Bernoulli model can be extended in testing if there is a difference between the two population proportions (Figure 2.4). We will assume there are p_i proportion of images activated out of n_i images in the i-th group. The total number of activated images are modeled as $Binomial(n_i, p_i)$. The null and alternate hypotheses for the two-sample test is then

$$H_0 : p_1 = p_2 \text{ vs. } H_1 : p_1 > p_2.$$

Note that

$$\mathbb{E}(\widehat{p}_1 - \widehat{p}_2) = \frac{p_1}{n_1} - \frac{p_2}{n_2}.$$

$$\mathbb{V}\widehat{p} = \frac{p_1(1-p_1)}{n_1} + \frac{p_2(1-p_2)}{n_2} \qquad (2.6)$$

Under null, the test statistic is similarly given by the Z-statistic as

$$Z_n = \frac{\widehat{p}_1 - \widehat{p}_2}{\sqrt{\frac{p_1(1-p_1)}{n_1} + \frac{p_2(1-p_2)}{n_2}}}.$$

The p-value is then given by

$$p\text{-value} = P\left(Z > \frac{\widehat{p}_1 - \widehat{p}_2}{\sqrt{\frac{p_1(1-p_1)}{n_1} + \frac{p_2(1-p_2)}{n_2}}}\right),$$

where p_1 and p_2 are observed proportions. There are many different test procedures for testing the equality of proportions. This is probably the most simple procedure among them.

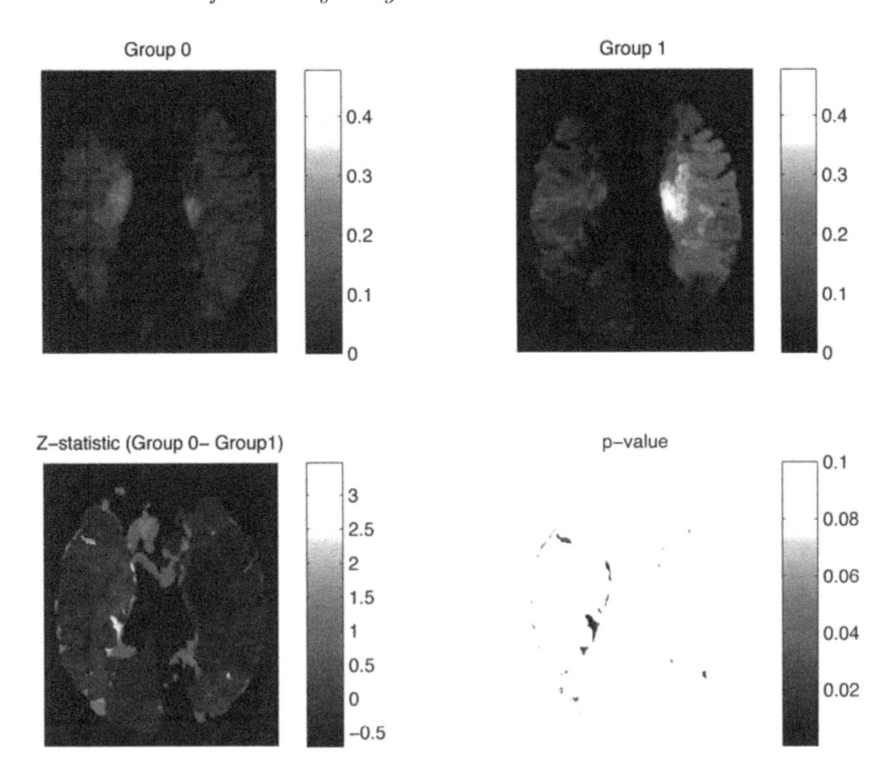

FIGURE 2.4
Two sample test result showing almost no region of group difference although there seems like a region of group difference where Z-values are larger than 3. This high Z-statistic value is most likely caused by the discretization error.

2.3 MATLAB Implementation

Images are stored in the bitmap format and files have a different length of file names. The first and the second group images are respectively stored in the directories `/stroke/0` and `/stroke/1`. We read the sequence of file names into structured arrays `g0` and `g1`. The first file name in the first group is then given by `g0(1).name`. The following code reads all the bitmap images in the directory.

```
g0 = dir('*.bmp')
n0=size(g0,1); % number of images in gorup 0
group0=zeros(370,301,n0);

for i=1:n0
    fname= g0(i).name;
```

```
    f=imread(fname,'bmp');
    group0(:,:,i) = double(f(:,:,1)/255);
end;

g1 = dir('*.bmp')
n1=size(g1,1); % number of images in gorup 1
group1=zeros(370,301,n1);
for i=1:n1
    fname= g0(i).name;
    f=imread(fname,'bmp');
    group1(:,:,i) = double(f(:,:,1)/255);
end;
```

`n0` and `n1` are the sample sizes. Each image slice of dimension 370×301 is loaded into the 3D matrices `group0` and `group1` for the two groups. The sample proportions are computed as

```
p0=mean(group0, 3);
p1=mean(group1,3);
```

The Z-statistic for one sample is computed as

```
 Zvalue = (p0-pc)./sqrt( pc.*(1-pc)/nc)
```

for some specific value `pc`. On the other hand, the Z-statistic for the two samples is computed as

```
 Zvalue = (p0-p1)./sqrt( p0.*(1-p0)/n0 + p1.*(1-p1)).
```

The p-value is then computed using `normcdf`. The whole routine is written as a function below. If the second argument is empty. i.e. `[]`, the function performs the one sample test.

```
[pvalue Zvalue] = stat_proportion(group0, []);
[pvalue Zvalue] = stat_proportion(group0, group1);
```

The resulting outputs are Figure 2.3 for the one sample test and Figure 2.4 for the two sample test. One sample test result shows many regions of statistical significance while the two sample test result shows almost no region of significance at 0.05.

3

General Linear Models

In this chapter, we introduce *general linear models* (GLM) that have been widely used in brain imaging applications. The GLM is a very flexible and general statistical framework encompassing a wide variety of fixed effect models such as the multiple regressions, the analysis of variance (ANOVA), the multivariate analysis of variance (MANOVA), the analysis of covariance (ANCOVA) and the multivariate analysis of covariance (MANCOVA) [366]. Note that the term *linear* is misleading in a sense that the model can also include mathematically nonlinear model terms such as the higher degree polynomials. This chapter is mainly based on [69].

The GLM provides a framework for testing various associations and hypotheses while accounting for nuisance covariates in the model in a straightforward fashion. The effect of age, sex, brain size and possibly IQ can have severe confounding effects on the final outcome of many anatomical and functional imaging studies. Older population's reduced functional activation could be the consequence of age-related atrophy of neural systems [244]. Brain volumes are significantly larger for children with autism 12 years old and younger compared with normally developing children [20]. Therefore, it is desirable to account for various confounding factors such as age and sex in the model. This can be done using GLM. The parameters of the GLM are mainly estimated by the least squares estimation and has been implemented in many statistical packages such as R (`www.r-project.org`) or Splus [281] and brain imaging packages such as SPM (`www.fil.ion.ucl.ac.uk/spm`) and fMRISTAT (`www.math.mcgill.ca/keith/fmristat`).

3.1 General Linear Models

Let y_i be the response variable, which is mainly coming from images and $\mathbf{x}_i = (x_{i1}, \cdots, x_{ip})$ to be the variables of interest and $\mathbf{z}_i = (z_{i1}, \cdots, z_{ik})$ to be nuisance variables corresponding to the i-th subject. We assume there are n subjects, i.e. $i = 1, \cdots, n$. We are interested in testing the significance of variables $\mathbf{x}_i$ while accounting for nuisance covariates $\mathbf{z}_i$. Then we set up the following GLM

$$y_i = \mathbf{z}_i \boldsymbol{\lambda} + \mathbf{x}_i \boldsymbol{\beta} + \epsilon_i$$

where $\boldsymbol{\lambda} = (\lambda_1, \cdots, \lambda_k)'$ and $\boldsymbol{\beta} = (\beta_1, \cdots, \beta_p)'$ are unknown parameter vectors to be estimated. We assume ϵ to be the usual zero mean Gaussian noise although the distributional assumption is not necessary for the least squares estimation.

The significance of the variable of interests $\mathbf{x}_i$ is determined by testing the null hypothesis

$$H_0 : \boldsymbol{\beta} = 0 \text{ vs. } H_1 : \boldsymbol{\beta} \neq 0.$$

The fit of the reduced model corresponding to $\beta = 0$, i.e.

$$y_i = \mathbf{z}_i \boldsymbol{\lambda}, \tag{3.1}$$

is measured by the sum of the squared errors (SSE):

$$\text{SSE}_0 = \sum_{i=1}^{n} (y_i - \mathbf{z}_i \widehat{\boldsymbol{\lambda}}_0)^2,$$

where $\widehat{\boldsymbol{\lambda}}_0$ is the least squares estimation obtained from the reduced model. The reduced model (3.1) can be written in a matrix form

$$\underbrace{\begin{pmatrix} y_1 \\ \vdots \\ y_n \end{pmatrix}}_{\mathbf{y}} = \underbrace{\begin{pmatrix} z_{11} & \cdots & z_{1k} \\ \vdots & \ddots & \vdots \\ z_{n1} & \cdots & z_{nk} \end{pmatrix}}_{\mathbf{Z}} \underbrace{\begin{pmatrix} \lambda_1 \\ \vdots \\ \lambda_n \end{pmatrix}}_{\boldsymbol{\lambda}}. \tag{3.2}$$

By multiplying $\mathbf{Z}'$ on the both sides, we obtain

$$\mathbf{Z}'\mathbf{y} = \mathbf{Z}'\mathbf{Z}\boldsymbol{\lambda}.$$

Now the matrix $\mathbf{Z}'\mathbf{Z}$ is a full rank and can be invertible if $n \geq k$, which is the usual case in brain imaging. Therefore, the matrix equation can be solved by performing a matrix inversion

$$\widehat{\boldsymbol{\lambda}}_0 = (\mathbf{Z}'\mathbf{Z})^{-1}\mathbf{Z}'\mathbf{y}.$$

Similarly the fit of the full model corresponding to $\boldsymbol{\beta} \neq 0$, i.e.

$$y_i = \mathbf{z}_i \boldsymbol{\lambda} + \mathbf{x}_i \boldsymbol{\beta} \tag{3.3}$$

is measured by

$$\text{SSE}_1 = \sum_{i=1}^{n} (y_i - \mathbf{z}_i \widehat{\boldsymbol{\lambda}}_1 - \mathbf{x}_i \widehat{\boldsymbol{\beta}}_1)^2,$$

where $\widehat{\boldsymbol{\lambda}}_1$ and $\widehat{\boldsymbol{\beta}}_1$ are the least squares estimation from the full model. The full model can be written in a matrix form by concatenating the row vectors

$\mathbf{z}_i$ and $\mathbf{x}_i$ into a larger row vector $(\mathbf{z}_i, \mathbf{x}_i)$, and the column vectors $\boldsymbol{\lambda}$ and $\boldsymbol{\beta}$ into a larger column vector $(\boldsymbol{\lambda}', \boldsymbol{\beta}')'$, i.e.

$$y_i = (\mathbf{z}_i, \mathbf{x}_i)(\boldsymbol{\lambda}', \boldsymbol{\beta}')'.$$

The full model can be also written in a matrix form and similarly solved by the matrix inversion.

Note that

$$\begin{aligned} \text{SSE}_1 &= \min_{\boldsymbol{\lambda}_1, \boldsymbol{\beta}_1} \sum_{i=1}^{n} (y_i - \mathbf{z}_i \boldsymbol{\lambda}_1 - \mathbf{x}_i \boldsymbol{\beta}_1)^2 \\ &\leq \min_{\boldsymbol{\lambda}_0} \sum_{i=1}^{n} (y_i - \mathbf{z}_i \boldsymbol{\lambda}_0)^2 = \text{SSE}_0. \end{aligned}$$

So the larger the value of $\text{SSE}_0 - \text{SSE}_1$, the more significant the contribution of the coefficients $\boldsymbol{\beta}$ is. Under the assumption of the null hypothesis H_0, the test statistic is the ratio

$$F = \frac{(\text{SSE}_0 - \text{SSE}_1)/p}{\text{SSE}_0/(n-p-k)} \sim F_{p,n-p-k}. \tag{3.4}$$

The larger the F value, it is more unlikely to accept H_0.

When $p = 1$, the test statistic F is distributed as $F_{1,n-1-k}$, which is the square of the student t-distribution with $n-1-k$ degrees of freedom, i.e. t^2_{n-1-k}. In this particular case, it is better to use the t-statistic. The advantage of using the t-statistic is that unlike the F-statistic, it has two sides so we can actually use it to test for one sided alternative hypothesis $H_1 : \beta_1 \geq 0$ or $H_1 : \beta_1 \leq 0$. Therefore, the t-statistic map can provide the direction of the group difference that the F-statistic map cannot provide.

3.1.1 R-square

The R-square of a model explains the proportion of variability in measurement that is accounted for by the model. Sometimes R-square is called the coefficient of determination and it is given as the square of a correlation coefficient for a very simple model. For a linear model involving the response variable y_i, the total sum of squares (SST) measures total variation in response y_i and is defined as

$$\text{SST} = \sum_{i=1}^{n} (y_i - \bar{y})^2,$$

where $\bar{y}$ is the sample mean of y_i.

On the other hand, SSE measures the amount of variability in y_i that is not explained by the model. Note that SSE is the minimum of the sum of squared residual of any linear model, SSE is always smaller than SST. Therefore, the amount of variability explained by the model is SST-SSE. The proportion of variability explained by the model is then

$$R^2 = \frac{\text{SST} - \text{SSE}}{\text{SST}},$$

which is the coefficient of determination. The R-square ranges between 0 and 1 and the value larger than 0.5 is usually considered as significant.

3.1.2 GLM for Whole Brain Images

In brain imaging, the linear model of type (3.2) is usually fitted in each voxel separately. If the image dimension is of size $100 \times 100 \times 200$ for instance, we need to fit 2 million linear models, which causes a serious computational bottleneck. So what we need is to reformulate the problem such that we fit all linear models simultaneously in a slice so that we only need to perform the least squares estimation 200 times.

Let $\mathbf{y}_j$ be the measurement vector at the j-th voxel in a slice. Assume there are m voxels in a slice. We have the same design matrix $\mathbf{Z}$ for all m voxels. Then we need to estimate the parameter vector $\boldsymbol{\lambda}_j$ in

$$\mathbf{y}_j = \mathbf{Z}\boldsymbol{\lambda}_j. \tag{3.5}$$

each j. Instead of solving (3.5) separately, we combine all of them together so that we have matrix equation

$$\underbrace{[\mathbf{y}_1, \cdots, \mathbf{y}_m]}_{\mathbf{Y}} = \mathbf{Z}\underbrace{[\boldsymbol{\lambda}_1, \cdots, \boldsymbol{\lambda}_m]}_{\boldsymbol{\Lambda}}. \tag{3.6}$$

The least squares estimation of the parameter matrix $\boldsymbol{\Lambda}$ proceeds similarly and given by

$$\widehat{\boldsymbol{\Lambda}} = (\mathbf{Z}'\mathbf{Z})^{-1}\mathbf{Z}'\mathbf{Y}.$$

The least squares estimation technique does not work for sparsely sampled data where $n \ll k$. In this case, $\mathbf{Z}'\mathbf{Z}$ is size $k \times k$ but only of rank n. So we can't invert $\mathbf{Z}'\mathbf{Z}$ directly and the method breaks down. The generalized inverse can be used instead of the usual matrix inverse for slightly underdetermined systems but for significantly underdetermined systems, we need to regularize using the l_1-norm penalty.

3.2 Voxel-Based Morphometry

GLM has often been used in *voxel-based morphometry* (VBM). Let us review the basic VBM that is needed to understand how GLM is used in VBM.

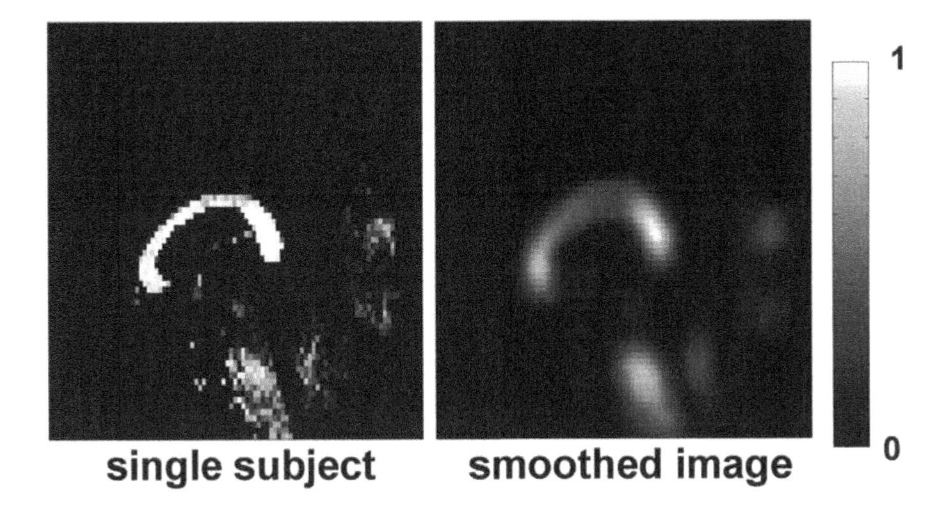

FIGURE 3.1
Left: White matter density map in the corpus callosum obtained through the SPM package. Right: Gaussian kernel smoothing on the white matter density map.

VBM involves a voxel-wise comparison of the local concentration of gray or white matters between populations [14]. It requires spatially normalizing images from all the subjects in the study to a template. This is followed by segmenting the gray and white matters and cerebrospinal fluid (CSF) from the spatially normalized images and smoothing out the segmented images. The binary segmented images are referred to as tissue densities (Figure 3.1). Statistical inference is subsequently done at each voxel level on the tissue densities while accounting for multiple comparisons.

The tissue segmentation is based on a Gaussian mixture model that assumes the image intensity values to follow the mixture of three independent Gaussians and the unknown parameters of Gaussian distributions are estimated by maximizing the likelihood function using the expectation maximization (EM) algorithm. Figure 3.2 shows the characteristic three-components Gaussian mixture observed in brain MRI. The widely used Statistical Parametric Mapping (SPM) package (Wellcome Department of Cognitive Neurology, London, UK, `www.fil.ion.ucl.ac.uk/spm`) is based on a Bayesian formulation of the Gaussian mixture model with a prior probability image obtained by averaging an already segmented large number of brain images [18, 14]. Based on the prior probability of each voxel belong to a specific tissue type, the Bayesian framework is used to get the posterior probability. This Bayesian update of the probability is iterated many times until the probability converges. The resulting probability map is interpreted as the tissue density. This is not physical density so it should be interpreted probabilistically.

The Bayesian segmentation framework utilizes the Bayes theorem in estimating the posterior probability of a voxel belonging to a particular tissue type from a given prior probability. Let C be the event of a voxel belong to a particular class. We may assume there are three classes corresponding to gray and white matters and CSF. The prior probability $P(C)$ is obtained by averaging a large sample of normalized binary segmentation and dividing the average by the total number of sample. Let T be the event that a voxel has a particular image intensity value. This is that we usually observe in T_1-weighted MRI. We wish to obtain the *conditional probability* $P(C|T)$ of the voxel belonging to the class C given that we have observed T:

$$P(C|T) = \frac{P(C \cap T)}{P(T)}. \tag{3.7}$$

$P(C|T)$ is interpreted as the probability of the voxel belonging to a specific class when the voxel has a particular intensity value. This is what we wish to determine in Bayesian segmentation and it is termed as tissue density in VBM. The numerator can be written as $P(C \cap T) = P(T|C)P(C)$ while, from the law of total probability, the total probability $P(T)$ can be decomposed as

$$P(T) = \sum_C P(T \cap C) = \sum_C P(T|C)P(C).$$

The conditional probability (3.7) can be written in terms of the prior probability as

$$P(C|T) = \frac{P(T|C)P(C)}{\sum_C P(T|C)P(C)}. \tag{3.8}$$

The likelihood term $P(T|C)$ is interpreted as the probability of a voxel obtaining a particular intensity value given the voxel belonging to a particular tissue type, and it can be estimated from mixture models. The likelihood term is given by evaluating the probability density for the class C at each voxel intensity value [14].

3.2.1 Mixture Models

The likelihood term is estimated using mixture models and the expectation-maximization (EM) algorithm. Mixture models have been widely used for segmenting images. This section will closely follow the description given in [65].

The image intensity value at a given voxel is assumed to come from different tissue classes with specific proportions p_j. We assume $0 < p_j < 1$ and $\sum_j p_j = 1$. We may assume that image intensity values for each class to follow a certain distribution f_j. This is the likelihood term $P(T|C)$. Then the k-component mixture model on image intensity values assumes image intensity values Y to come from k different distributions $f_1, \cdots, f_k$ with proportions

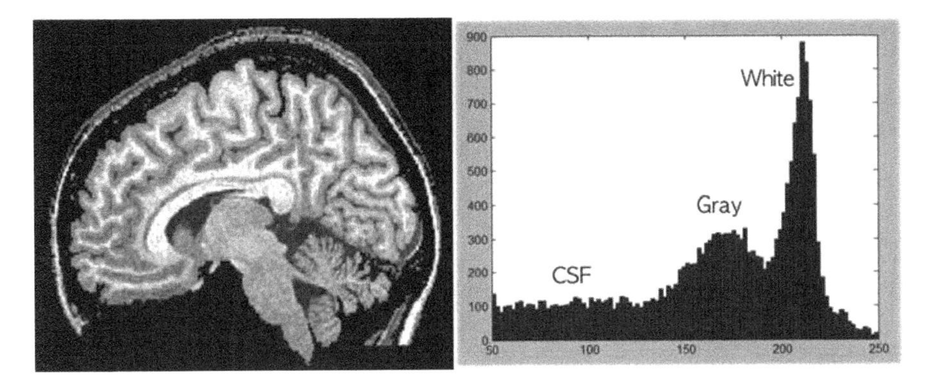

FIGURE 3.2
Image intensity of MRI slice showing the characteristic three-components Gaussian mixture. EM-algorithm can be used to determine the parameters of the Gaussian mixture.

$p_1, \cdots, p_k$. Another way of saying this is that the k-component mixture model can be obtained by mixing samples obtained from distributions f_j with p_j proportions. This can be modeled by conditioning on a multinomial distribution.

Let X_j be an indicator variable for the j-th class such that $P(X_j = 1) = p_j$ and $P(X_j = 0) = 1 - p_j$. X_j is a Bernoulli random variable. The collection of variables $X = (X_1, \cdots, X_k)$ form a multinomial distribution with parameters $(p_1, \cdots, p_k)$ if we have the additional constraint $X_1 + \cdots + X_k = 1$. The probability mass function of X is given by

$$f(x_1, \cdots, x_k) = P(X_1 = x_1, \cdots, X_k = x_k) = p_1^{x_1} \cdots p_k^{x_k}.$$

Define a random variable Y condition on the event $X_j = 1$ such that $Y \sim f_j$ if $X_j = 1$. The conditional density $f(y|x_j = 1) = f_j$ is the distribution for the j-th class. The joint density between X_j and Y is then given by $f(x_j = 1, y) = p_j f_j(y)$, which can be compactly written as

$$f(x, y) = [p_1 f_1(y)]^{x_1} \cdots [p_k f_k(y)]^{x_k}.$$

The marginal density of Y is subsequently given as

$$f(y) = \sum_x f(x, y) = \sum_{i=1}^{k} p_i f_i(y). \tag{3.9}$$

The collection of unknown parameters in (3.9) is denoted as Θ. The unknown parameters include the mixing proportions p_j as well as parameters of the distribution f_i. Then we write the *k-component mixture model* as

$$f(y|\Theta) = \sum_{i=1}^{k} p_i f_i(y). \tag{3.10}$$

to indicate the dependence of the model on the parameters Θ. The parameters Θ in (3.10) are estimated by the *maximum likelihood estimation* (MLE). Suppose we have a sample $Y = \{Y_1, \cdots, Y_n\}$ drawn from the distribution $f(y|\Theta)$. The likelihood estimation of Θ is given by maximizing the loglikelihood:

$$\widehat{\Theta} = \arg\max_{\Theta} \prod_{i=1}^{n} f(y_i|\Theta) = \arg\max_{\Theta} \sum_{i=1}^{n} \ln f(y_i|\Theta).$$

The loglikelihood function is then subsequently maximized using an iterative approximation technique called the expectation maximization (EM) algorithm.

3.2.2 EM-Algorithm

The expectation maximization (EM) algorithm was first introduced by [105]. For the introductory overview on the algorithm, see [302] and [122]. The EM-algorithm proceeds as follows.

Following the argument in [302], we augment the observed data Y with latent (unobserved or missing) data Y^m such that the complete data $Y^c = (Y, Y^m)$. The latent data is introduced as an artifice to make the problem tractable. The probability density of the complete data Y^c is denoted as $f(y^c) = f(y, y^m)$. The conditional density for the latent data Y^m, condition on observation Y, is

$$f(y^m|y, \Theta) = \frac{f(y, y^m|\Theta)}{f(y|\Theta)}.$$

Again we introduced Θ to indicate the dependence of the probability on the parameters. Taking the logarithm on both sides, we get the loglikelihood for the observed data

$$\ln f(Y|\Theta) = \ln f(Y^c|\Theta) - \ln f(Y^m|Y, \Theta).$$

Since the logarithm is a strictly increasing function, the value that maximizes $f(Y|\Theta)$ also maximizes $\ln f(Y|\Theta)$. Now taking the expectation with respect to $f(y^m|y, \Theta_0)$ for some fixed Θ_0 on the both sides, we have

$$\mathbb{E}[\ln f(Y|\Theta)|Y, \Theta_0] = \mathbb{E}[\ln f(Y^c|\Theta)|Y, \Theta_0] - \mathbb{E}[\ln f(Y^m|Y, \Theta)|Y, \Theta_0]. \quad (3.11)$$

Now denote the expected loglikelihood for the complete data as

$$Q(\Theta|\Theta_0, Y) = \mathbb{E}[\ln f(Y^c|\Theta)|Y, \Theta_0].$$

We maximize the likelihood in iterative two-steps:

(1) E-step: compute the expectation $Q(\Theta|\widehat{\Theta}_{j-1}, Y)$.

(2) M-step: maximize $Q(\Theta|\widehat{\Theta}_{j-1}, Y)$ and take

$$\widehat{\Theta}_j = \arg\max_{\Theta} Q(\Theta|\widehat{\Theta}_{j-1}, Y). \tag{3.12}$$

Starting with the initial estimate $\widehat{\Theta}_0$, we have a sequence of estimators $\widehat{\Theta}_1, \widehat{\Theta}_2, \cdots$ and it can be shown to converges to the true MLE $\widehat{\Theta}$. However, the proof is beyond the scope of the book and we will only show that the Q function monotonically increases. The argument is as follows. By the definition (3.12), we have

$$Q(\widehat{\Theta}_j|\widehat{\Theta}_j, y) \leq Q(\widehat{\Theta}_{j+1}|\widehat{\Theta}_j, y).$$

Now let $R(\Theta|\Theta_0, Y) = \mathbb{E}[\ln f(Y^m|Y, \Theta)|Y, \Theta_0]$. This is the second term in (3.11). From the Jensen's inequality, we can show that

$$R(\Theta|\Theta_0, y) - R(\Theta_0|\Theta_0, y) \leq \ln \int \frac{f(y^m|y, \Theta)}{f(y^m|y, \Theta_0)} f(y^m|y, \Theta_0)\, dy^m = 0.$$

Hence we have $R(\widehat{\Theta}_{j+1}|\widehat{\Theta}_j, y) \leq R(\widehat{\Theta}_j|\widehat{\Theta}_j, y)$. Consequently

$$\begin{aligned}
\ln f(y|\widehat{\Theta}_j) &= Q(\widehat{\Theta}_j|\widehat{\Theta}_j, y) - R(\widehat{\Theta}_j|\widehat{\Theta}_j, y) \\
&\leq Q(\widehat{\Theta}_{j+1}|\widehat{\Theta}_j, y) - R(\widehat{\Theta}_{j+1}|\widehat{\Theta}_j, y) \\
&\leq \ln f(y|\widehat{\Theta}_{j+1}).
\end{aligned}$$

The inequality guarantees the sequence of estimators $\widehat{\Theta}_j$ monotonically increases the likelihood function. Further, since the monotonically increasing sequence is bounded, i.e. $\ln f(y|\widehat{\Theta}_j) \leq \ln f(y|\widehat{\Theta})$, where $\widehat{\Theta}$ is the MLE, the sequence must be converging to a constant, but it is not clear if the limit is in fact $\ln f(y|\widehat{\Theta})$. To guarantee that the limit converges to the true maximum likelihood estimator, additional conditions are needed [47, 397].

The difficulty of implementing the EM-algorithm is at the E-step where we need to compute the conditional expectation $Q(\Theta|\widehat{\Theta}_{j-1}, y)$. The Monte Carlo version of the EM-algorithm overcome this problem by simulating the missing data Y^m from the conditional density $f(y^m|y, \Theta)$ so that

$$\widehat{Q}(\Theta|\Theta_0, y) = \frac{1}{k}\sum_{j=1}^{k} \ln f(Y, Y^m|\Theta).$$

3.2.3 Two-Components Gaussian Mixture

As an illustration, two-components Gaussian mixture model will be fitted using the EM-algorithm. Extension of three-components mixture model is similar. The image intensity will be modeled as a Gaussian mixture of the form

$$f(y) = p_1 f_1(y) + p_2 f_2(y)$$

where $p_1+p_2 = 1$ and $f_1 \sim N(\mu_1, \sigma_1^2)$ and $f_2 \sim N(\mu_2, \sigma_2^2)$ are all known. There are 5 unknown parameters $\Theta = \{p_1, \mu_1, \mu_2, \sigma_1^2, \sigma_2^2\}$ to be estimated. Once p_1 is estimated, p_2 is automatically given as $1 - p_1$. The likelihood function is given by

$$f(\Theta|y) = \prod_{i=1}^{n} \Big[p_1 f_1(y_i) + p_2 f_2(y_i)\Big].$$

The loglikelihood is

$$L(\Theta|y) = \sum_{i=1}^{n} \ln \Big[p_1 f_1(y_i) + p_2 f_2(y_i)\Big].$$

The loglikelihood is maximized by solving

$$\frac{\partial L(\Theta|y)}{\partial p_i} = 0, \quad \frac{\partial L(\Theta|y)}{\partial \mu_i} = 0, \quad \frac{\partial L(\Theta|y)}{\partial \sigma_i^2} = 0$$

but this is not tractable. So we argument the data with the latent data and apply the EM-algorithm.

Let X be a Bernoulli random variable with $P(X = 1) = p$ and $P(X = 0) = q = 1 - p$. This choice of latent random variable makes the subsequent EM-algorithm to be tractable. Now define the conditional distribution $Y \sim f_1$ if $X = 1$ and $Y \sim f_2$ if $X = 0$. This defines the conditional density $f(y|x)$. The joint density $f(x, y)$ is $f(1, y) = pf_1(y)$ and $f(0, y) = qf_2(y)$. This can be compactly written as

$$f(x, y) = \Big[pf_1(y)\Big]^x \Big[qf_2(y)\Big]^{1-x}.$$

The marginal density of Y is obviously

$$f(y) = \sum_{x=0,1} f(x, y) = pf_1(y) + qf_2(y).$$

The conditional density of X given Y is then

$$f(x|y) = \frac{[pf_1(y)]^x [qf_2(y)]^{1-x}}{pf_1(y) + qf_2(y)}.$$

The conditional expectation of X with respect to $f(x|y)$ is then

$$\mathbb{E}(X|y, p) = \frac{pf_1(y)}{pf_1(y) + qf_2(y)}. \tag{3.13}$$

The likelihood for the complete data (x, y) is

$$f(\Theta|x, y) = \prod_{i=1}^{n} \big[pf_1(y_i)\big]^{x_i} \big[qf_2(y_i)\big]^{1-x_i}$$

and the corresponding loglikelihood is given by

$$L(\Theta|x,y) = \sum_{i=1}^{n} x_i \ln\left[\frac{pf_1(y_i)}{qf_2(y_i)}\right] + \ln\left[qf_2(y_i)\right].$$

Take the expectation with respect to the latent variable X to get the Q-function

$$\begin{aligned} Q(\Theta|\Theta_0,y) &= \mathbb{E}\big[\ln L(\Theta|X,Y)\big|y,\Theta_0\big] \\ &= \sum_{i=1}^{n} \mathbb{E}(X_i|y,\Theta_0) \ln\left[\frac{pf_1(y_i)}{qf_2(y_i)}\right] + \ln\left[qf_2(y_i)\right]. \end{aligned} \quad (3.14)$$

From (3.13), we have

$$\mathbb{E}(X_i|y,\Theta_0) = \frac{p_0 f_1(y_i)}{p_0 f_1(y_i) + q_0 f_2(y_i)} = \pi_{1i}$$

is the posterior probability of the i-th observation coming from the first class. Hence the expression (3.14) can be written as

$$Q(\Theta|\Theta_0,y) = \sum_{i=1}^{n} \pi_i \ln\left[\frac{pf_1(y_i)}{qf_2(y_i)}\right] + \ln\left[qf_2(y_i)\right]. \quad (3.15)$$

Maximizing Q with respect to p by solving $\partial Q/\partial p = 0$, we obtain

$$p = \frac{1}{n}\sum_{i=1}^{n} \pi_{1i}. \quad (3.16)$$

(3.16) states that the prior probability for the 1st class is estimated as the average of the posterior probabilities in the 1st class. Note that we did not use the explicit forms for f_1 and f_2 so this result is general for any type of mixture distribution. Based on (3.16), we set up the iteration

$$\widehat{p}_{j+1} = \frac{1}{n}\sum_{i=1}^{n} \frac{\widehat{p}_j f_1(y_i)}{\widehat{p}_j f_1(y_i) + (1-\widehat{p}_j) f_2(y_i)}$$

with any arbitrary initial $\widehat{p}_0 \in (0,1)$. For other parameters, we obtain similar iterative formulas:

$$\mu_j = \frac{\sum_i \pi_{ji} y_i}{\sum_i \pi_{ji}}$$

and

$$\sigma_j^2 = \frac{\sum_i \pi_{ji}(y_i - \mu_j)^2}{\sum_i \pi_{ji}}.$$

3.3 Case Study: VBM in Corpus Callosum

VBM is applied to characterizing white matter differences in corpus callosum in autism. Autism is a neurodevelopmental disorder of brain function that has begun to attract *in vivo* structural magnetic resonance imaging (MRI) studies in the region of the corpus callosum [117, 165, 241, 282, 283]. The corpus callosum is a white matter structure that can be used as an index of neural connectivity between brain regions [165] (Figure 3.3). There is little understanding about the link between the functional deficit and the underlying abnormal anatomy in autism, which provides motivation for our study. These studies use the Witelson partition or a similar partition scheme of the corpus callosum [383]. Witelson partitioned the midsagittal cross-sectional images of the corpus callosum along the maximum anterior-posterior line [350] and defined the region of the genu, rostrum, midbodies, isthmus and splenium from the anterior to posterior direction. Based on the Witelson partition, there has been a consistent finding in abnormal reduction in anterior, midbody and posterior of the corpus callosum [48].

Piven *et al.* (1997) compared 35 autistic individuals with 36 normal control subjects controlling for total brain volume, gender and IQ and detected a statistically significant smaller midbody and posterior regions of the corpus callosum in the autistic group [283]. Manes *et al.* (1999) compared 27 low functioning autistic individuals with 17 normal controls adjusting for the total brain volume [241]. They found a smaller corpus callosum compared to the control group in genu, rostrum, anterior midbody, posterior midbody and isthmus but did not find statistically significant differences in the rostrum and the splenium although the sample mean of the rostrum and splenium size are smaller than that of the control group. Hardan *et al.* (2000) compared 22 high functioning autistic to 22 individually matched control subjects and showed smaller genu and rostrum of the corpus callosum adjusting for the total brain volume based on the Witelson partition [165]. The smaller corpus callosum size was considered as an indication of a decrease in interhemispheric connectivity. They did not detect other regions of significant size difference. For an extensive review of structural MRI studies for autism that have been published between 1966 and 2003, one may refer to Brambilla *et al.* (2003) [48].

The shortcoming of the Witelson partition is the artificial partitioning. The Witelson partition may dilute the power of detection if the anatomical difference occurs near the partition boundary. Alternative voxel-wise approaches that avoid predefined regions of interests (ROI) have begun to be used in structural autism studies. Vidal *et al.* (2003) used the tensor-based morphometry (TBM) to show reduced callosal thickness in the genu, midbody and splenium in autistic children [371]. Hoffmann *et al.* (2004) used a similar TBM to show curvature difference in the midbody [173]. Abell *et al.*

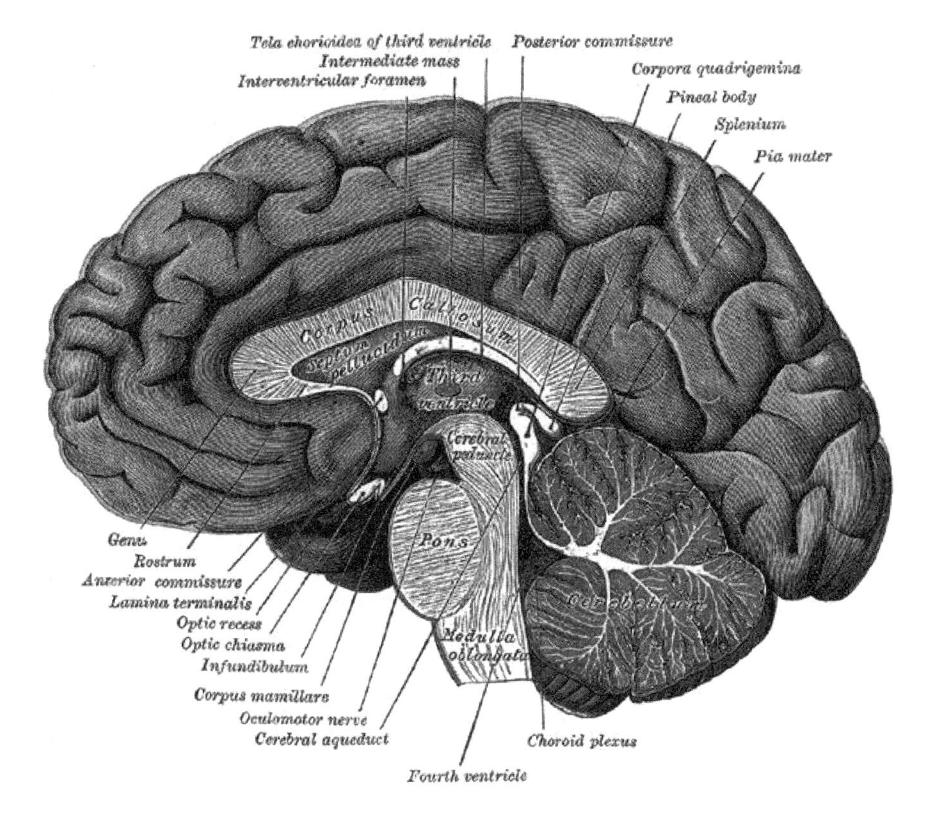

FIGURE 3.3
Midsagittal cross section of brain. Corpus callosum is the collection of neural fibers that connects the left and the right cerebral hemispheres. The posterior part of the corpus callosum is called the splenium; the anterior part is called the genu. Between them is the midbody. The illustration is from Gray's Anatomy [149].

(1999) used voxel-based morphometry (VBM) in high functioning autism to show decreased gray matter volume in the right paracingulate sulcus, the left occipito-temporal cortex and increased amygdala and periamygdaloid cortex [1].

The advantage of the VBM framework over the Witelson partition approach is that it is completely automated and does not require artificial partitioning of the corpus callosum that introduces undesirable bias. Further it is not restricted to *a priori* ROIs enabling us to perform the statistical analysis at each voxel level and to pinpoint the exact location of the anatomical differences within ROI even if there are no ROI size differences. Although VBM was originally developed for whole brain 3D morphometry, we will concentrate on the midsagittal cross sectional corpus callosum regions to be able to compare the result with the previous 2D Witelson partition studies [165, 241, 283].

3.3.1 White Matter Density Maps

Gender and handedness affect the corpus callosum anatomy [382, 383] so all the subjects used in the data set are right-handed males. Sixteen autistic subjects were recruited for this study from a list of individuals with a diagnosis of high functioning autism in the Madison and Milwaukee area maintained for research purposes by the Waisman center at the University of Wisconsin-Madison. Diagnoses were confirmed with the Autism Diagnostic Interview - Revised (ADI-R) or clinical interview administered by a trained and certified psychologist at the Waisman center. All participants met DSM-IV criteria for autism or Asperger's pervasive developmental disorder. Twelve healthy, typically developing males with no current or past psychological diagnoses served as a control group. The average age for the control subject is 17.1 ± 2.8 and the autistic subjects is 16.1 ± 4.5 which is in a compatible age range.

High resolution anatomical MRI scans were obtained using a 3-Tesla GE SIGNA (General Electric Medical Systems, Waukesha, WI) scanner with a quadrature head RF coil. A three-dimensional, spoiled gradient-echo (SPGR) pulse sequence was used to generate T1-weighted images. The imaging parameters were TR/TE 21/8 ms, flip angle 30°, 240 mm field of view, 256x192 in-plane acquisition matrix (interpolated on the scanner to 256x256), and 128 axial slices (1.2 mm thick) covering the whole brain. Then the midsagittal cross-sections of the white matter are segmented using the SPM-package (Figure 3.4).

3.3.2 Manipulating Density Maps

The segmented imaging data is stored in the directory `/CCdensity` as text files with the file extension `*.txt`. For instance, `CCautism12.txt` is the white matter density for the 12th autistic subject while `CCcontrol03.txt` is the white matter density for the 3rd control subject. This is the data set published in [69] where white matter density in the corpus callosum was analyzed. Figure 3.4 shows the images of white matter density at corpus callosum. MATLAB codes below are given in `http://brainimaging.waisman.wisc.edu/~chung/BIA`. Subject identifiers are usually given in numbers:

```
c1=[01 02 05 08 09]';
c2=[10 11 12 13 14 15 16]';

a1=[01 02 03 04 05 06 07 08 09]'
a2=[10 11 13 14 16 17 18]';
```

The file names are then given with subject identifiers as strings:

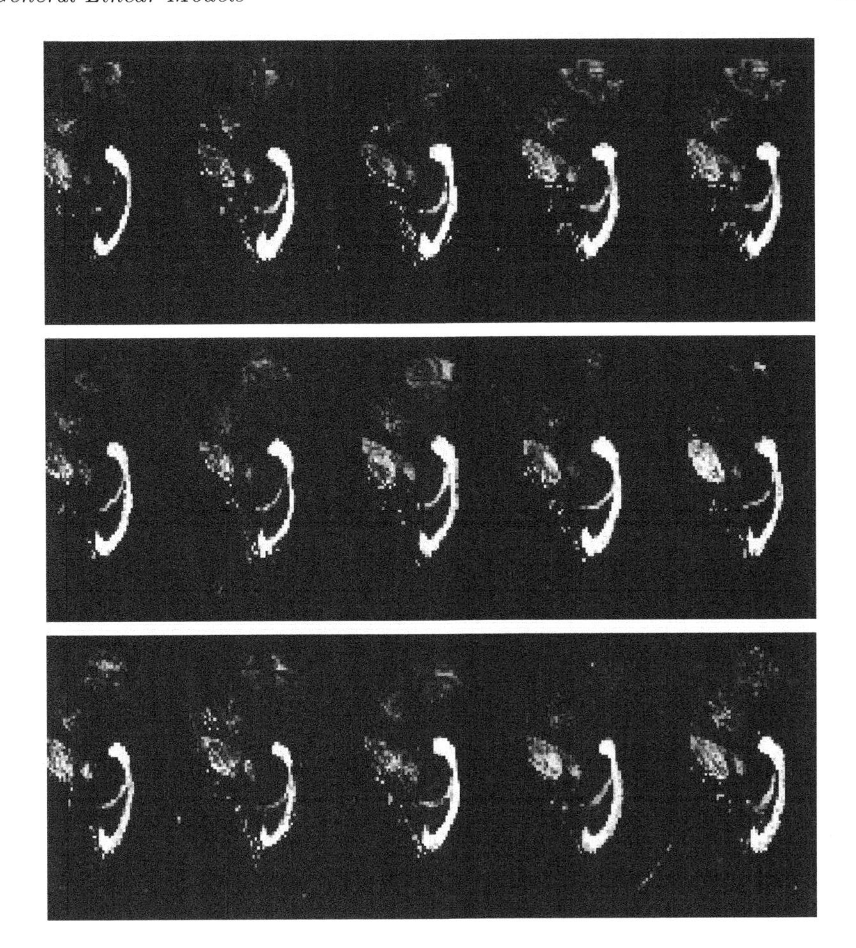

FIGURE 3.4
Midsagittal cross-section images of corpus callosum for 15 subjects. We are interested in determining if there is any group difference in white matter.

```
file_c1=strcat('CCcontrol0',int2str(c1),'.txt');
file_c2=strcat('CCcontrol',int2str(c2),'.txt');
file_c=[file_c1; file_c2]

file_a1=strcat('CCautism0',int2str(a1),'.txt');
file_a2=strcat('CCautism',int2str(a2),'.txt');
file_a=[file_a1; file_a2]
```

This produces the list of file names:

```
file_c =
```

```
CCcontrol01.txt
CCcontrol02.txt
CCcontrol05.txt
.
.
.
```

We need to concatenate in two different ways since `int2str(08)` does not produce 08 but 8. The total number of subjects in each group is

```
n_c=size(file_c,1)
n_a=size(file_a,1)

n_c =
    12

n_a =
    16
```

The dimension of images is 95×68. So we let

```
d1=95; d2=68;
```

Once we have the list of files, we load them sequentially into matrices:

```
density_a=zeros(n_c,d1,d2);
density_a=zeros(n_a,d1,d2);

for i=1:n_c
    temp=load(file_c(i,:));
    temp=reshape(temp,d1,d2);
    density_c(i,:,:)=temp;
end;

for i=1:n_a
    temp=load(file_a(i,:));
    temp=reshape(temp,d1,d2);
    density_a(i,:,:)=temp;
end;
```

All 28 subject images can be visualized using

```
img=[];
for i=1:n_c
    img = [img, squeeze(density_c(i,:,:))];
end;
figure; imagesc(img); colormap('bone'); colorbar;
```

```
img=[];
for i=1:n_a
    img = [img, squeeze(density_a(i,:,:))];
end;
figure; imagesc(img); colormap('bone'); colorbar
```

The resulting density maps are shown in Figure 3.4. The first 12 images are controls and the next 16 images are autistic subjects. The additional command `set(gcf,'Color','w')` will set the background of an image white. To save the image as a file, `print('-dtiff', '-r300', 'CC')` can be used.

3.3.3 Numerical Implementation

A simple example of GLM is the usual two-sample t-test setting. Given two groups, we are interested in testing the significance of group difference on tissue density. So we consider the following GLM:

$$\texttt{density}_i = \lambda_1 + \beta_1 \cdot \texttt{group}_i + \epsilon, \tag{3.17}$$

where the dummy variable `group` is 1 for autism and 0 for control. This is the case for $k = 1$, $z_{i1} = 1$ and $p = 1$. Another more complicated example is the case of liner regression for two groups, which can be combined into a single GLM:

$$\texttt{density}_i = \lambda_1 + \lambda_2 \cdot \texttt{age}_i + \beta_1 \cdot \texttt{group}_i \tag{3.18}$$

This is the case for $k = 2$ and $p = 1$ (Figure 3.7). We will implement (3.18) in `MATLAB` and estimate the parameters using the least squares method.

Once we loaded images, we set up a general linear model and estimate the parameters in a least squares fashion. The combined age information of both groups is stored in `age` which is a column vector of size 28×1. The design matrix `X` consists of a column of ones and vector `age`.

```
age_c = [15 18 18 16 15 13 18 15 21 17 16 23]'
age_a = [15 20 17 13 12 15 25 14 15 14 24 18 10 12 22 12]'
age=[age_c;age_a];

const=ones(n_c+n_a,1);
X=[const, age];
```

The density map for controls `density_c` is of size $12 \times 95 \times 68$. We reshape it into a 2-dimensional matrix of size $12 \times (95 \cdot 68)$.

```
p_c=reshape(density_c,12,d1*d2);
p_a=reshape(density_a,16,d1*d2);
p=[p_c; p_a];
```

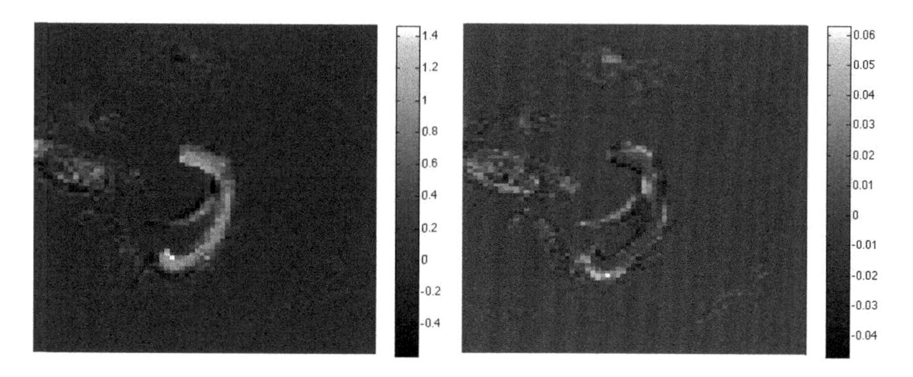

FIGURE 3.5
The least squares estimation of the parameters `lambda0` and `lambda1`.

Then we estimate the parameters in the linear model

$$\texttt{density} = \texttt{lambda0} + \texttt{lambda1} \cdot \texttt{age} \tag{3.19}$$

using the pseudoinverse operation `pinv` (Figure 3.5). The estimated parameter maps are given in Figure 3.5.

```
lambda=zeros(2,d1*d2);
for i=1:(d1*d2)
    lambda(:,i)=pinv(X)*p(:,i);
end

lambda0=reshape(lambda(1,:),d1,d2);
lambda1=reshape(lambda(2,:),d1,d2);

figure;imagesc(lambda0);colorbar; colormap('bone')
figure;imagesc(lambda1);colorbar; colormap('bone')
```

The sum of the residual at each pixel is given by

```
SSE0 = sum((p - (X*lambda)).^2);
SSE0=reshape(SSE0,d1,d2);
```

We add the additional term `group` in (3.19):

$$\texttt{density} = \texttt{lambda0} + \texttt{lambda1} \cdot \texttt{age} + \texttt{lambda2} \cdot \texttt{group}.$$

For this, we have another column in the design matrix `X`:

```
group = [zeros(12,1); ones(16,1)]
X=[const, age, group];
```

The design matrix now looks like

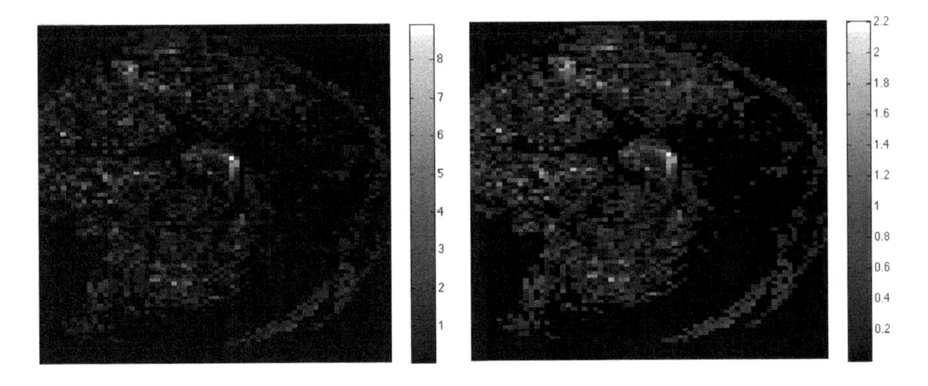

FIGURE 3.6
The F-statistic of testing the significance of the term `group` and the corresponding p-value in $\log_{10}$ scale. The signal is detected in the genu of the corpus callosum. Since we didn't smooth the density maps, we are getting a lot of noise.

```
...
1     16     0
1     23     0
1     15     1
1     20     1
...
```

The parameters are again estimated by the pseudoinverse operation:

```
lambda=zeros(3, d1*d2);
for i=1:(d1*d2)
    lambda(:,i) = pinv(X)*p(:,i);
end
```

The total sum of squared error is then

```
SSE1 = sum((p - (X*lambda)).^2);
SSE1=reshape(SSE1,d1,d2);
```

The statistic of testing the significance of `group` is based on the ratio of the sum of squared errors between `SSE0` and `SSE1` (Figure 3.6).

```
F=25*(SSE0-SSE1)./SSE0;
imagesc(F);colorbar; colormap('bone')
pvalue=1 - fcdf(F,1,25);
figure; imagesc(-log10(pvalue)); colormap('bone')
```

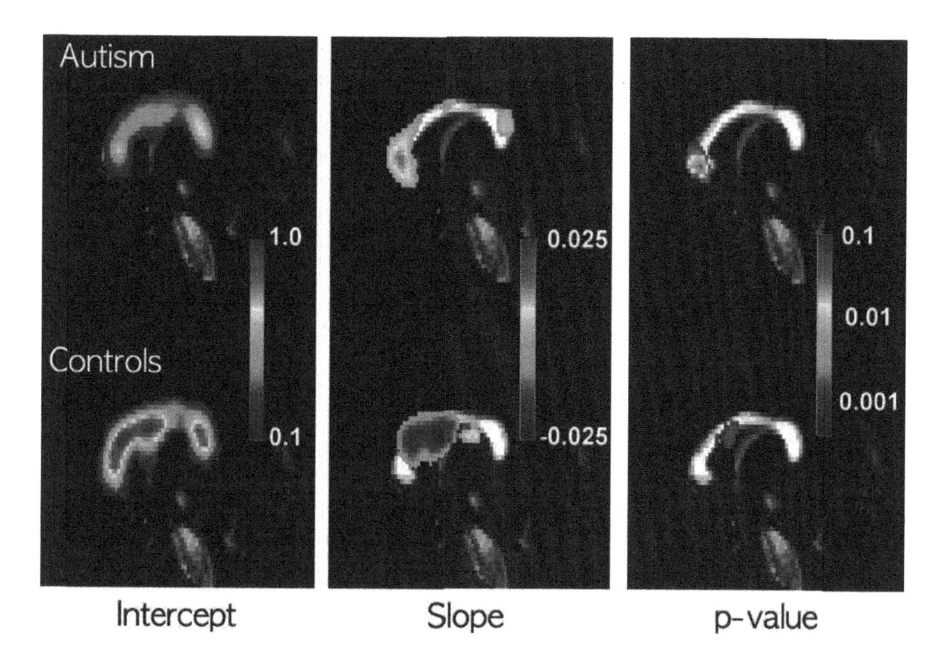

FIGURE 3.7
White matter density is linearly fitted over age and group at each voxel. The intercept and slopes of the linear regression is for each group. The autistic group shows lower white matter density compared to the control at lower age but gains white matter over time while the control group shows decreasing white matter density with age [69]. (See color insert.)

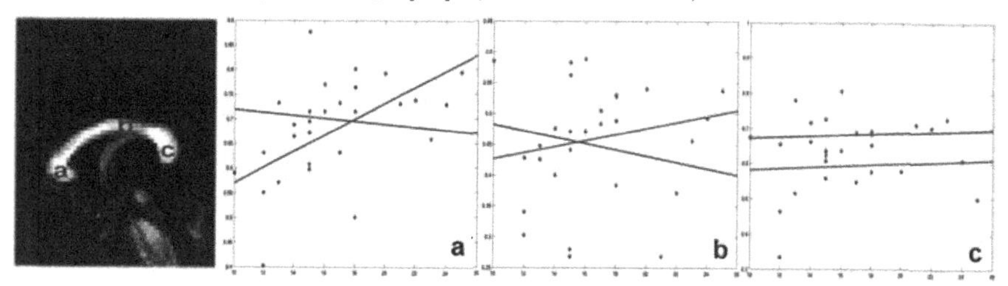

FIGURE 3.8
Linear regression fit for each group (red=autism, blue = control) (a) Genu (b) Midbody and (c) splenium of the corpus callosum. (See color insert.)

3.4 Testing Interactions

Let us cover a few other issues that are left out in the previous sections. As shown in Figures 3.7 and 3.8, each group has different growth rate, the change of white matter density over age. The question is if the growth rate difference is statistically significant. This can be tested within the GLM framework. Note the GLM incorporates two-way ANOVA and ANCOVA. Assume there are total n subjects. Let $\texttt{density}_i$ be the white matter density for the i-th subject. Let $\texttt{age}_i$ be the age of the i-th subject. Let $\texttt{group}_i$ be the group variable of the i-th subject taking value 1 for autistic and 0 for controls. Then we have the following GLM:

$$\texttt{density}_i = \beta_0 + \beta_1 \texttt{age}_i + \beta_2 \texttt{group}_i + \beta_3 \texttt{age}_i \cdot \texttt{group}_i + \epsilon_i. \tag{3.20}$$

From (3.20), we have two separate models for each group. For controls ($\texttt{group}_i = 0$),

$$\texttt{density}_i = \beta_0 + \beta_1 \texttt{age}_i + \epsilon_i.$$

For autistic subjects ($\texttt{group}_i = 1$), we have

$$\texttt{density}_i = (\beta_0 + \beta_2) + (\beta_1 + \beta_3)\texttt{age}_i + \epsilon_i.$$

Testing the equality of the growth rate is equivalent to testing

$$H_0 : \beta_1 = \beta_1 + \beta_3$$

or equivalently

$$H_0 : \beta_3 = 0.$$

We need to test for the significance of the interaction term β_3 in the model.

The fit of model is measured by the sum of squared errors (SSE). Let $\texttt{SSE}_0$ and $\texttt{SSE}_1$ be the SSE for the reduced (when $\beta_3 = 0$)and the full model (when $\beta_3 \neq 0$) respectively. Then

$$\texttt{SSE}_0 = \sum_{i=1}^{m+n} (\texttt{density}_i - \widehat{\beta}_0 + \widehat{\beta}_1 \texttt{age}_i - \widehat{\beta}_2 \texttt{group}_i)^2,$$

where $\widehat{\beta}_i$ are the estimated regression coefficients from the reduced model. Similarly for the full model,

$$\texttt{SSE}_1 = \sum_{i=1}^{m+n} (\texttt{density}_i - \widehat{\gamma}_0 + \widehat{\gamma}_1 \texttt{age}_i - \widehat{\gamma}_2 \texttt{group}_i - \widehat{\gamma}_3 \texttt{age}_i \cdot \texttt{group}_i)^2,$$

where $\widehat{\gamma}_i$ are the estimated regression coefficients from the full model. Then the F-statistic is given by the ratio of SSE:

$$F = \frac{(\texttt{SSE}_0 - \texttt{SSE}_1)/1}{\texttt{SSE}_0/(n-1-3)} \sim F_{1,n-1-3}.$$

4

Gaussian Kernel Smoothing

Image acquisition and segmentation are likely to introduce noise. Further image processing such as image registration and parameterization can introduce additional noise. It is thus imperative to reduce noise measurements and boost signal. In order to increase the signal-to-noise ratio (SNR) and smoothness of data required for the subsequent random field theory based statistical inference, some type of smoothing is necessary [203]. Among many image smoothing methods, *Gaussian kernel smoothing* has emerged as a de facto smoothing technique among brain imaging researchers due to its simplicity in numerical implementation [212, 275]. Gaussian kernel smoothing also increases statistical sensitivity and statistical power as well as Gausianness. Gaussian kernel smoothing can be viewed as weighted averaging of voxel values. Then from the central limit theorem, the weighted average should be more Gaussian.

4.1 Kernel Smoothing

Kernel smoothing is the most widely used image smoothing technique in brain image analysis. Consider the integral transform

$$Y(t) = \int K(t,s)X(s)\,ds,$$

where K is the *kernel* of the integral. Given the input signal X, Y represents the output signal. The smoothness of the output depends on the smoothness of the kernel. We assume the kernel to be unimodal and isotropic. When the kernel is isotropic, it has radial symmetry and should be invariant under rotation. So it has the form

$$K(t,s) = f(\|t-s\|)$$

for some smooth function f. Since the kernel only depends on the difference of the arguments, with the abuse of notation, we can simply write K as

$$K(t,s) = K(t-s).$$

We may further assume the kernel is normalized such that

$$\int K(t)\,dt = 1.$$

With this specific form of kernel K, (4.1) can be written as

$$Y(t) = K * X(t) = \int K(t-s)X(s)\,ds.$$

The integral transform (4.1) is called *kernel smoothing*. We may assume further that K is dependent on *bandwidth* σ. The bandwidth will determine the spread of kernel weights such that

$$\lim_{\sigma \to \infty} K(t,s;\sigma) = 1 \quad (4.1)$$

$$\lim_{\sigma \to 0} K(t,s;\sigma) = \delta(t-s), \quad (4.2)$$

where δ is the Dirac-delta function [37, 106]. The Dirac-delta function is a special case of *generalized functions* [133, 339]. The Dirac-delta function is usually traditionally defined as

$$\delta(t) = 0 \text{ if } t \neq 0,\ \delta(t)\,dt = 1. \quad (4.3)$$

This is also referred to as the *impulse function* in literature. Figure 4.1 illustrates 1D Gaussian kernel smoothing in a simple example.

4.2 Gaussian Kernel Smoothing

All brain images are inherently noisy due to errors associated with image acquisition. Compounding the image acquisition errors, there are errors caused by image registration and segmentation. So it is necessary to smooth out the segmented images before any statistical analysis is performed to boost statistical power. Among many possible kernel smoothing methods [212, 275], *Gaussian kernel smoothing* has emerged as a de facto smoothing technique in brain imaging.

The *Gaussian kernel* in 1D is defined as

$$K(t) = \frac{1}{\sqrt{2\pi}} e^{t^2/2}.$$

Let's scale the Gaussian kernel K by the bandwidth σ:

$$K_\sigma(t) = \frac{1}{\sigma} K\Big(\frac{t}{\sigma}\Big).$$

This is the density function of the normal distribution with mean 0 and variance σ^2.

The n-dimensional isotropic Gaussian kernel is defined as the product of n

1D kernels. Let $t = (t_1, \cdots, t_n)' \in \mathbb{R}^n$. Then the n-dimensional kernel is given by

$$\begin{aligned} K_\sigma(t) &= K_\sigma(t_1)K_\sigma(t_2)\cdots K_\sigma(t_n) \\ &= \frac{1}{(2\pi)^{n/2}\sigma^n} \exp\Big(\frac{1}{2\sigma^2}\sum_{i=1}^n t_i^2\Big). \end{aligned}$$

Subsequently, n-dimensional isotropic Gaussian kernel smoothing $K_\sigma * X$ can be done by applying 1-dimensional smoothing n times in each direction by factorizing the kernel as

$$\begin{aligned} K_\sigma * X(t) &= \int K_\sigma(t-s)X(s)\,ds \\ &= \int K_\sigma(t_1-s_1)K_\sigma(t_2-s_2)\cdots K_\sigma(t_n-s_n)X(s)\,ds \\ &= \int K_\sigma(t_1-s_1)\cdots K_\sigma(t_{n-1}-s_{n-1})\,ds_1\cdots ds_{n-1} \\ &\quad \times \int K_\sigma(t_n-s_n)X(s)\,ds_n. \end{aligned}$$

Note that $K_\sigma * X$ is the scale-space representation of image X first introduced in [384]. Each $K_\sigma * X$ for different values of σ produces a blurred copy of its original. The resulting scale-space representation from coarse to fine resolution can be used in multiscale approaches such as hierarchical searches and image segmentation [230, 287, 288, 394, 393].

4.2.1 Fullwidth at Half Maximum

The *bandwidth* σ defines the spread of kernel. In brain imaging, the spread of the kernel is usually measured in terms of the *full width at the half maximum* (FWHM) of Gaussian kernel K_σ, which is given by $2\sqrt{2\ln 2}\sigma$. This can be easily seen by representing the kernel using the radius $r^2 = x_1^2 + \cdots x_n^2$:

$$K_\sigma(r) = \frac{1}{(2\pi)^{n/2}\sigma^n} \exp\Big(\frac{r^2}{2\sigma^2}\Big).$$

The peak of the kernel is $K_\sigma(0)$. Then the *full width at half maximum* (FWHM) of the peak is given by

$$\text{FWHM} = 2\sqrt{2\ln 2}\sigma.$$

The FWHM increases linearly as σ increases in Euclidean space.

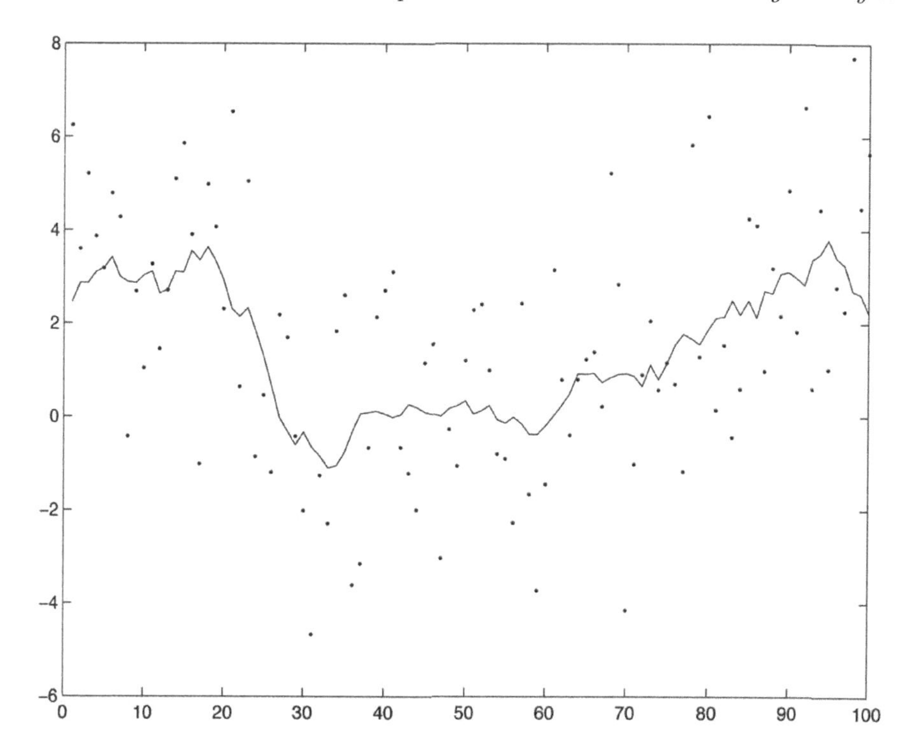

FIGURE 4.1
Simulated noise $N(0, 2^2)$ is added to signal $\mu(t) = (t-50)^2/500$. Gaussian kernel smoothing is applied with bandwidth 10.

4.3 Numerical Implementation

We present 1D and 2D Gaussian kernel smoothing here as illustrations.

4.3.1 Smoothing Scalar Functions

Consider a noise 1D functional signal

$$Y(t) = (t-50)^2/500 + \epsilon(t),$$

where $\epsilon(t)$ is distributed as $N(0, 2^2)$ at each point t. At each integer point between 0 and 100, we have noisy measurement (Figure 4.1). We are interested in smoothing out the noisy measurement and estimating the underlying smooth signal. Using `inline` function, we define kernel `K` and kernel smoothing is performed by convolution.

```
K=inline('exp(-(x.^2)/2/sigma^2)');
```

```
dx= -5:5;

sum(K(10,dx))
weight=K(10,dx)/sum(K(10,dx))
sum(weight)

t=1:100;
mu=(t-50).^2/500;
noise= normrnd(0, 2, 1,100);
Y=mu + noise
figure; plot(Y, '.');

smooth=conv(Y,weight,'same');
hold on; plot(smooth, 'r');
```

4.3.2 Smoothing Image Slices

Suppose that noisy observation is obtained at each grid point $(\frac{i}{100}, \frac{j}{100}) \in [0,1]^2$. This basically forms a 2D image. The signal μ is assumed to be

$$\mu(t_1, t_2) = \cos(10t_1) + \sin(10t_2).$$

The signal is contaminated by Gaussian white noise $\epsilon \sim N(0, 0.4^2)$. Using `meshgrid` command, we generate 101×101 2D grid points. The grid coordinates are stored in `px` and `py`, which are 101×101 matrix. Figure 4.2 shows the simulated image $Y = \mu + \epsilon$.

```
[px,py] = meshgrid([0:0.01:1]);

>>px(1:3,1:3)

ans =
         0    0.0100    0.0200
         0    0.0100    0.0200
         0    0.0100    0.0200
         ...

>>py(1:3,1:3)

ans =
         0         0         0
    0.0100    0.0100    0.0100
    0.0200    0.0200    0.0200
    ...

mu=cos(10*px)+sin(8*py);
```

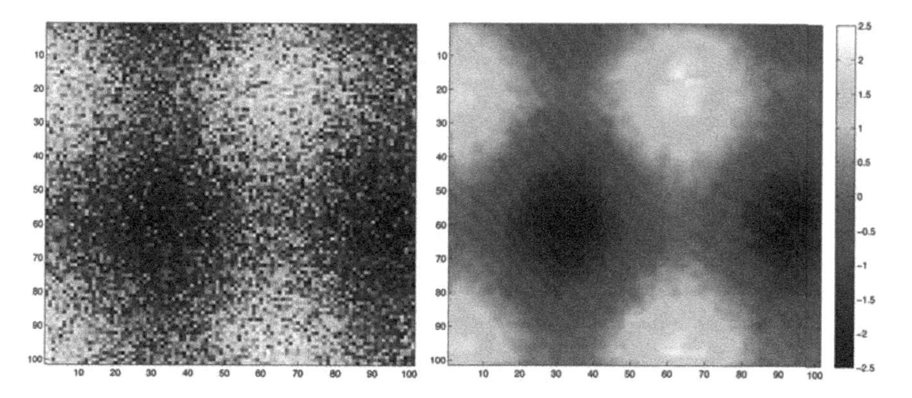

FIGURE 4.2
Left: simulated noise $N(0, 0.4^2)$ is added to signal $\mu(t_1, t_2) = \cos(10t_1) + \sin(10t_2)$. Right: Gaussian kernel smoothing is applied with bandwidth 1.

```
e=normrnd(0,0.4,101,101);
Y=mu+e;
figure; imagesc(Y); colorbar;
```

The simulated image Y is fairly noisy. Gaussian kernel smoothing with bandwidth 1 is applied to Y to increase smoothness. The Gaussian kernel is constructed using `inline` function.

```
>>K=inline('exp(-(x.^2+y.^2)/2/sig^2)')

K =
     Inline function:
     K(sig,x,y) = exp(-(x.^2+y.^2)/2/sig^2)
```

The inline function `K` has three arguments: two coordinates `x` and `y` and bandwidth `sigma`. The kernel is then constructed discretely using 5×5 grid and then renormalizing it such that the kernel sums up to 1. For small `sig`, the discrete kernel weights are focused in the center of the 5×5 window while for large `sig`, the kernel weights are more dispersed. Error will increase if a smaller window is used. Smoothing is done by the convolution `conv2`. The smoothed image is in Figure 4.2.

```
[dx,dy]=meshgrid([-2:2]);
>>weight=K(0.5,dx,dy)/sum(sum(K(0.5,dx,dy)))

weight =
0.0000 0.0000 0.0002 0.0000 0.0000
0.0000 0.0113 0.0837 0.0113 0.0000
0.0002 0.0837 0.6187 0.0837 0.0002
```

```
0.0000 0.0113 0.0837 0.0113 0.0000
0.0000 0.0000 0.0002 0.0000 0.0000

>>weight=K(1,dx,dy)/sum(sum(K(1,dx,dy)))

weight =
0.0030 0.0133 0.0219 0.0133 0.0030
0.0133 0.0596 0.0983 0.0596 0.0133
0.0219 0.0983 0.1621 0.0983 0.0219
0.0133 0.0596 0.0983 0.0596 0.0133
0.0030 0.0133 0.0219 0.0133 0.0030

Ysmooth=conv2(Y,weight,'same');
figure; imagesc(Ysmooth); colorbar;
```

4.4 Case Study: Smoothing of DWI Stroke Lesions

We will use the same stroke lesion data introduced in Section 2.2. One slice of sample T1 MRI is stored as bitmap image format as `DWI.bmp`, which can be read using the built-in `MATLAB` function `imread`. Stroke lesions in DWI are segmented by a human expert and stored as
`DWI-segmentation.bmp`.

```
DWimg= 'DWI';
DWI=imread(DWimg,'bmp');

segmentation = 'DWI-segmentation';
f=imread(segmentation,'bmp');
```

Once we read the image files, we are interested in smoothing the binary segmentation using the Gaussian kernel smoothing procedure. Bitmap files consist of a matrix of three columns representing red, green and blue colors. So we simply take out the first column and normalize it to have pixel values 0 or 1. The normalized segmentation is then smoothed using Gaussian kernel smoothing, which is implemented as `gaussblur2D.m`. The first argument to function `gaussblur2D` is the 2D array to smooth out and the second argument is the bandwidth given in terms of full width at the half maximum (FWHM) of the kernel.

```
binarize = double(f(:,:,1)/255);
fblur= gaussblur2D(binarize,10);

n=size(fblur);
```

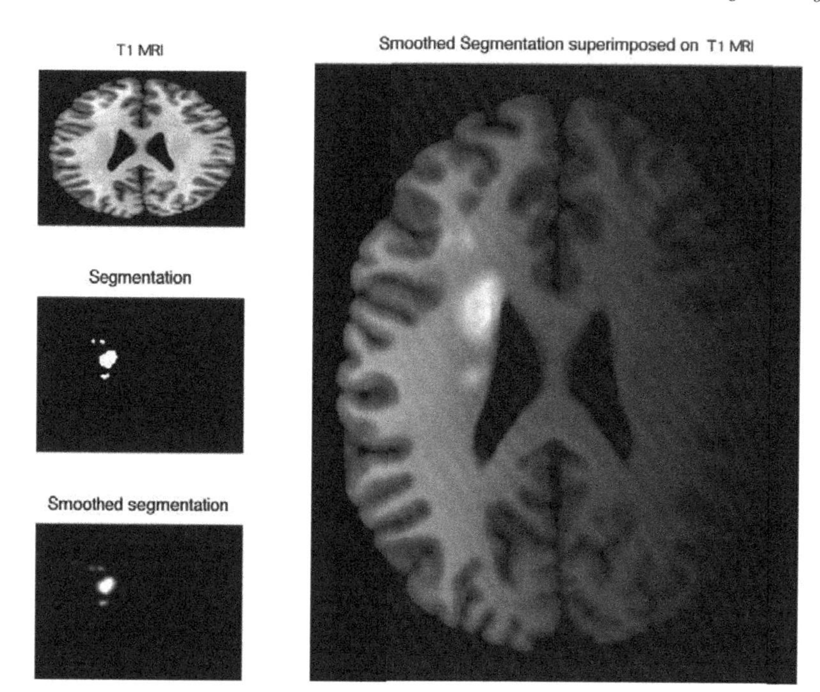

FIGURE 4.3
T1 MRI and the segmentation of ischemic regions from diffusion-weighted image (DWI). The segmentation is done manually using DWI and smoothed with Gaussian kernel. Smoothing is necessary before any group analysis in reducing the regions of false positives.

```
fblurout = zeros(n(1), n(2), 3);
fblurout(:,:,1)=fblur;
fblurout(:,:,3)=fblur;
```

For visualization, we use `subplot` consisting of 3×3 image grid. Four different images are displayed by partitioning the 3×3 image grid. We have superimposed the smoothed segmentation on top of MRI by specifying the transparency of each pixel using `set` command with `'AlphaData'` option. Figure 4.3 is obtained by running the codes below.

```
figure; subplot(3,3,1);
imagesc(DWI);
axis off
title('DWI')

subplot(3,3,4);
imagesc(f);
```

```
axis off
title('Segmentation')

subplot(3,3,7); imagesc(fblurout)
axis off
title('Smoothed segmentation')

subplot(3,3,[2:3, 8:9]);
imagesc(DWI);
axis off
hold on

imgAlpha = repmat(0:1/n(2):1-1/n(2),n(1),1);
img = imagesc(fblurout);
set(img,'AlphaData',imgAlpha);
title('Smoothed Segmentation superimposed on DWI')
```

4.5 Effective FWHM

Once we smooth the image with kernel K_σ, the smoothness of the image changes. The smoothness of image after smoothing can be measured in terms of *effective FWHM*. The unbiased estimator of eFWHM is first introduced in [390], where it is estimated along edges in the lattice. Label the two voxels at the end of an edge by 1 and 2. Let the length of edge be Δx. Suppose there are n images in a group. Let r_{ij} denote the residual for the i-th image at voxel j. The normalized residuals at the two ends are

$$u_{ij} = \frac{r_{ij}}{\sqrt{\sum_{i=1}^{n} r_{ij}^2}}.$$

The roughness of the noise is defined as the standard deviation of the derivative of the noise divided by the standard deviation of the noise itself. Let

$$\Delta u = \sqrt{\sum_{i=1}^{n} (u_{i1} - u_{i2})^2}.$$

Then an unbiased estimator of the roughness is given by

$$\lambda = \frac{\Delta u}{\Delta x}.$$

Then the eFWHM along the edge is given by

$$e\text{FWHM} = \frac{\sqrt{4ln2}}{\lambda}.$$

The effective-FWHM is often used in the random field theory based on multiple comparisons correction through the `fMRISTAT` package [396].

4.6 Checking Gaussianness

In this section, we will explain various methods for checking Gaussianness of imaging measurements. Since many statistical models assume normality, checking if imaging data follows normality is fairly important. Also image smoothing not only increases the smoothness of underlying image intensity values, but it also increases the normality of data. In paper submissions to various imaging journals, this is an often asked question by reviewers.

4.6.1 Quantile-Quantile Plots

Checking the normality of imaging data is fairly important when the underlying statistical model assumes the normality of data. But how do we know the data will follow normality? This is easily checked visually using the *quantile-quantile (QQ) plot* first introduced by Wilk and Gnanadesikan [381]. The QQ-plot is a graphical method for comparing two distributions by plotting their quantiles against each other. A special case of QQ-plot is the *normal probability plot* where the quantiles from an empirical distribution are plotted on the vertical axis while the theoretical quantiles from a Gaussian distribution are plotted on the horizontal axis. It is used to check graphically if the empirical distribution follows the theoretical Gaussian distribution. If the data follows Gaussian, the normal probability plot should be close to a straight line.

4.6.2 Quantiles

Definition 1 *The quantile point q for random variable X is a point that satisfies*

$$P(X \leq q) = F_X(q) = p,$$

where F_X is the cumulative distribution function (CDF) of X.

Assuming we can find the inverse of CDF, the quantile is given by

$$q = F_X^{-1}(p).$$

This function is mainly referred to as a quantile function. The quantile-quantile (QQ) plot of two random variables X and Y is then defined to be a parametric curve $\mathcal{C}(p)$ parameterized by $p \in [0, 1]$:

$$\mathcal{C}(p) = \left(F_X^{-1}(p), F_Y^{-1}(p)\right).$$

4.6.3 Empirical Distribution

The CDF $F_X(q)$ measures the proportion of random variable X less than given value q. So by counting the number of measurements less than q, we can empirically estimate the CDF. Let $X_1, \cdots, X_n$ be a random sample of size n. Then order them in increasing order:

$$\min(X_1, \cdots, X_n) = X_{(1)} \leq X_{(2)} \leq \cdots \leq X_{(n)} = \max(X_1, \cdots, X_n).$$

Suppose $X_{(j)} \leq q < X_{(j+1)}$. This implies that there are j samples that are smaller than q. So we approximate the CDF as

$$\widehat{F_X}(q) = \frac{j}{n}.$$

The j/n-th *sample quantile* is then $X_{(j)}$. Some authors define the sample quantile as the $(j-0.5)/n$-th sample quantile. The factor 0.5 is introduced to account for the discretization error.

In numerical implementation, it is easier to implement the empirical distribution using the *step function* $\mathcal{I}_q(x)$ which is implemented as $\mathcal{I}_q(x) = 1$ if $x \leq q$ and $\mathcal{I}_q(x) = 0$ if $x > q$. Then the CDF is estimated as

$$\widehat{F_X}(q) = \frac{1}{n}\sum_{i=1}^{n} \mathcal{I}_q(X_i),$$

where $\mathcal{I}_q(X_i)$ counts if X_i is less than q. A different possibly more sophisticated estimation can be found in [128].

4.6.4 Normal Probability Plots

The QQ-plot for two Gaussian distributions is a straight line. This can be easily proved as follows. Suppose

$$X \sim N(\mu_1, \sigma_1^2) \text{ and } Y \sim N(\mu_2, \sigma_2^2).$$

Let $Z \sim N(0,1)$ and $\Phi(z) = P(Z \leq z)$, the CDF of the standard normal distribution. If we denote q_1 and q_2 to be the p-th quantiles for X and Y respectively, we have

$$p = P(X \leq q_1) = P\Big(\frac{X-\mu_1}{\sigma_1} \leq \frac{q_1-\mu_1}{\sigma_1}\Big) = \Phi\Big(\frac{q-\mu_1}{\sigma_1}\Big).$$

Hence the parameterized QQ-plot is given by

$$\begin{aligned} q_1(p) &= \mu_1 + \sigma_1 \Phi^{-1}(p), \\ q_2(p) &= \mu_2 + \sigma_2 \Phi^{-1}(p). \end{aligned}$$

This is a parametric form of the QQ-plot. The QQ-plot without the parameter p is then trivially given by

$$\frac{q_1-\mu_1}{\sigma_1} = \frac{q_2-\mu_2}{\sigma_2},$$

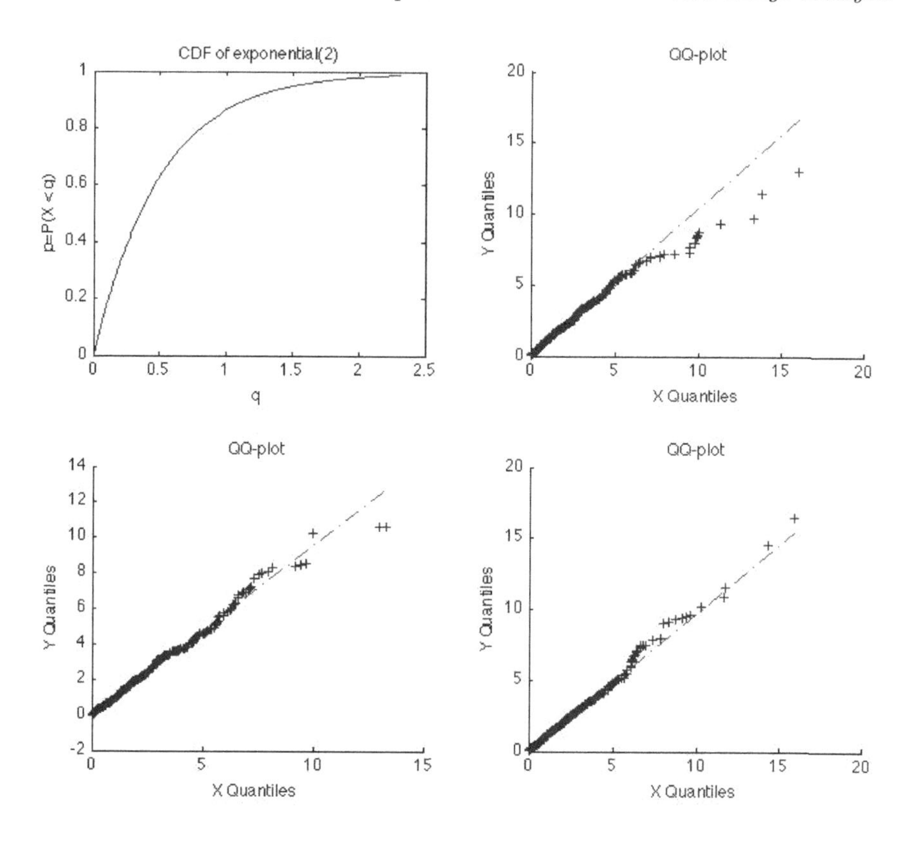

FIGURE 4.4
Top left: CDF of a exponential random variable with parameter 2. QQ-plots are obtained by randomly generating 500 exponential random variables. Each time, we may have slightly different QQ-plots at extreme values. In interpreting QQ-plots, we should not focus attention on extreme outlying points.

the equation for a line. This shows the QQ-plot of two normal distributions is a straight line. This idea can be used to determine the normality of a given sample. We can check how closely the sample quantiles correspond to the normal distribution by plotting the QQ-plot of the sample quantiles vs. the corresponding quantiles of a normal distribution. In normal probability plot, we plot the QQ-plot of the sample against the standard normal distribution $N(0,1)$.

4.6.5 `MATLAB` Implementation

As an example, consider the problem of plotting the quantile function for the exponential random variable X with parameter $\lambda = 2$, i.e. $X \sim exp(2)$. It can be shown that

$$F_X^{-1}(p) = -\frac{1}{2}\ln(1-p).$$

The actual CDF can be plotted using the `inline` function, which can define a function quickly without writing a separate function file.

```
p=[1:99]/100;
q=inline('-log(1-p)/2');
plot(q(p),p);
xlabel('x')
ylabel('P(X < q)
```

When we generate the QQ plot of $X \sim exp(2)$ and $Y \sim exp(2)$, since they are identical distributions, you expect the straight line $y = x$ as the QQ plot (Figure 4.4). This can be done by the exponential random number generator `exprnd`.

```
X=exprnd(2, 500,1);
Y=exprnd(2, 500,1);
subplot(2,2,2); qqplot(X,Y)
title('QQ-plot')
```

4.7 Effect of Gaussianness on Kernel Smoothing

Gaussian kernel smoothing can increase the Gaussianness of multiple images. As an example, consider a key shaped binary image `toy-key.tif` (Figure 4.5). Gaussian noise with large variance is added to the binary image to mask the signal.

```
signal= imread('toy-key.tif');
signal=(double(signal)-219)/36;
figure; subplot(2,2,1); imagesc(signal);
colormap('bone'); colorbar
title('Signal')

noise= normrnd(0, 5, 596, 368);
f = signal + noise;
subplot(2,2,2); imagesc(f); colormap('bone'); colorbar
title('Signal + N(0,5)')
```

To recover the signal, we performed Gaussian kernel smoothing with the bandwidths 1 and 10. For a sufficiently large bandwidth of 10, we are able to recover the underlying key shaped object. If properly used, Gaussian kernel smoothing can recover the underlying signal pretty well.

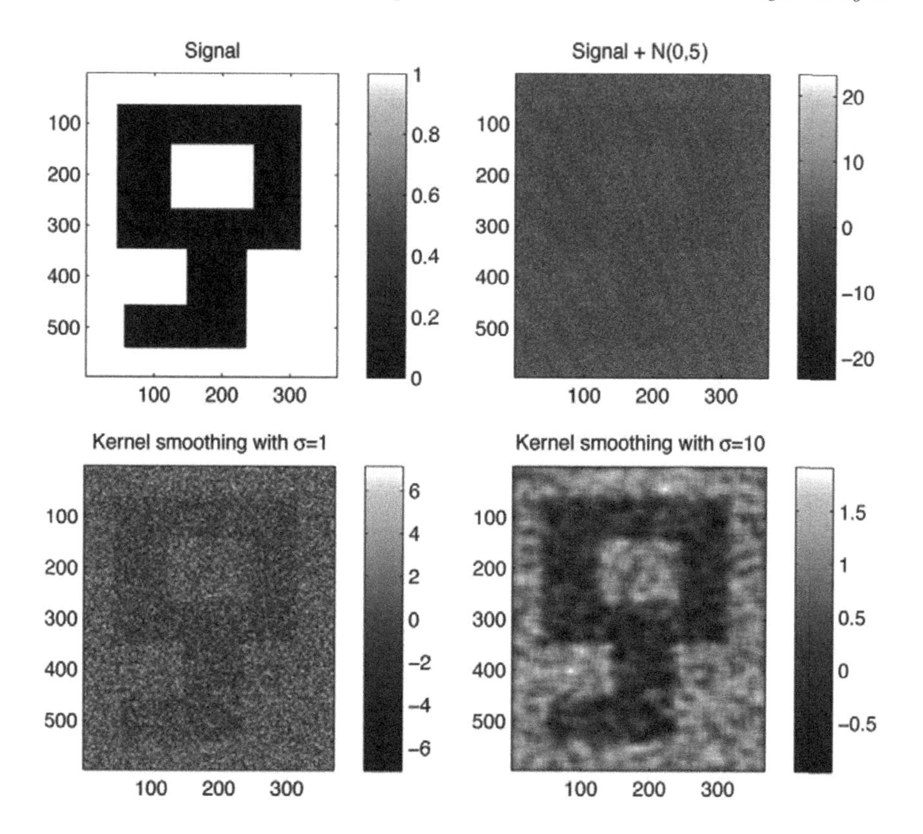

FIGURE 4.5
If properly used, Gaussian kernel smoothing can be used in detecting a hidden signal. The key shaped signal is contaminated with $N(0,5)$. Since the noise variability is so huge, the underlying signal is not clearly visible. By performing kernel smoothing with increasing bandwidths, it is possible to recover the original signal.

```
K=inline('exp(-(x.^2+y.^2)/2/sigma^2)');
[dx,dy]=meshgrid([-10:10]);

sigma=100;
weight=K(sigma,dx,dy)/sum(sum(K(sigma,dx,dy)));

weight=K(1,dx,dy)/sum(sum(K(1,dx,dy)));
smooth=conv2(f,weight,'same');
subplot(2,2,3); imagesc(smooth); colormap('bone'); colorbar
title('Kernel smoothing with \sigma=1')

weight=K(10,dx,dy)/sum(sum(K(10,dx,dy)));
```

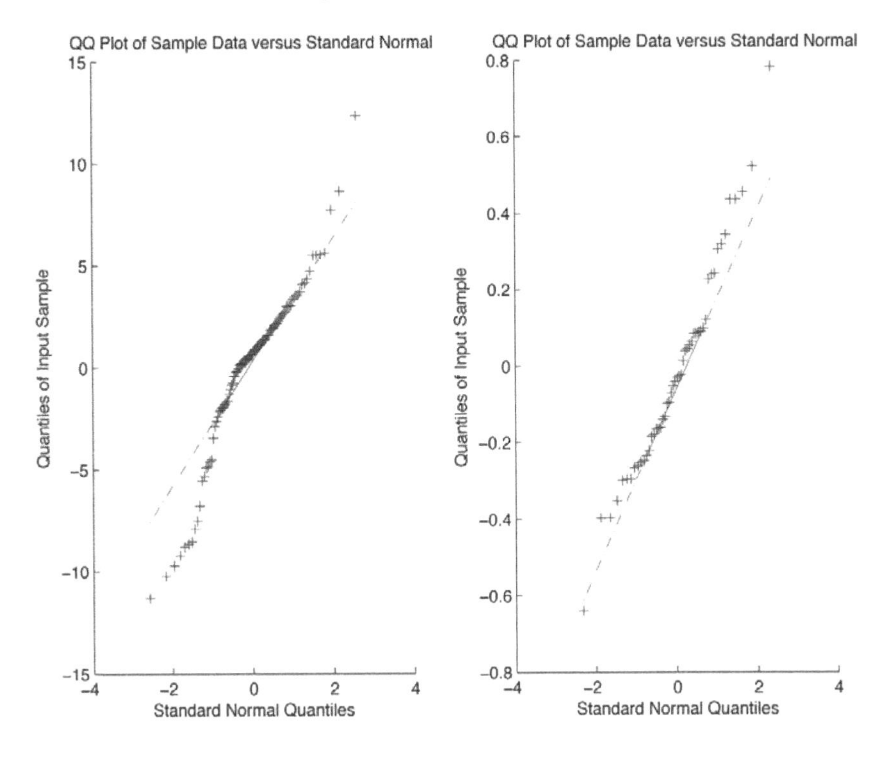

FIGURE 4.6
Image intensity value at a particular pixel. After Gaussian kernel smoothing with bandwidth 100, Gaussianness has increased as expected.

```
smooth=conv2(f,weight,'same');
subplot(2,2,4); imagesc(smooth); colormap('bone'); colorbar
title('Kernel smoothing with \sigma=10')
```

To show kernel smoothing can increase Gaussianness, we selected a pixel at $(314, 150)$ which is at the edge of the binary object. For 50 measurements at the pixel, we plotted the QQ-plot. As shown in the left in Figure 4.6, the pixel values are not showing Gaussianness. However, after smoothing with bandwidth 100, the Gaussianness has been increased.

```
for i=1:50
    noise= normrnd(0, 5, 596, 368);
    weight=K(100,dx,dy)/sum(sum(K(100,dx,dy)));
    f=signal+noise;
    smooth=conv2(f,weight,'same');
    pixelvalue(i)=f(314, 150);
    pixelvalues(i)=smooth(314, 150);
end;
```

```
figure; subplot(1,2,1); qqplot(pixelvalue)
subplot(1,2,2); qqplot(pixelvalues)
```

5

Random Fields Theory

So far we have studied general linear models (GLM) that are constructed at each voxel. In this chapter, we explore the *multiple comparisons* issue that is necessary to properly threshold statistical maps for the whole image. The multiple comparisons are crucial in determining overall statistical significance in correlated test statistics over the whole brain. In practice, t- or F-statistics in adjacent voxels are correlated. So there is the problem of multiple comparisons, which we have simply neglected up to now. For multiple comparisons that account for spatially correlated test statistics, various methods are proposed: Bonferroni correction, random field theory [388, 394], false discovery rates [35, 36, 134] and permutation tests [261].

In brain imaging studies, it is necessary to model measurements at each voxel as a random field. For instance, in the deformation-based morphometry (DBM), deformation fields are usually modeled as continuous random fields [81]. The generalization of a continuous stochastic process defined in $\mathbb{R}$ to a higher dimensional abstract space is called a *random field*. For an introduction to random fields, see [6], [109] and [399]. In the random field theory as used in [388, 394], measurement Y at voxel position $x \in \mathcal{M}$ is modeled as

$$Y(x) = \mu(x) + \epsilon(x),$$

where μ is the unknown signal to be estimated and ϵ is the measurement error. The measurement error at each fixed x can be modeled as a random variable. Then the collection of random variables $\{\epsilon(x) : x \in \mathcal{M}\}$ is called a *stochastic process* or *random field*. The more precise measure-theoretic definition can be found in [6]. Random field modeling can be done beyond the usual Euclidean space to curved cortical and subcortical manifolds [194, 83]. This chapter will closely follow the description given in [65].

5.1 Random Fields

It is necessary to define various mathematical terminologies at least at the conceptual level.

Definition 2 *Given a probability space, a random field $T(x)$ defined in $\mathbb{R}^n$ is* *a function such that for every fixed $x \in \mathbb{R}^n$, $T(x)$ is a random variable on the probability space.*

Definition 3 *The covariance function $R(x, y)$ of a random field T is defined as*

$$R(x, y) = \mathbb{E}\big[T(x) - \mathbb{E}T(x)\big]\big[T(y) - \mathbb{E}T(y)\big].$$

Consider a random field T. If the joint distribution

$$F_{x_1,\cdots,x_m}(z_1, \cdots, z_m) = P\big[T(x_1) \leq z_1, \cdots, T(x_m) \leq z_m\big]$$

is invariant under the translation

$$(x_1, \cdots, x_m) \to (x_1 + \tau, \cdots, x_m + \tau),$$

T is said to be stationary or homogeneous. For a stationary random field T, we can show

$$\mathbb{E}T(x) = \mathbb{E}T(0)$$

and subsequently

$$R(x, y) = f(x - y)$$

for some function f. Although the converse is not always true, such a case is not often encountered in practical applications [399] so we may equate the stationarity with the condition

$$\mathbb{E}T(x) = \mathbb{E}T(0), \ R(x, y) = f(x - y).$$

A special case of stationary fields is an isotropic field which requires the covariance function to be rotation invariant, i.e.

$$R(x, y) = f(\|x - y\|)$$

for some function f. $\| \cdot \|$ is the geodesic distance in the underlying manifold.

5.1.1 Gaussian Fields

The most important class of random fields is Gaussian fields. A more rigorous treatment can be found in [6]. Let us start defining a multivariate normal distribution from a Gaussian random variable.

Definition 4 *A random vector $T = (T_1, \cdots, T_m)$ is multivariate normal if $\sum_{i=1}^m c_i T_i$ is Gaussian for every possible $c_i \in \mathbb{R}$.*

Then a Gaussian random field can be defined from a multivariate normal distribution.

Definition 5 *A random field T is a* Gaussian random field *if $T(x_1), \cdots, T(x_m)$ are multivariate normal for every $(x_1, \cdots, x_m) \in \mathbb{R}^m$.*

An equivalent definition to Definition 5 is as follows.

Definition 6 *T is a Gaussian random field if the finite joint distribution $F_{x_1,\cdots,x_m}(z_1,\cdots,z_m)$ is a multivariate normal for every $(x_1,\cdots,x_m)$.*

T is a mean zero Gaussian field if $\mathbb{E}T(x) = 0$ for all x. Because any mean zero multivariate normal distribution can be completely characterized by its covariance matrix, a mean zero Gaussian random field T can be similarly determined by its covariance function R. Two fields T and S are independent if $T(x)$ and $S(y)$ are independent for every x and y. For mean zero Gaussian fields T and S, they are independent if and only if the cross-covariance function

$$R(x,y) = \mathbb{E}\big[T(x)T(y)\big]$$

vanishes for all x and y.

The Gaussian white noise is a Gaussian random field with the Dirac-delta function δ as the covariance function. Note the Dirac delta function is defined as $\delta(x) = \infty, x = 0$, $\delta(x) = 0 x \neq 0$ and $\int \delta(x) = 1$. Numerically we can simulate the Dirac delta function as the limit of the sequence of Gaussian kernel K_σ when $\sigma \to \infty$. The Gaussian white noise is simulated as an independent and identical Gaussian random variable at each voxel.

5.1.2 Derivative of Gaussian Fields

Any linear operation on Gaussian fields is again Gaussian fields. Suppose $\mathcal{G}$ be a collection of Gaussian random fields. For given $X, Y \in \mathcal{G}$, we have $c_1 X + c_2 Y \in \mathcal{G}$ again for all c_1 and c_2. Therefore, $\mathcal{G}$ forms an infinite-dimensional vector space. Not only the linear combination of Gaussian fields is again Gaussian but also the derivatives of Gaussian fields are Gaussian. To see this, we define mean-square convergence.

Definition 7 *A sequence of random fields T_h, indexed by h converges to T as $h \to 0$ in mean-square if*

$$\lim_{h\to 0} \mathbb{E}\big|T_h - T\big|^2 = 0.$$

We will denote the mean-square convergence using the usual limit notation:

$$\lim_{h\to 0} T_h = T.$$

The convergence in mean-square implies the convergence in mean. This can be seen from

$$\mathbb{E}\big|T_h - T\big|^2 = \mathbb{V}\big[T_h - T\big]^2 + \big(\mathbb{E}|T_h - T|\big)^2.$$

Now let $T_h \to T$ in mean square. Each term in the right hand side should also converges to zero proving the statement.

Now we define the derivative of field in mean square sense as

$$\frac{dT(x)}{dx} = \lim_{h \to 0} \frac{T(x+h) - T(x)}{h}.$$

Note that if $T(x)$ and $T(x+h)$ are Gaussian random fields, $T(x+h) - T(x)$ is again Gaussian, and hence the limit on the right hand side is again Gaussian. If R is the covariance function of the mean zero Gaussian field T, the covariance function of its derivative field is given by

$$\mathbb{E}\Big[\frac{dT(x)}{dx}\frac{dT(y)}{dy}\Big] = \frac{\partial^2 R(x,y)}{\partial x \partial y}.$$

5.1.3 Integration of Gaussian Fields

The integration of Gaussian fields is also Gaussian. To see this, define the integration of a random field as the limit of Riemann sum. Let $\cup_{i=1}^n \mathcal{M}_i$ be a partition of $\mathcal{M}$, i.e.

$$\mathcal{M} = \cup_{i=1}^n \mathcal{M}_i \text{ and } \mathcal{M}_i \cap \mathcal{M}_j = \emptyset \text{ if } i \neq j.$$

Let $x_i \in \mathcal{M}_i$ and $\mu(\mathcal{M}_i)$ be the volume of $\mathcal{M}_i$. Then we define the integration of field T as

$$\int_{\mathcal{M}} T(x)\, dx = \lim \sum_{i=1}^n T(x_i)\mu(\mathcal{M}_i),$$

where the limit is taken as $\mu(\mathcal{M}_j) \to 0$ for all j. When we integrate a Gaussian field, it is the limit of a linear combination of Gaussian random variables so it is again a Gaussian random variable. In general, any linear operation on Gaussian fields will result in Gaussian fields.

5.1.4 t, F and χ^2 Fields

We can use i.i.d. Gaussian fields to construct χ^2-, t-, F-fields, all of which are extensively studied in [56, 396, 394, 388]. The χ^2-field with m degrees of freedom is defined as

$$T(x) = \sum_{i=1}^m X_i^2(x),$$

where $X_1, \cdots, X_m$ are independent, identically distributed Gaussian fields with zero mean and unit variance. Similarly, we can define t and F fields as well as Hotelling's T^2 field. The Hotelling's T^2-statistic has been widely used in detecting morphological changes in deformation-based morphometry [56, 86, 131, 194, 359]. In particular, [56] derived the excursion probability of the Hotelling's T^2-field and applied it to detect gender specific morphological differences.

5.2 Simulating Gaussian Fields

For the random field theory based multiple correction to work, it is necessary to have smooth images. In this section, we show how to simulate smooth Gaussian fields by performing Gaussian kernel smoothing on white noise. This is the easiest way of simulating Gaussian fields.

White noise is defined as a random field whose covariance function is proportional to the Dirac-delta function δ, i.e.

$$R(x, y) \propto \delta(x - y).$$

For instance, me may take $R(x, y) = \lim_{\sigma \to 0} K_\sigma(\|x-y\|)$, the limit of the usual isotropic Gaussian kernel. White noise is usually characterized via generalized functions.

One example of white noise is the generalized derivative of Brownian motion (Wiener process) called white Gaussian noise.

Definition 8 *Brownian motion (Wiener process) $B(x), x \in \mathbb{R}^+$ is zero mean Gaussian field with covariance function*

$$R_B(x, y) = \min(x, y).$$

Following Definition 8, we can show $\mathbf{Var} B(x) = x$ by taking $x = y$ in the covariance function and $B(0) = 0$ by letting $x = 0$ in the variance. The increments of Wiener processes in nonoverlapping intervals are independent identically distributed (iid) Gaussian. Further the paths of the Wiener process is continuous in the Kolmogorov sense while it is not differentiable. For a different but identical canonical construction of Brownian motion, see [268]. Higher dimensional Brownian motion can be generalized by taking each component of vector fields to be i.i.d. Brownian motion.

Although the path of Wiener process is not differentiable, we can define the generalized derivative via integration by parts with a smooth function f called a test function in the following way

$$f(x)B(x) = \int_0^x f(y) \frac{dB(y)}{dy}\, dy + \int_0^x \frac{f(y)}{dy} B(y)\, dy.$$

Taking the expectation on both sides we have

$$\int_0^x f(y) \mathbb{E} \frac{dB(y)}{dy}\, dy = 0.$$

It should be true for all smooth f so $\mathbb{E}\frac{dB(y)}{dy} = 0$. Further it can be shown that the covariance function of process $dB(y)/dy \propto \delta(x - y)$.

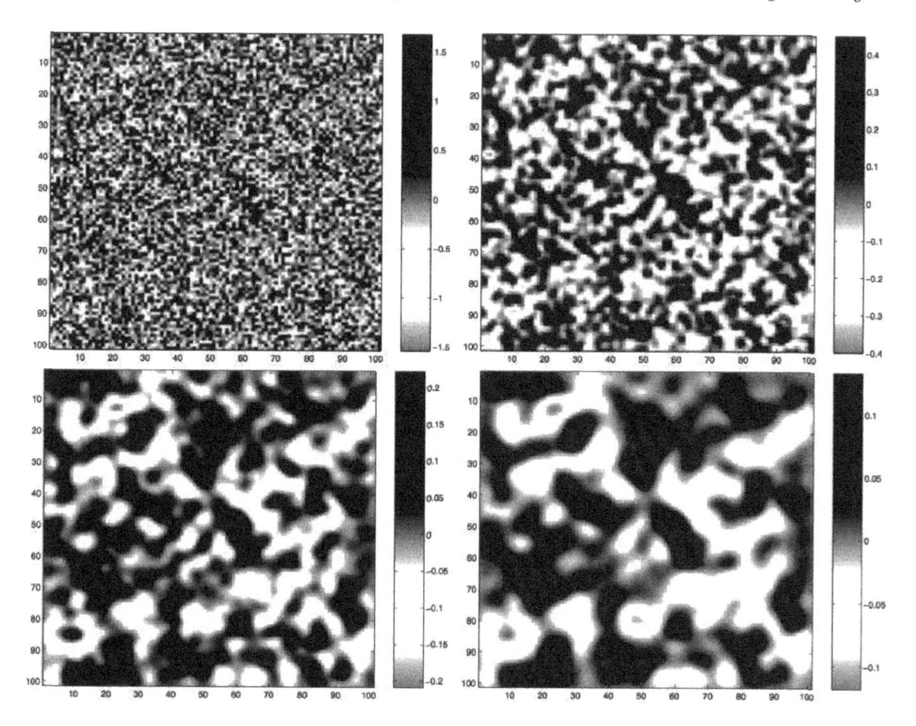

FIGURE 5.1
Random fields simulation via iterated Gaussian kernel smoothing with $\sigma = 0.4$. $N(0, 0.4^2)$. White noise, 1, 4 and 9 iterations in sequence.

The Gaussian white noise can be used to construct smooth Gaussian random fields of the form

$$X(x) = K * W(x) = K * \frac{dB(x)}{dx},$$

where K is a Gaussian kernel. Since Brownian motion is zero mean Gaussian process, $X(x)$ is obviously zero mean field with the covariance function

$$R_X(x, y) = \mathbb{E}[K * W(x) K * W(y)] \tag{5.1}$$

$$\propto \int K(x - z)K(y - z)\, dz. \tag{5.2}$$

The case when K is an isotropic Gaussian kernel was investigated by D.O. Siegmund and K.J. Worsley with respect to optimal filtering in scale space theory [327].

In numerical implementation, we use the discrete white Gaussian noise which is simply a Gaussian random variable.

Example 2 *Let w be a discrete version of white Gaussian noise given by*

$$w(x) = \sum_{i=1}^{m} Z_i \delta(x - x_i),$$

where i.i.d. $Z_i \sim N(0, \sigma_w^2)$. Note that

$$K * w(x) = \sum_{i=1}^{m} Z_i K(x - x_i). \tag{5.3}$$

The collection of random variables $K * w(y_1), \cdots, K * w(y_l)$ forms a multivariate normal at arbitrary points $y_1, \cdots, y_l$. Hence the field $K * w(x)$ is a Gaussian field.

The covariance function of the field 5.3 is given by

$$R(x, y) = \sum_{i,j=1}^{m} \mathbb{E}(Z_i Z_j) K(x - x_i) K(y - x_j) \tag{5.4}$$

$$= \sum_{i=1}^{m} \sigma_w^2 K(x - x_i) K(y - x_i). \tag{5.5}$$

As usual we may take K to be a Gaussian kernel. Let us simulate some Gaussian fields.

Example 3 *The unknown signal is assumed to be $\mu(x, y) = \cos(10x) + \sin(8y), (x, y) \in [0, 1]^2$ and white noise error $w \sim N(0, 0.4^2)$ which is shown in the top left of Figure 5.1. Then iteratively more smooth version of Gaussian random fields are constructed by*

```
w=normrnd(0,0.4,101,101);
smooth_w=w;
for i=1:10
  smooth_w=conv2(smooth_w,K,'same');
  figure;imagesc(smooth_w);colorbar;
end;
```

where the kernel weight `K` *is constructed following the method given in Section 4.3.*

5.3 Statistical Inference on Fields

Given functional measurement Y, we have model

$$Y(x) = \mu(x) + \epsilon(x),$$

where μ is unknown signal and ϵ is a zero mean unit variance Gaussian field. We assume $x \in \mathcal{M} \subset \mathbb{R}^n$. In brain imaging, one of the most important problems is that of signal detection, which can be stated as the problem of identifying the regions of statistical significance. So it can be formulated as an inference problem

$$H_0 : \mu(x) = 0 \text{ for all } x \in \mathcal{M} \text{ vs. } H_1 : \mu(x) > 0 \text{ for some } x \in \mathcal{M}.$$

Let

$$H_0(x) : \mu(x) = 0$$

at a fixed point x. Then the null hypothesis H_0 is a collection of multiple hypotheses $H_0(x)$ over all x. Therefore, we have

$$H_0 = \bigcap_{x \in \mathcal{M}} H_0(x).$$

We may assume that $\mathcal{M}$ is the region of interest consisting of the finite number of voxels. We also have the corresponding point-wise alternate hypothesis

$$H_1(x) : \mu(x) > 0$$

and the alternate hypothesis H_1 is constructed as

$$H_1 = \bigcup_{x \in \mathcal{M}} H_0(x).$$

If we use Z-statistic as a test statistic, for instance, we will reject each $H_0(x)$ if $Z > h$ for some threshold h. So at each fixed x, for level $\alpha = 0.05$ test, we need to have $h = 1.64$. However, if we threshold at $\alpha = 0.05$, 5% of observations are false positives. Note that the false positives are pixels where we are incorrectly rejecting $H_0(x)$ when it is actually true. However, these are the false positives related to testing $H_0(x)$. For determining the true false positives associated with testing H_0, we need to account for multiple comparisons.

Definition 9 *The type-I error is the probability of rejecting the null hypothesis (there is no signal) when the alternate hypothesis (there is a signal) is true.*

The type-I error is also called the *family-wise error rate* (FWER) and given by

$$\begin{aligned} \alpha &= P(\text{ reject } H_0 \mid H_0 \text{ true }) \\ &= P(\text{ reject some } H_0(x) \mid H_0 \text{ true }) \\ &= P\Big(\bigcup_{x \in \mathcal{M}} \{Y(x) > h\} \Big| \mathbb{E}Y = 0\Big). \end{aligned} \quad (5.6)$$

Unfortunately, $Y(x)$ is correlated over x and it makes the computation of type-I error almost intractable for random fields other than Gaussian.

5.3.1 Bonferroni Correction

One standard method for dealing with multiple comparisons is to use the Bonferroni correction. Note that the probability measure is additive so that for any event E_j, we have

$$P\Big(\bigcup_{j=1}^{\infty} E_j\Big) \leq \sum_{j=1}^{\infty} P(E_j).$$

This inequality is called Bonferroni inequality and it has been used in the construction of simultaneous confidence intervals and multiple comparisons when the number of hypotheses are small. From (5.6), we have

$$\alpha = P\Big(\bigcup_{x \in \mathcal{M}} \{Y(x) > h\} \;\Big|\; \mathbb{E}Y = 0\Big) \quad (5.7)$$

$$\leq \sum_{x \in \mathcal{M}} P\big(Y(x_j) > h \mid \mathbb{E}Y = 0\big) \quad (5.8)$$

So by controlling each type-I error separately at

$$P\big(Y(x_j) > h \mid \mathbb{E}Y = 0\big) < \frac{\alpha}{\#\mathcal{M}}$$

we can construct the correct level α test. Here $\#\mathcal{M}$ is the number of voxels.

The problem with the Bonferroni correction is that it is too conservative. The Bonferroni inequality (5.8) becomes exact when the measurements across voxels are all independent, which is unrealistic. Since the measurements are expected to be strongly correlated across voxels, we have highly correlated statistics. So in a sense, we have a less number of comparisons to make.

Let us illustrate the Bonferroni correction procedure using MATLAB. Consider 100×100 image Y of standard normal random variables (Figure 5.2). The threshold corresponding to the significance $\alpha = 0.05$ is 1.64.

```
Y=normrnd(0,1,100,100);
figure; imagesc(Y); colorbar; colormap('hot')

norminv(0.95,0,1)

ans =

    1.6449

[Yl, Yh] = threshold_image(Y, 1.64);
figure; imagesc(Yh); colormap('hot'); colorbar;
```

By thresholding the image at 1.64, we obtain approximately about 5% of pixels as false positives. To account for the false positives, we perform the

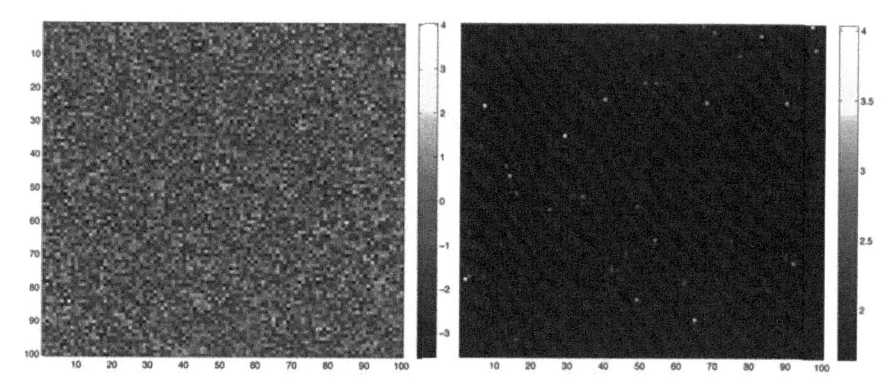

FIGURE 5.2
Image consisting of $N(0,1)$ noise. At the thresholding 1.64 corresponding to the significance level $\alpha = 0.05$, 5% of all pixels are false positives.

Bonferroni correction. For image of size 100×100, there are 10000 pixels. Therefore, $\alpha/\#\mathcal{M} = 0.05/10000 = 0.000005$ is the corresponding point-wise p-value and the corresponding threshold is 4.42. In this example, there is no pixel that is higher than 4.42 so we are not detecting any false positives as expected.

```
n=100*100;
size(find(reshape(Y,n,1)>=1.64),1)/n
norminv(1-0.05/10000,0,1)

ans =

    4.4172
```

5.3.2 Rice Formula

We can obtain a less conservative estimate for (5.6) using the random field theory. Assuming $\mathbb{E}Y = 0$, we have

$$\begin{aligned}
\alpha(h) &= P\Big(\bigcup_{x\in\mathcal{M}} \{Y(x) > h\}\Big) \\
&= 1 - P\Big(\bigcap_{x\in\mathcal{M}} \{Y(x) \le h\}\Big) \\
&= 1 - P\Big(\sup_{x\in\mathcal{M}} Y(x) \le h\Big) \\
&= P\Big(\sup_{x\in\mathcal{M}} Y(x) > h\Big). \qquad (5.9)
\end{aligned}$$

In order to construct the α-level test corresponding to H_0, we need to know the distribution of the supremum of the field Y. The corresponding p-value based on the supremum of the field, i.e. $\sup_{x \in \mathcal{M}} Y$, is called the *corrected p-value* to distinguish it from the usual p-value obtained from the statistic Y. Note that the p-value is the smallest α-level at which the null hypothesis H_0 is rejected.

Analytically computing the exact distribution of the supremum of random fields is hard. If we denote $Z = \sup_{x \in \mathcal{M}} Y(x)$ and F_Z to be the cumulative distribution of Z, for the given $\alpha = 0.05$, we can compute $h = 1 - F_Z^{-1}(\alpha)$. Then the region of statistically significant signal is localized as $\{x \in \mathcal{M} : Y(x) > h\}$.

The distribution of supremum of Brownian motion is somewhat simple due to its independent increment properties. However, for smooth random fields, it is not so straightforward. Read [4] for an overview of computing the distribution of the supremum of smooth fields.

Consider 1D smooth stationary Gaussian random process $Y(x), x \in \mathcal{M} = [0, 1] \subset \mathbb{R}$. Let N_h to be the number of times Y crosses over h from below (called upcrossing) in $[0, 1]$. Then we have

$$\begin{aligned} P\Big(\sup_{x \in [0,1]} Y(x) > h\Big) &= P(N_h \geq 1 \text{ or } Y(0) > h) \\ &\leq P(N_h \geq 1) + P(Y(0) > h) \\ &\leq \mathbb{E}N_h + P(Y(0) > h). \end{aligned}$$

If R is the covariance function of the field Y, we have

$$R(0) = \sigma^2 = \mathbb{E}Y^2(x).$$

It can be shown that from Rice formula [5, 299],

$$\mathbb{E}N_h = \frac{1}{\pi}\Big(\frac{-R''(0)}{R(0)}\Big)^{1/2} \exp\Big(\frac{h^2}{2\sigma^2}\Big).$$

Also note that

$$P(Y(0) > h) = 1 - \Phi\Big(\frac{h}{\sigma}\Big)$$

where Φ is the cumulative distribution function of the standard normal. Then from the inequality that bounds the cumulative distribution of the standard normal [119], we have

$$\Big(1 - \frac{\sigma^2}{h^2}\Big)\frac{\sigma}{\sqrt{2\pi}h} e^{-h^2/2\sigma^2} \leq 1 - \Phi\Big(\frac{h}{\sigma}\Big) \leq \frac{\sigma}{\sqrt{2\pi}h} e^{-h^2/2\sigma^2}$$

So

$$P\Big(\sup_{x \in [0,1]} Y(x) > h\Big) \leq \Big[c_1 + \frac{c_2}{\sqrt{2\pi}h}\Big] e^{-h^2/2\sigma^2}$$

for some c_1 and c_2. In fact we can show that

$$P\Big(\sup_{x\in[0,1]} Y(x) > h\Big) = \Big[c_1 + \frac{c_2}{h} + O(h^{-2})\Big]e^{-h^2/2\sigma^2}.$$

5.3.3 Poisson Clumping Heuristic

To extend the Rice formula to a higher dimension, we need a different mathematical machinery. For this method to work, the random field Y needs to be sufficiently smooth and isotropic. The smoothness of a random field corresponds to the random field being differentiable. There are very few cases for which exact formulas for the excursion probability (5.9) are known [3]. For this reason, approximating the excursion probability is necessary for most cases.

From the *Poisson clumping heuristic* [7],

$$P\Big(\sup_{x\in\mathcal{M}} Y(x) < h\Big) \approx \exp\Big(-\frac{\|\mathcal{M}\|}{\mathbb{E}\|A_h\|}P\big(Y(x) \geq h\big)\Big),$$

where $\|\cdot\|$ is the Lebesgue measure of a set and the random set

$$A_h = \{x \in \mathcal{M} : Y(x) > h\}$$

is called the *excursion set* above the threshold h. This approximation involves unknown $\mathbb{E}\|A_h\|$, which is the mean clump size of the excursion set. The distribution of $\|A_h\|$ has been estimated for the case of Gaussian [7], χ^2, t and F fields [55] but for general random fields, no approximation is available yet.

5.4 Expected Euler Characteristics

An alternate approximation to the supremum distribution based on the expected Euler characteristic (EC) of A_h is also available. The Euler characteristic approach reformulates the geometric problem as a topological problem. Read [2, 57, 354, 389] for an overview of the Euler characteristic method.

For sufficiently high threshold h, it is known that

$$P\Big(\sup_{x\in\mathcal{M}} Y(x) > h\Big) \approx \mathbb{E}\chi(A_h) = \sum_{d=0}^{N} \mu_d(\mathcal{M})\rho_d(h) \quad (5.10)$$

where $\mu_d(\mathcal{M})$ is the d-th Minkowski functional or *intrinsic volume* of $\mathcal{M}$ and ρ_d is the d-th Euler characteristic (EC) density of Y [391]. For details on intrinsic volume, read [315]. The expansion (5.10) also holds for non-isotropic fields but we will not pursue it any further. Compared to other approximation methods such as the Poisson clump heuristic and the tube formulae, the advantage of using the Euler characteristic formulation is that a simple exact expression can be found for $\mathbb{E}\ \chi(A_h)$. Figure 5.3 and Figure 5.4 show how $\chi(A_h)$ and $\mathbb{E}\ \chi(A_h)$ change as the threshold h increases for a simple binary object with a hole.

5.4.1 Intrinsic Volumes

The d-th intrinsic volume of $\mathcal{M}$ is a generalization of d-dimensional volume. Note that $\mu_0(\mathcal{M})$ is the Euler characteristic of $\mathcal{M}$. $\mu_N(\mathcal{M})$ is the volume of $\mathcal{M}$ while $\mu_{N-1}(\mathcal{M})$ is half the surface area of $\mathcal{M}$. There are various techniques for computing the intrinsic volume [354]. The methods depend on the smoothness of the underlying manifold $\mathcal{M}$. For a solid sphere with radius r, the intrinsic volumes are

$$\mu_0 = 1, \mu_1 = 4r, \mu_2 = 2\pi r^2, \mu_3 = \frac{4}{3}\pi r^3.$$

For a 3D box of size $a \times b \times c$, the intrinsic volumes are

$$\mu_0 = 1, \mu_1 = a + b + c, \mu_2 = ab + bc + ac, \mu_3 = abc.$$

In general, the intrinsic volume can be given in terms of a curvature matrix. Let $K_{\partial\mathcal{M}}$ be the curvature matrix of $\partial\mathcal{M}$ and $\mathrm{detr}_d(K_{\partial\mathcal{M}})$ be the sum of the determinant of all $d \times d$ principal minors of $K_{\partial\mathcal{M}}$. For $d = 0, \cdots, N-1$ the Minkowski functional $\mu_d(\mathcal{M})$ is defined as

$$\mu_d(\mathcal{M}) = \frac{\Gamma(\frac{N-i}{2})}{2\pi^{\frac{N-i}{2}}} \int_{\partial\mathcal{M}} \mathrm{detr}_{N-1-d}(K_{\partial\mathcal{M}})\, dA,$$

and $\mu_N(\mathcal{M}) = \|\mathcal{M}\|$, the Lebesgue measure of $\mathcal{M}$.

For irregular jagged shapes such as the 2D corpus callosum shape $\mathcal{M}$, the intrinsic volume can be estimated in the following fashion [393, 69]. Treating pixels inside $\mathcal{M}$ as points on a lattice, let V be the number of vertices that forms the corners of pixels, E be the number of edges connecting each adjacent lattice points and F be the number of faces formed by four connected edges. We assume the distance between the adjacent lattice points is δ in all directions. Then

$$\mu_0 = V - E + F, \mu_1 = (E - 2F)\delta, \mu_2 = F\delta^2.$$

To find the number of edges and pixels contained in $\mathcal{M}$, we start from an initial face (pixel) somewhere in the corpus callosum and add one face at a time while counting the additional edges and faces. In this fashion, we can grow a graph that will eventually contains all the pixels that form the corpus callosum. A numerical method for computing the intrinsic volume for jagged irregular shapes has been implemented in FMRISTAT package (`www.math.mcgill.ca/keith/fmristat`).

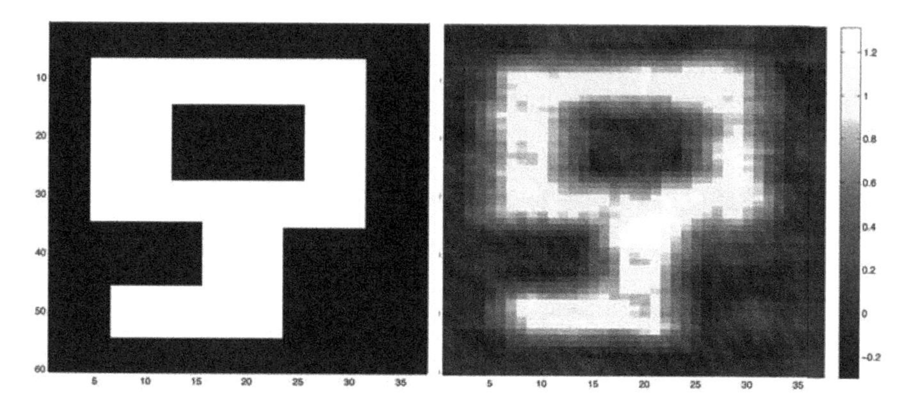

FIGURE 5.3
Gaussian white noise is added and smoothed to the key shaped binary object. The Euler characteristic of an object with a hole is 0.

5.4.2 Euler Characteristic Density

The d-th EC-density is given by

$$\rho_d(h) = \mathbb{E}\big[(Y > h)\det(-\ddot{Y}_d)|\dot{Y}_d = 0\big]P(\dot{Y}_d = 0),$$

where dot notation indicates partial differentiation with respect to the first d components. The subscript d represents the first d components of Y. Computation of the conditional expectation is nontrivial other than for Gaussian fields. For zero mean and unit variance Gaussian field Y, we have for instance

$$\begin{aligned}
\rho_0 &= P(Y > h) = 1 - \Phi(h) \\
\rho_1 &= \lambda^{1/2}\frac{e^{-h^2/2}}{2\pi} \\
\rho_2 &= \lambda h\frac{e^{-h^2/2}}{(2\pi)^{3/2}} \\
\rho_3 &= \lambda^{3/2}(h^2 - 1)\frac{e^{-h^2/2}}{(2\pi)^2},
\end{aligned}$$

where λ measures the smoothness of fields, defined as the variance of the derivative of component of Y. The exact expression for the EC density ρ_d is available for other random fields such as t, χ^2, F fields [388], Hotelling's T^2 fields [56] and scale-space random fields [327]. In each case, the EC density ρ_d is proportional to $\lambda^{\frac{d}{2}}$ and it changes depending on the smoothness of the field.

If $X_1, \cdots, X_\alpha, Y_1, \cdots, Y_\beta$ are i.i.d. stationary zero mean unit variance

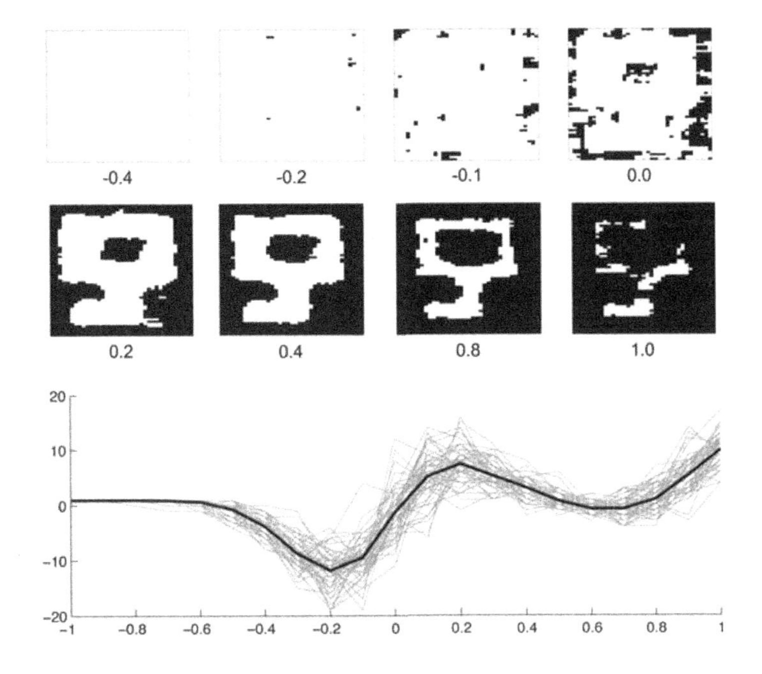

FIGURE 5.4
The mean Euler characteristic of the excursion sets of the shaped object shown in Figure 5.3. The number below each object is the threshold.

Gaussian fields. Then F-field with α and β degrees of freedom is given by

$$F(x) = \frac{\sum_{j=1}^{\alpha} X_j^2(x)/\alpha}{\sum_{j=1}^{\beta} Y_j^2(x)/\beta}.$$

To avoid singularity, we need to assume the total degrees of freedom $\alpha+\beta \gg N$ to be sufficiently larger than the dimension of space [388]. The EC-density for F-field is then given by

$$\begin{aligned}
\rho_0 &= \int_h^{\infty} \frac{\Gamma(\frac{\alpha+\beta}{2})}{\Gamma(\frac{\alpha}{2})\Gamma(\frac{\beta}{2})} \frac{\alpha}{\beta} \left(\frac{\alpha x}{\beta}\right)^{\frac{(\alpha-2)}{2}} \left(1+\frac{\alpha x}{\beta}\right)^{-\frac{(\alpha+\beta)}{2}} dx, \\
\rho_1 &= \lambda^{1/2} \frac{\Gamma(\frac{\alpha+\beta-1}{2}) 2^{\frac{1}{2}}}{\Gamma(\frac{\alpha}{2})\Gamma(\frac{\beta}{2})} \left(\frac{\alpha h}{\beta}\right)^{\frac{(\alpha-1)}{2}} \left(1+\frac{\alpha h}{\beta}\right)^{-\frac{(\alpha+\beta-2)}{2}}, \\
\rho_2 &= \frac{\lambda}{2\pi} \frac{\Gamma(\frac{\alpha+\beta-2}{2})}{\Gamma(\frac{\alpha}{2})\Gamma(\frac{\beta}{2})} \left(\frac{\alpha h}{\beta}\right)^{\frac{(\alpha-2)}{2}} \left(1+\frac{\alpha h}{\beta}\right)^{-\frac{(\alpha+\beta-2)}{2}} \\
&\times \left[(\beta-1)\frac{\alpha h}{\beta} - (\alpha-1)\right].
\end{aligned}$$

If the random field Y is given as the convolution of a smooth kernel $K_h(x) = K(x/h)/h^N$ with a white Gaussian noise [327, 392], the covariance matrix of $\dot{Y} = dY/dx$ is given by

$$\mathbf{Var}(\dot{Y}) = \frac{\int_{\mathbb{R}^N} \dot{K}(\frac{x}{h})\dot{K}^t(\frac{x}{h})\,dx}{h^2 \int_{\mathbb{R}^N} K^2(\frac{x}{h})\,dx}.$$

Applying it to a Gaussian kernel $K(x) = (2\pi)^{-n/2}e^{-\|x\|^2/2}$ gives

$$\lambda = \mathbf{Var}(\dot{Y}_1) = 1/(2h^2).$$

In terms of FWHM of the kernel K_h,

$$\lambda = 4\ln 2/\text{FWHM}^2.$$

5.4.3 Numerical Implementation of Euler Characteristics

In this section, we show how to compute the expected Euler characteristic in MATLAB. The presented routine can be used in estimating the excursion probability numerically. Consider a 2D binary object (Figure 5.3), which is stored as a 2D image `toy-key.tif`. After loading the image using `imread`, we perform scaling on image intensity values so that it becomes a binary object. The Euler characteristic of the binary object is then computed using `bweuler`.

```
I=imread('toy-key.tif');
I= imresize(I,.1, 'nearest');
I=(max(max(I))-I);
I=I/max(max(I));
I=double(I);
figure;imagesc(I); colormap('hot')
eul = bweuler(I)

eul =

     0
```

Since there is a hole in the object, the Euler characteristic is 0. We will add Gaussian white noise $N(0, 0.5^2)$ to the binary object and smooth out with FWHM of 10 using `gaussblur`. The resulting image is a Gaussian random field with sufficient smoothness. The smoothed image is stored as `smooth` and displayed in Figure 5.3.

```
e=normrnd(0,1, 60, 37);
Y=I + e;
figure;imagesc(Y); colorbar; colormap('hot')
smooth = gaussblur(Y,10);
figure;imagesc(smooth);colorbar; colormap('hot');
```

At each threshold `h` between -1 and 1, we threshold `smooth` and store it as a new variable `excursion`. Then compute the Euler characteristic of `excursion`. For computing the mean of the Euler characteristic, we simulated Gaussian random fields 50 times using the `for`-loop.

```
figure;
eulsum=zeros(1,21);
for k=1:50
    Y=I+ normrnd(0,1,60,37);
    smooth = gaussblur(Y,10);
    eul=[];
    j=1;
    for h=-1:0.1:1
        [Yl, Yh] = threshold_image(smooth, h);
        Yh=reshape(Yh,60*37,1);
        excursion=zeros(60*37,1);
        excursion(find(Yh>h))=1;
        excursion=reshape(excursion, 60, 37);
        eul(j) = bweuler(excursion);
        j=j+1;
    end;
    hold on; plot(-1:0.1:1, eul, 'Color', [0.7 0.7 0.7])
    eulsum=eulsum+eul;
end;
hold on; plot(-1:0.1:1, eulsum/50, 'Color', 'k', 'LineWidth',2)
```

6

Anisotropic Kernel Smoothing

In Chapter 4, we explained how to perform isotropic kernel smoothing where the kernel weights are isotropic. However, it is often necessary to smooth images anisotropically by shaping the kernel. In this chapter, we introduce anisotropic kernel smoothing and its application in DTI. Figure 6.1 illustrates a situation where spatially adaptive anisotropic smoothing is applied along the principal direction of water diffusion in DTI.

As a main application of anisotropic kernel smoothing, we present a novel probabilistic approach of representing the connectivity of the brain white fiber in diffusion tensor imaging via anisotropic Gaussian kernel smoothing. This approach is simpler than solving a diffusion equation, which has been often used in the probabilistic representation of white matter connectivity. The constructed connectivity metric is deterministic in a sense that it avoids using Monte-Carlo random walk simulation in constructing the transition probability so the resulting connectivity maps do not change from one computational run to another. The same computational framework can also be also used in smoothing functional and structural signals along the white fiber tracks. This chapter is based on the technical report [74].

6.1 Anisotropic Gaussian Kernel Smoothing

Assume $H = H'$. The isotropic kernel under linear transform $x \to H^{1/2}x$ changes the shape of the kernel to an *anisotropic kernel*

$$K_H(x) = \frac{1}{(2\pi)^{n/2} \det H^{1/2}} K(x'H^{-1}x/2). \tag{6.1}$$

$\det H^{1/2}$ is the Jacobian determinant of the transformation that normalizes the kernel. Note that this is the density of n-dimensional multivariate normal with the covariance matrix HH', i.e. $N(0, HH')$. The matrix H is called the *bandwidth matrix* and it measures the amount of smoothing. It can be shown that K_H is the Dirac-delta function when all the eigenvalues $\lambda_1, \cdots, \lambda_n$ of H go to zero, i.e.

$$\lim_{\lambda_1,\cdots,\lambda_n \to 0} K_H(x) = \delta(x).$$

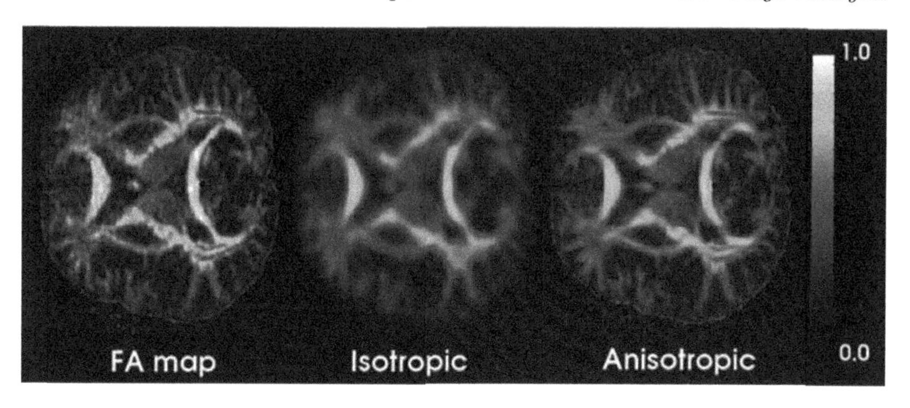

FIGURE 6.1
Smoothing applied to FA map. Isotropic smoothing blurs anatomical detail while anisotropic smoothing does not.

From now on we will denote $H \to 0$ if all $\lambda_i \to 0$ and $H \to \infty$ if all $\lambda_i \to \infty$.

Given noise observation

$$X(t) = \mu(t) + \epsilon(t),$$

where ϵ is a mean zero random field and μ is unknown signal, the kernel smoothing estimator of μ is given by

$$\widehat{\mu}(t) = K_H * X(t) = \int K_H(t-s)X(s)\, ds$$

The important properties of the estimator are

(1) $\lim_{H \to 0} \widehat{\mu}(t) = \int \delta(t-s)X(s)\, ds = X(t)$.

(2) $\lim_{H \to \infty} \widehat{\mu}(t) = 0$.

(3) $\mathbb{E}\widehat{\mu}(t) = K_H * \mu(t) \to \mu(t)$ as $H \to 0$. The kernel estimator becomes more unbiased as $H \to 0$.

(4) Assuming the mean signal is bounded, i.e. $|\mu| < \infty$,

$$\mathbb{E}\widehat{\mu}(t) \leq \int K_H(t) \sup \mu(t)\, dt \leq \sup_t \mu(t).$$

Similarly we can bound from below so that

$$\inf_t \mu(t) \leq \mathbb{E}\widehat{\mu}(t) \leq \sup \mu(t).$$

(5) Another interesting property is

$$\begin{aligned} \int K_H * X(t)\, dt &= \int_t \int_s K_H(t-s) X(s)\, ds\, dt \\ &= \int X(s)\, ds. \end{aligned}$$

This implies that the integral mean signal is preserved after kernel smoothing.

6.1.1 Truncated Gaussian Kernel

In general, it is cumbersome to manipulate kernels with infinite support numerically. We usually truncate kernel K_H outside some compact subset $\mathcal{M}$. We define a truncated and normalized kernel

$$\widetilde{K}_H(x) = \frac{K_H(x)\mathbf{1}_{\mathcal{M}}(x)}{\int_{\mathcal{M}} K_H(x)\, dx},$$

where $\mathbf{1}_{\mathcal{M}}$ is an indicator function that is zero everywhere except $\mathcal{M}$ where it is one. The kernel smoothing estimator is then given by

$$\widehat{\mu}(x) = \widetilde{K}_H * X(x).$$

Even though functional signal X is continuous, it will be observed at n discrete points x_i so the signal can be written as

$$X(x) = \sum_{i=1}^{n} X(x_i)\delta(x - x_i).$$

From the property of the Dirac-delta function, we have

$$\widehat{\mu}(\mathbf{x}) = \sum_{i=1}^{n} \widetilde{K}_H(\mathbf{x} - \mathbf{x}_i) X(\mathbf{x}_i)$$

where $\widetilde{K}_H$ is now a discrete normalized kernel

$$\widetilde{K}_H(x - x_i) = \frac{K_H(x - x_i)}{\sum_{x_j \in \mathcal{M}} K_H(x - x_j)}.$$

The discrete normalized kernel estimator in the nonparametric regression setting is usually called *Nadaraya-Watson kernel estimator* [258, 379].

6.2 Probabilistic Connectivity in DTI

Diffusion tensor imaging (DTI) is a new technique that provides the directional information of water diffusion in the white matter of the brain. The directional information of water diffusion is usually represented as a symmetric positive definite 3×3 matrix $D = (d_{ij})$ which is usually termed as the *diffusion tensor* or *diffusion coefficients*. The principal eigenvector $\mathbf{V}$ of D determines the main direction of the water diffusion.

There are various probabilistic and stochastic models for tracing fibers [31, 33, 34, 159]. Tench et al. introduced a hybrid streamline-based tractography where the direction of principal eigenvector is modeled stochastically to overcome the shortcomings of DTI [356]. Koch et al. introduced a Monte-Carlo random walk simulation that uses a different transition probability [210]. This algorithm has certain restrictions built in the random walk so that it was only allowed to jump in a direction within 90 degrees from the previous jump direction, which restricts the jump to a very small number of voxels in the neighborhood. Furthermore they considered the voxels with the FA-values [31] and sum of the eigenvectors bigger than certain thresholds. Then based on the Monte-Carlo simulation of 4000 random walks, they computed the probabilistic connectivity measure.

Hagmann et al. used a hybrid approach combining Monte-Carlo random walk simulation with information about the white fiber track curvature function in the corpus callosum [159] . Then assuming bivariate normal distribution of the random walk hitting a vertical plane at some distance apart, they estimated the covariance matrix and performed a statistical hypothesis testing of the homogeneity of covariance matrix in the different regions of the corpus callosum.

Batchelor et al. (2001) solved an anisotropic heat equation where the diffusion coefficients of the heat equation are the diffusion coefficients of DTI [33]. To get the probabilistic measure of the connectivity, the diffusion equation is solved with the initial condition where every vertex is zero except a seed region where it is given the value one:

$$\frac{\partial f}{\partial t}(p,t) = \nabla \cdot D \nabla f(p,t) \tag{6.2}$$

$$f(p,t=0) = 1 \text{ at } p = p_0 \text{ and } 0 \text{ elsewhere.} \tag{6.3}$$

The value 1 is diffused though the white matter and the numerical values between 0 and 1 are taken as a probability of white matter connectivity. Mathematically it is equivalent as the Monte-Carlo random walk simulation without restriction. The boundary condition $D\nabla f \cdot \mathbf{n} = 0$, i.e. the boundary is insulated and no heat diffuses out of the boundary, can be also enforced. A Crank-Nicholson scheme with Galerkin finite element discretization in space, and finite difference in time was then used to solve (6.3) [21]. Instead of solving the diffusion equation (6.3) directly, we can perform an equivalent iterative anisotropic kernel smoothing scheme [74, 403]. Recently, fast marching based tractography has been popularized [271, 182, 337]. Unlike probabilistic tractography, the fast marching methods do not present a computational burden.

The white fiber tracking is prone to cumulative acquisition noise and partial volume effect so the estimated white fiber tracks might possibly be erroneous in some cases [30, 356]. So it is crucial to develop a connectivity metric that is robust under the effect of acquisition noise and partial voluming. Such a robust connectivity metric can be used in VBM types of voxelwise inference on connectivity difference between two populations [14]. In the classical VBM, the gray and white matter densities are computed and used for inference on tissue concentration at each voxel. In DTI, instead of the tissue densities, we can use the connectivity metric that measures the strength of how two regions of the brain are connected via the white fiber tracts.

6.3 Riemannian Metric Tensors

Let D_0 be raw diffusion coefficients. D_0 is represented as $n \times n$ matrix in $\mathbb{R}^n$. We normalize it by $D = D_0/\text{tr}D_0$ (Figure 6.2). This normalization guarantees that the sum of eigenvalues of D to be $\text{tr}D = \sum_{j=1}^n \lambda_j = 1$. Consider a vector field $\mathbf{V} = (V_1, \cdots, V_n)'$ which is the principal eigenvector of D. Now suppose that we would like to smooth signals along the vector fields such that we smooth more along the larger vector fields. Suppose that the stream line or flow $\mathbf{x} = \psi(t)$ corresponding to the vector field is given by

$$\frac{d\psi}{dt} = \mathbf{V}(\psi(t)).$$

This ordinary differential equation gives a family of integral curves whose tangent vector is $\mathbf{V}$ [40]. The line element is

$$d\psi^2 = V_1^2\, dx_1^2 + \cdots + V_n^2 dx_n^2.$$

So $g_{ij} = V_i^2 \delta_{ij}$. We want to smooth more along the larger metric distance so we let $HH' = 2tG, G = (g_{ij})$, i.e.

$$HH' = 2t\text{Diag}(V_1^2, V_2^2, \cdots, V_n^2).$$

By introducing the scaling parameter t, we left room for adjusting the amount of smoothing. For the above choice of the covariance matrix, the anisotropic kernel is given by

$$K_t(\mathbf{x}) = (4\pi t)^{-n/2} \prod_{j=1}^{n} \frac{1}{|V_j|} \exp\Big(-\frac{x_j^2}{4tV_j^2}\Big). \tag{6.4}$$

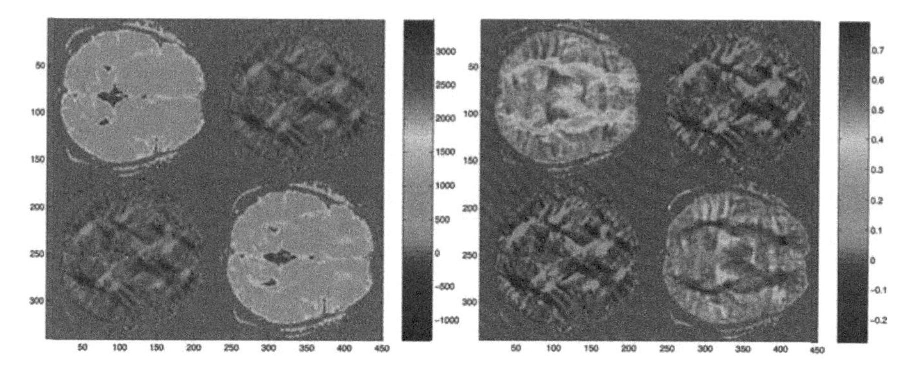

FIGURE 6.2
Left: the original diffusion tensor D_0 for a 2D slice. Right: normalized diffusion tensor $D = D_0/\mathrm{tr}D_0$ for 2D. The diffusion tensor D gives a natural Riemannian metric tensor in DTI. (See color insert.)

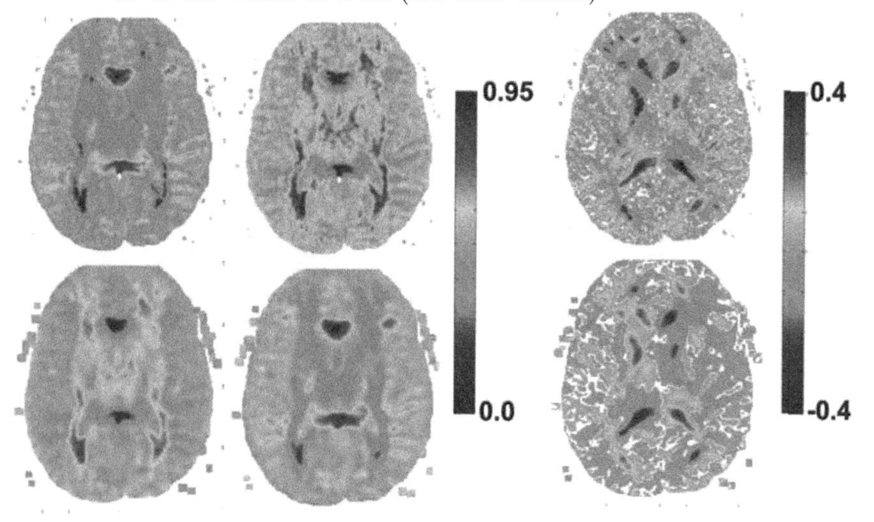

FIGURE 6.3
Top: the original diffusion tensors d_{11}, d_{22} and d_{12}. Bottom: the inverse of the diffusion tensor, d^{11}, d^{22} and d^{12}, which is needed for computing the anisotropic kernel. It has been smoothed and normalized by the trace of the inverse. (See color insert.)

In the case $V_j = 0$ while $|\mathbf{V}| \neq 0$ in (6.4), the kernel diverges. To avoid the divergence in numerical computation, we can introduce very small ϵ and use $\mathbf{V} + \epsilon$ instead of $\mathbf{V}$ in computing the kernel. In the numerical computation, we used $\epsilon = 10^{-10}$. The advantage of using only the principal eigenvectors would be the simplicity of the implementation while the drawback is that the computation of the principal eigenvectors takes a fair amount of time in `MATLAB` as well as the fact that the method does not completely utilize other eigenvectors.

Any reasonable functional relationship between HH' and G such as $HH' = G^2$ or $HH' = (I+G)^2$ can be used for the covariance matrix depending on how one wants to smooth signals in images. The Riemannian metric structure can come naturally from the diffusion tensor D. The inverse of the diffusion tensor D gives a natural Riemannian metric tensor along the white matter fibers in the brain, i.e. $G = D$. For the rigorous mathematical justification, see [102]. The same idea of matching the diffusion tensor to the Riemannian metric tensor has been introduced in [265], where they used it to compute the tensor-warped distances in the white fibers. Based on the natural Riemannian metric tensor D is for the diffusion, our anisotropic kernel is given by

$$K_t(\mathbf{x}) = \frac{\exp(-\mathbf{x}'D^{-1}\mathbf{x}/4t)}{(4\pi t)^{n/2}(\det D)^{1/2}}.$$

Assuming D is constant everywhere, it can be shown that $K_t * f(\mathbf{x})$ is a solution to an anisotropic diffusion equation

$$\frac{\partial g}{\partial t} = \nabla \cdot (D\nabla g) \tag{6.5}$$

with the initial condition $g(\mathbf{x}, 0) = f(\mathbf{x})$ after time t. If D is not constant, $K_t * f(\mathbf{x})$ is an approximate solution to (6.5) in the small neighborhood of $\mathbf{x}$ where D can be considered as constant. The exact solution to the equation (6.5) with the initial condition $g(\mathbf{x}, 0) = \delta(\mathbf{x})$ has been used as the probabilistic representation of white fiber track connectivity [33]. Note the conservation of total probability

$$\int_{\mathbb{R}^n} g(\mathbf{x}, t)\, d\mathbf{x} = 1 \text{ for all } t.$$

So in the diffusion equation approach, at each iteration step, the connectivity probability will always sum up to one and this will be also true for our kernel approach.

6.4 Chapman-Kolmogorov Equation

In this section, as an application of anisotropic kernel smoothing, a probabilistic connectivity based on the transition probability of diffusion is presented in detail [74, 403]. Let $P_t(p, q)$ be the *transition density* of a particle going from p to q under diffusion. This is the conditional probability density of the particle hitting q at time t when the particle is at p at time 0. Similarly the *transition probability* of going from point p to another region of interest Q is

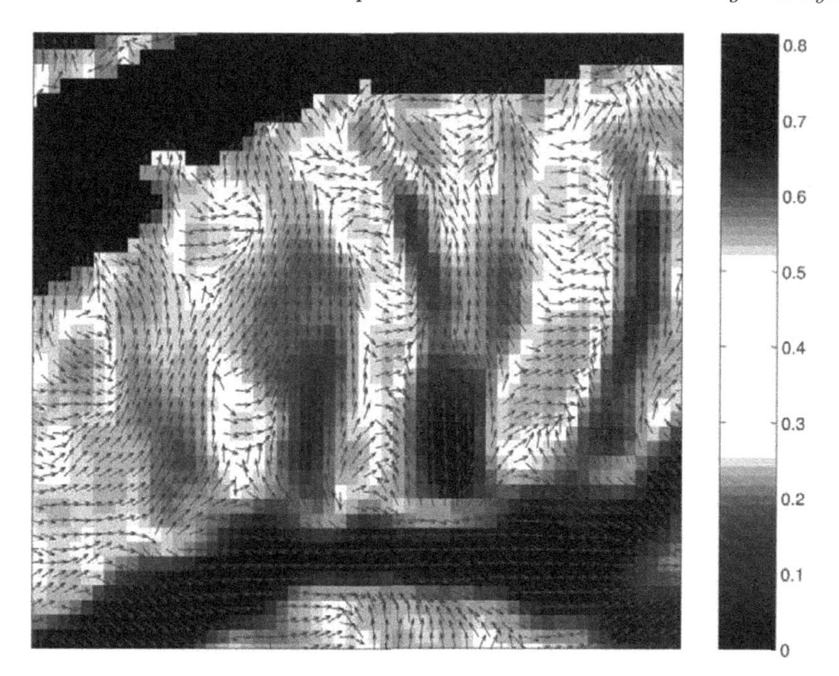

FIGURE 6.4
The principal eigenvalues and eigenvectors of the diffusion tensor. The arrows are the principal eigenvectors. The anisotropic kernel technique smooths along the direction of the principal eigenvectors and the amount of the smoothing is related to the principal eigenvalues.

given by

$$P_t(p, Q) = \int_Q P_t(p, x)\, dx.$$

Note that

$$P_t(p, \mathbb{R}^n) = \int_{\mathbb{R}^n} P_t(p, x)\, dx = 1.$$

The region Q can be a collection of voxels and it may possibly be consisting of a single voxel p. So we will interchangeably use $P_t(p, q)$ as either transition probability density or transition probability if there is no ambiguity. The transition probability is the most natural probabilistic measure associated with diffusion process and the connectivity measure based on the transition probability will be presented in this section.

If the diffusion coefficient D is constant in $\mathbb{R}^n$, it can be shown that

$$P_t(p, q) = K_t(q - p),$$

the Gaussian kernel with bandwidth t [341]. Since D is varying over the brain regions, it is only valid when p and q are a short distance apart and we may take $D(x)$ to be constant in the neighborhood of voxel position x.

The transition probability of a particle going from p to any arbitrary q is the total sum of the probabilities of going from p to q through all possible intermediate points $x \in \mathbb{R}^n$. Therefore,

$$P_t(p, q) = \int_{\mathbb{R}^n} P_s(p, x) P_{t-s}(x, q)\, dx \tag{6.6}$$

for any $0 < s < t$. It is traditionally called the Chapman-Kolmogorov equation [272]. The equation still holds in the case when s is either 0 or t, since in that case one of the probabilities in the integral becomes the Dirac-delta function and, in turn, the integral collapses to the probability on the left side.

Note that the probability $P(p, x)$ decreases exponentially as the distance between p and x increases so we approximate (6.6) in a small region B_p centered around p. For any point $x \in B_p$,

$$P_s(p, x) \doteq K_s(p - x).$$

Then for any arbitrary points p and q,

$$P_t(p, q) \doteq \frac{\int_{B_p} K_s(p - x) P_{t-s}(x, q)\, dx}{\int_{B_p} K_s(p - x)\, dx}. \tag{6.7}$$

When $s \to 0$, the approximation becomes exact since all the weights of the kernel will be in B_p. The denominator is a correction term for compensating the underestimation in the numerator. Note that this is the integral version of Gaussian kernel smoothing of data $P_{t-s}(x, q)$ for given q. Comparing with the formulation of Gaussian kernel smoothing, we rewrite (6.7) as

$$P_t(p, q) \doteq \tilde{K}_s * P_{t-s}(p, q), \tag{6.8}$$

where the convolution is with respect to the first argument p and $\tilde{K}_s$ is the truncated Gaussian kernel normalized by $\int_{B_p} K_s(p - x)\, dx$. Note that when $s \to 0$, the equation becomes exact. Figure 6.5 shows the computed transition probability in the 8-neighbors scheme.

The kernel smoothing formulation (6.8) is mainly valid when s is small. For large s, we borrow the iterative smoothing framework [76, 77]. We discretize t into N equal time intervals $t = N\Delta t$ and let $s = \Delta t$. Then (6.8) can be written as

$$F_j(q) = \tilde{K}_{\Delta t} * F_{j-1}(q), \tag{6.9}$$

where $F_j(q) = P_{j\Delta t}(p, q)$ for a given p and the initial condition

$$F_0(q) = P_0(p, q) = \delta(p - q).$$

The reason we get the Dirac-delta function is that the transition probability of a particle at p hitting any point q instantaneously is zero except when $q = p$.

One important property of our iterative procedure is the conservation of the total probability at each iteration. From (6.9), we have

$$\begin{aligned}\int_{\mathbb{R}^n} F_{j+1}(x)\,dx &= \int_{B_x} \tilde{K}_{\Delta t}(x-y)\,dy \int_{\mathbb{R}^n} F_j(x)\,dx \\ &= \int_{\mathbb{R}^n} F_j(x)\,dx.\end{aligned}$$

Since

$$F_1(\mathbf{q}) = \tilde{K}_{\Delta t} * \delta(q) = \tilde{K}_{\Delta t}(q),$$

F_1 is a probability function and it will integrate to one so

$$\int_{\mathbb{R}^n} F_j(x)\,dx = 1.$$

Hence F_j is also a probability function at each iteration. As the number of iteration increases, the total probability will be dispersed over all region of white matter from the seed (Figure 6.6).

If there are one million voxels within the brain, each voxel will have the connection probability of one over a million, which is extremely small. So even though the connectivity measure based on the transition probability is a mathematically sound one, it may not be a good one for visualization. So what we need is the log-scale of the transition probability, i.e. $\rho = \ln P_t(p,q)$ and we propose this as a probabilistic metric for measuring the strength of the anatomical connectivity. We will refer to this metric as the *log-transition probability*. For simplicity we may let $p = 0$ and let $\rho(q) = \ln P_t(0,q)$ for fixed t. If the diffusion coefficient is constant, the log-transition probability can be represented in a simple formula

$$\rho(x) = -x'D^{-1}x - \sum_{i=1}^{n} \ln \lambda_i - \frac{n}{2}\log(4\pi t),$$

where λ_i are the eigenvalues of D. When $D = I$,

$$\rho(x) = -x'x - \frac{n}{2}\log(4\pi t).$$

For a region of interest Q, the log-transition probability of reaching Q would be

$$\rho(q) = \ln \int_Q P_t(\mathbf{0}, x)\,dx.$$

See Figure 6.6 for log-transition probability obtained by taking the splenium as the seed.

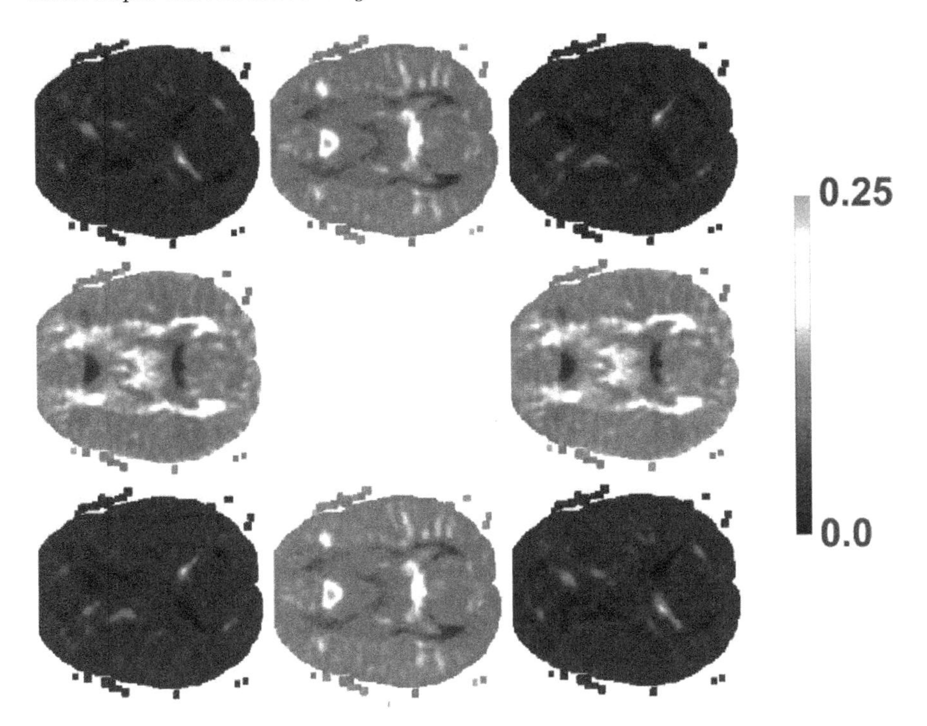

FIGURE 6.5
The transition probabilities from the center voxel to neighboring voxels. Using these probabilities as the weights, kernel smoothing is applied as the weighted averaging.

6.5 Cholesky Factorization of DTI

We show the robustness of the log-transition probability measure under simulated noise. We generate noise based on Cholesky factorization which guarantees the symmetric positive definite noise structure (Figures 6.7 and 6.8). Let $E = (e_{ij})$ be a random upper triangular matrix whose components $e_{ij}(\mathbf{x})$ are independent and identically distributed stationary Gaussian random fields. Further let $R = (r_{ij})$ be the upper triangular Cholesky factor such that $D = R'R$ [167]. Then we propose the following statistical model:

$$r_{ij}(\mathbf{x}) = \mu_{ij}(\mathbf{x}) + e_{ij}(\mathbf{x}), \tag{6.10}$$

where $\mu_{ij}(\mathbf{x})$ is the mean of the component of the Cholesky factor of D. In matrix form, equation (6.10) can be written as $D = (\mu + E)'(\mu + E)$, where $\mu = (\mu_{ij})$. A slightly different stochastic model would be

$$D(\mathbf{x}) = \mu'(\mathbf{x})\mu(\mathbf{x}) + E'(\mathbf{x})E(\mathbf{x}). \tag{6.11}$$

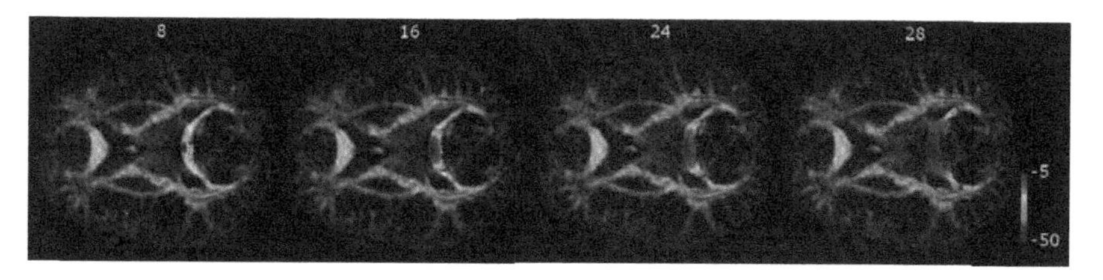

FIGURE 6.6
The transition probability in log scale from the seed at the splenium in the corpus callosum with $\Delta t = 0.1$ and $N = 8, 16, 24, 28$ iterations. (See color insert.)

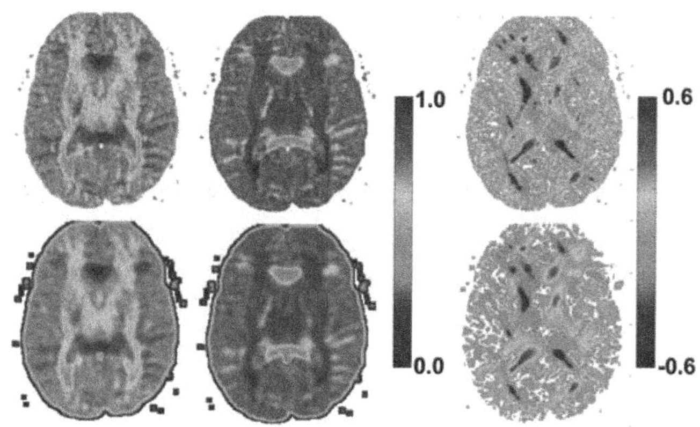

FIGURE 6.7
Top: Cholesky factors r_{11}, r_{22}, r_{12} of diffusion tensor D. Bottom: smoothed Cholesky factors with 8mm FWHM isotropic Gaussian kernel. (See color insert.)

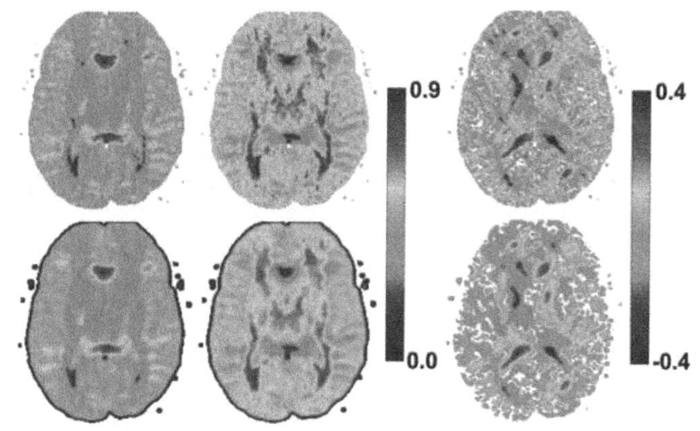

FIGURE 6.8
Top: normalized diffusion tensors d_{11}, d_{22}, d_{12} respectively. Bottom: smoothed diffusion tensors. We are smoothing the Cholesky factors r_{ij}. (See color insert.)

Note that the sum of symmetric positive definite matrix is again symmetric positive definite. The difference between model (6.10) and (6.11) is that in model (6.10), there are crossproduct terms between μ and E while there are none in (6.11); however, model (6.10) is simpler to manipulate component-wise for the simulation purpose. For our simulation, we have used (6.10).

In MATLAB, we generated the isotropic Gaussian random field by convolving white noise with isotropic Gaussian kernel. Let $w(\mathbf{x})$ be the white noise whose covariance function is of the form $\sigma_w^2 \delta(\mathbf{x} - \mathbf{y})$ for some constant σ_w^2 [109]. Let

$$K_t(\mathbf{x}) = \exp(-\mathbf{x}'\mathbf{x}/4t)/(4\pi t)^{n/2}$$

be the isotropic kernel. Then we generate a Gaussian field via $e(\mathbf{x}) = K_t * w(\mathbf{x})$. The covariance function of e is then

$$R_e(\mathbf{x}, \mathbf{y}) = \int_{\mathbb{R}^2} K_t(\mathbf{x} - \mathbf{x}')K_t(\mathbf{y} - \mathbf{y}')R_w(\mathbf{x}', \mathbf{y}')\, d\mathbf{x}'\, d\mathbf{y}'.$$

From the property of the Dirac-delta function, the variance of e can be shown to be

$$\sigma_e^2 = \sigma_w^2 \int_{\mathbb{R}} K_t^2(\mathbf{x} - \mathbf{x}')\, d\mathbf{x}'.$$

Since $K_t^2(\mathbf{x}) = K_{t/2}(\mathbf{x})/(8\pi t)^{n/2}$, by integrating the above integral we have $\sigma_e^2 = \sigma_w^2/(8\pi t)^{n/2}$. Based on this relationship, we simulated an isotropic Gaussian field with the specific variance σ_e^2 from the discrete version of white noise with variance σ_w^2. σ_e^2 is estimated from DTI. In all voxels where the diffusion tensor is positive definite, the sample variance for each e_{ij} ranged from 0.006 up to 0.009. For simulation we let $\sigma_e^2 = 0.01$ to give a little bit more noise than actual DTI to show the robustness of our connectivity metric.

6.6 Experimental Results

DTI of normal subjects were obtained using 1.5 Tesla SIGNA scanners. A conventional single-shot spin echo EPI pulse sequence was modified to obtain diffusion-weighted (DW) images from any arbitrary set of specified diffusion-weighting directions [217].

In discretizing the integral version of kernel smoothing, we assume the center of voxels in DTI to be on the integer lattice $\mathbb{Z}^n$. The affine transformation that maps DTI to the lattice $\mathbb{Z}^n$ without changing the maximal connectivity measure is used to simplify the computation. Take $B_{\mathbf{x}}$ to be the collection of voxels incident to voxel $\mathbf{x}$ and $\mathbf{x}$ itself. There are 3^n voxels in $B_{\mathbf{x}}$. The smaller the bandwidth parameter Δt, the more concentrated the weighting is to these incident voxels. The smaller support $B_{\mathbf{x}}$ is used to speed up the computational time. Let's illustrate how the method works in the case of the isotropic kernel.

Anisotropic kernel case is similar except that the kernel changes from voxel to voxel. The 2D isotropic kernel when $H = I$ for $\Delta t = 0.1$ is given by

$$\begin{pmatrix} 0.0054 & 0.0653 & 0.0054 \\ 0.0653 & 0.7958 & 0.0653 \\ 0.0054 & 0.0653 & 0.0054 \end{pmatrix}.$$

The contributions outside 9 voxels are less than 0.0001 and negligible so we use $\Delta t \leq 0.1$ and 9 voxel neighborhood for our computation. If a more accurate connection probability is desired, one may tempted to use more neighborhood; however, that is not necessary if Δt is very small. Note that

$$\sum_{\mathbf{x}_j \in B_{\mathbf{x}}} K_{\Delta t}(\mathbf{x}_j) = 1.0785$$

while

$$\sum_{\mathbf{x}_j \in \mathbb{Z}^2} K_{\Delta t}(\mathbf{x}_j) = 1.0787,$$

about a 0.0002% difference. So even after 500 iterations using the above truncated kernel, there will be only 0.1% difference to true value in the discrete lattice. Since the dimension of DTI is less than 500^3, the probability of connection can be computed in less than 500 iterations. Normalizing the above kernel, we get

$$\widetilde{K}_{\Delta t} = \begin{pmatrix} 0.0050 & 0.0606 & 0.0050 \\ 0.0606 & 0.7378 & 0.0606 \\ 0.0050 & 0.0606 & 0.0050 \end{pmatrix}.$$

The discrete version of the Dirac-delta function is the Kronecker's delta. So we let $F_0(\mathbf{q}) = 0$ everywhere except $F_0(\mathbf{p}) = 1$. Then using the discrete convolution, we compute

$$F_{j+1}(\mathbf{q}) = \widetilde{K}_{\Delta t} * F_j(\mathbf{q}).$$

At each stage of iteration the total probability is conserved, i.e. $\sum_{\mathbf{q} \in \mathbb{Z}^n} F_j(\mathbf{q}) = 1$.

6.7 Discussion

We introduced a novel approach of spatially adaptive anisotropic Gaussian kernel smoothing in representing the white fiber track connectivity via the concept of the transition probability of the stochastic process. Compared to the previous approaches of Monte-Carlo simulation or diffusion equation, our kernel method is very simple to implement but fast and robust enough to be used in large DTI data set. As an added bonus, our anisotropic kernel method can be used to smooth data along the white fiber tracks to get the continuous and smooth representation of data while preserving the directional characteristic of DTI. So it is hoped that the anisotropic kernel method can be further investigated in relation to the registration, segmentation and other image processing in DTI.

Our spatially adaptive anisotropic kernel smoothing is compatible with solving anisotropic diffusion equations in computational speed since both approaches can be viewed as iterative adaptive local weighted averaging. To speed up the kernel smoothing, we might use decomposition scheme of [137].

Our probabilistic connectivity metric can be used in VBM-like morphometry. The hypothesis of interest is

$$H_0 : \rho(\mathbf{x}) = \rho_0 \text{ vs. } H_1 : \rho(\mathbf{x}) > \rho_0$$

for fixed $\mathbf{x}$. An appropriate test statistic is

$$Z = \frac{\bar{\rho}(\mathbf{x}) - \rho_0}{S(\rho(\mathbf{x}))},$$

where $\bar{\rho}$ and $S(\rho)$ are the sample mean and standard deviation of a sample log-transition probability image $\rho_1, \cdots, \rho_m$ respectively. Unfortunately, our statistic Z will not be a T random field due to the nonlinearity of the kernel smoothing used. For the inference, we need to know the probability distribution of ρ from the distribution of D; however, it is hard to compute the exact distribution analytically. So we may estimate the distribution of ρ via bootstrapping.

7

Multivariate General Linear Models

Multivariate general linear models (MGLM) [9, 355, 396] generalize widely used univariate general linear models [394] by incorporating vector valued responses and explanatory variables. For instance, the surface coordinates or displacement vector fields can be taken as the response variable P (Figure 7.4). The displacement vector fields are mainly obtained from the deformation-based morphometry (DBM) [81]. This chapter is mainly based on [80]. For this chapter, it is necessary to review multivariate normal distributions,

7.1 Multivariate Normal Distributions

The multivariate normal distribution is a generalization of the univariate normal distribution to higher dimensions. So it can be defined from 1D normal distributions as follows.

Definition 10 *For independent and identically distributed standard normal distributions* $z_1, z_2 \sim N(0,1)$, *let* $z = (z_1, z_2)'$. *Then* $w = Hz$ *is distributed as a multivariate normal with the covariance matrix* HH', *i.e.*

$$w = Hz \sim N(0, HH').$$

The covariance matrix HH' is computed in the following way.

$$\mathbf{Var}(Hz) = \mathbb{E}\big[(Hz)(Hz)'\big] = HH'$$

since $\mathbb{E}zz' = I$. The covariance matrix is symmetric positive definite.

Theorem 1 *Any* $n \times n$ *symmetric positive definite matrix* $V = (v_{ij})$ *can be factored as* $V = HH'$, *where* $H = (h_{ij})$ *is a lower triangular matrix and* H' *is the upper triangular matrix. Further* $h_{11} = \sqrt{v_{11}}$ *and* $h_{i1} = v_{i1}/h_{ii}$ *for* $i = 2, \cdots, n$.

The decomposition given in Theorem 1 is called the Cholesky factorization. `MATLAB` implements it as `chol`. The `MATLAB` convention uses the upper triangular matrix as the Cholesky factor. Let us compute the Cholesky factor H' of $V = \begin{pmatrix} 2 & 1 \\ 1 & 2 \end{pmatrix}$.

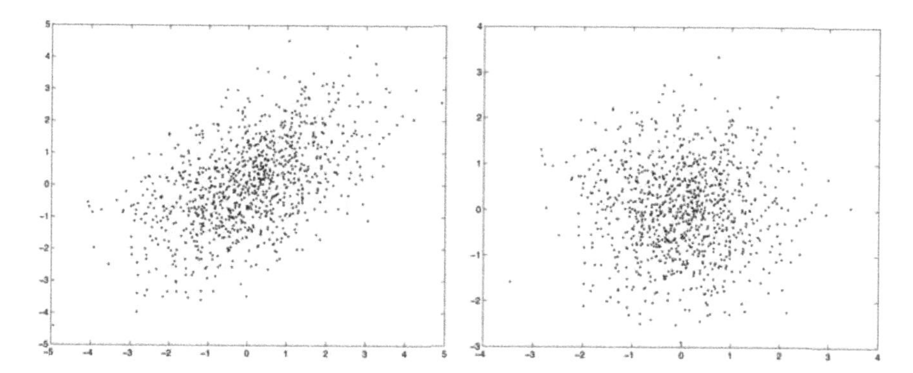

FIGURE 7.1
Left: 1000 bivariate normal vectors with a given covariance matrix V. The principal eigenvector of V correspond to the direction of data elongation. Right: Using the Cholesky factorization, we can uncorrelate the data.

```
V=[2 1
   1 2];

>>Htr=chol(V)
Htr =
    1.4142    0.7071
         0    1.2247

>> Htr'*Htr
ans =
    2.0000    1.0000
    1.0000    2.0000
```

Simulating 1000 multivariate normal random vectors with zero mean and covariance

$$V = \begin{pmatrix} 2 & 1 \\ 1 & 2 \end{pmatrix}$$

can be done via Cholesky factorization. We use the Gaussian random number generator `normrnd`, which generates univariate normal distribution.

```
z1 = normrnd(0,1,1000,1);
z2=normrnd(0,1,1000,1);
w=Htr'*[z1 z2]';
figure; plot(w(1,:), w(2,:),'.k')
```

The sample covariance is then given by

```
>> cov(w')
```

```
ans =
    2.0353    1.0426
    1.0426    1.9536
```

7.1.1 Checking Bivariate Normality of Data

For given bivariate data $w_i = (x_i, y_i)$, we are interested in testing if the data follows bivariate normal. You can use χ^2 goodness-of-fit test but a simpler approach would be to check if given bivariate data can be generated via linear transform

$$w = Hz + \mu, z \sim N(0, I).$$

for some unknown H and μ. Note that

$$w = Hz + \mu \sim N(\mu, HH')$$

where $z \sim N(0, I)$. Using the simulated 1000 w's, let us check if w's are distributed as a bivariate Gaussian. In this particular example, there is no need to estimate μ.

Let $\widehat{V} = \widehat{H}\widehat{H}'$ be the Cholesky decomposition. Then

$$z = \widehat{H}^{-1}w.$$

```
hatV = cov(w');

>> hatH=chol(hatV)'

hatH =
     1.4266         0
     0.7308    1.1914

>> inv(hatH)

ans =
     0.7010         0
    -0.4300    0.8393

newz=inv(hatH)*w;
figure; plot(newz(1,:), newz(2,:) ,'.k')
qqplot(newz(1,:));
qqplot(newz(2,:));
```

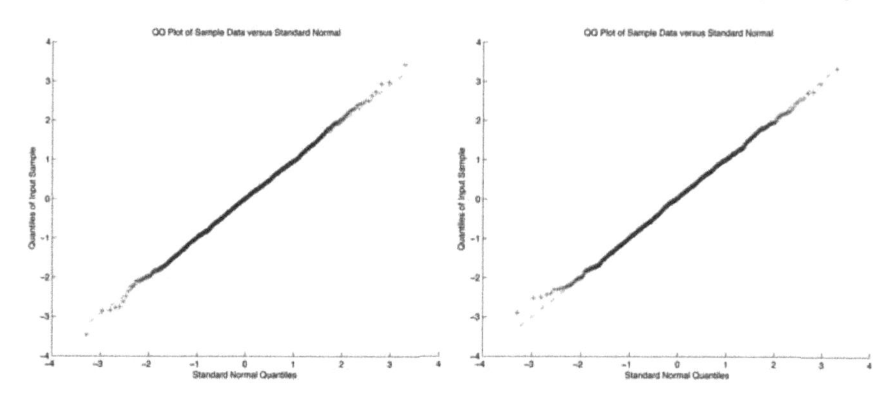

FIGURE 7.2
The QQ-plot of the components of z. Since the components of z are distributed as standard normals, we conclude that w is distributed as bivariate normal.

7.1.2 Covariance Matrix Factorization

Other factorization techniques for covariance matrix V are also available. Suppose V has positive eigenvalues λ_i and corresponding normalized eigenvectors e_i. Let $Q = (e_1, e_2, e_3)$. Then

$$QQ' = I$$

and

$$V = QDiag(\lambda_1, \lambda_2, \lambda_3)Q',$$

where $Diag$ is a diagonal matrix with λ_i in the diagonal terms. Then we have

$$V^{1/2} = QDiag(\lambda_1^{1/2}, \lambda_2^{1/2}, \lambda_3^{1/2})Q'.$$

For $z \sim N(0, I)$, $w = V^{1/2}z \sim N(0, V)$.

```
>>V=[2 1
     1 2]
>> [Q D]=eig(V)
Q = -0.7071    0.7071
     0.7071    0.7071
D =  1     0
     0     3
>> Q*D.^(1/2)*Q'
ans = 1.3660    0.3660
      0.3660    1.3660
>> V^(1/2)
ans = 1.3660    0.3660
      0.3660    1.3660
>>w=V^(1/2)*[z1 z2]';
```

It can be shown that generating bivariate normal this way gives the identical distribution as the Cholesky factorization.

Other factorization techniques exist for covariance matrix V. Suppose V has positive eigenvalues λ_i and corresponding normalized eigenvectors e_i. Consider the orthonormal matrix $Q = (e_1, e_2, e_3)$. Then $QQ' = I$ and $V = QDiag(\lambda_1, \lambda_2, \lambda_3)Q'$. So $V^{1/2} = QDiag(\lambda_1^{1/2}, \lambda_2^{1/2}, \lambda_3^{1/2})Q'$. For if we start with $z \sim N(0, I)$, $w = V^{1/2}z \sim N(0, V)$, the desired bivariate normal with the specified covariance matrix.

```
V=[2 1
   1 2]

>> [Q D]=eig(V)
Q = -0.7071     0.7071
     0.7071     0.7071
D =  1     0
     0     3

>> Q*D.^(1/2)*Q'
ans = 1.3660    0.3660
      0.3660    1.3660

>> V^(1/2)
ans = 1.3660    0.3660
      0.3660    1.3660

>>w=V^(1/2)*[z1 z2]';
```

The bivariate normal vector `w` will be similar to the left of Figure 7.1 in shape.

7.2 Deformation-Based Morphometry (DBM)

Morphological differences in the brain have been examined primarily by MRI-volumetry before the development of voxel-wise morphometries. The classical MRI-volumetry requires segmentation of anatomically corresponding regions of interest (ROI), either manually or automatically in MR images. Then the total volumes V_1 and V_2 of the homologous ROIs are calculated by counting the total number of voxels. Afterwards, the volume difference $\Delta V = V_2 - V_1$ is used as an index of morphological changes [141, 293, 300, 357]. On the other hands, *deformation-based morphometry* (DBM), which utilizes spatial position difference of corresponding voxels, does not require segmentation of *a priori* regions of interest [99, 14]. The advantage of DBM over the classical MRI-volumetry is that it does not require the *a priori* knowledge of the

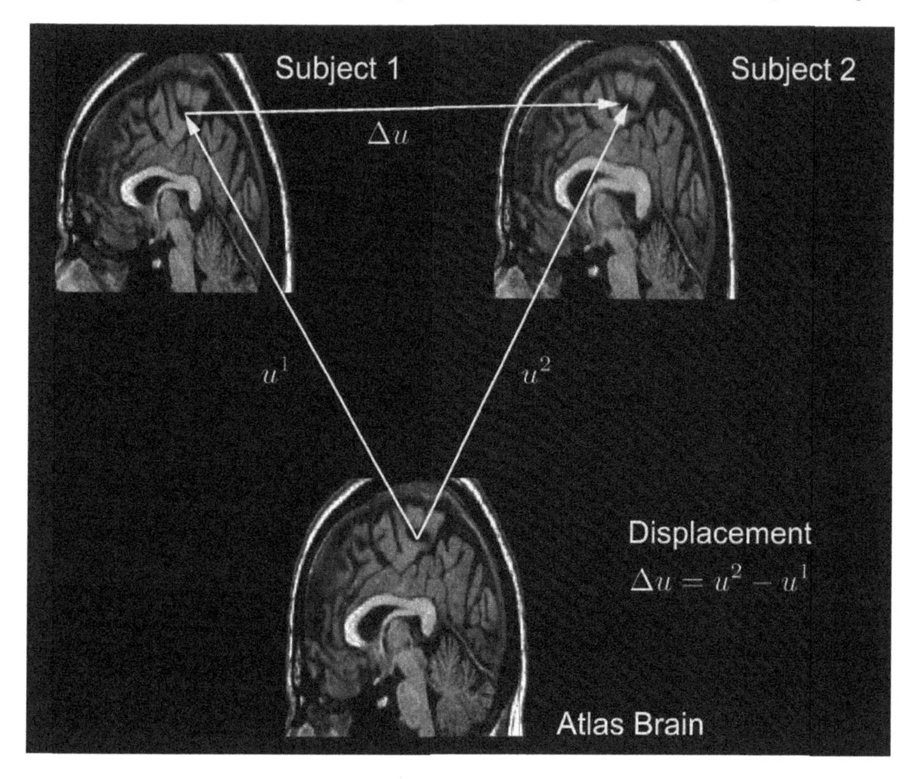

FIGURE 7.3
Schematic of registration for multiple brain images. If u^j is the displacement obtained from registering from the atlas brain to the j-th subject, the relative displacement Δu between the subjects 1 to 2 is $\Delta u = u^2 - u^1$.

ROI to perform the morphological analysis and structural differences can be detected at a voxel level. For example, using DBM, it is possible to detect local structural differences within the hippocampus and identify exactly what part of the hippocampus is responsible for the most anatomical variation in a group of subjects. The second advantage is that it does not require *a priori* knowledge of the ROI to perform the morphological analysis. Moreover, DBM improves the power of detecting the regions of volume change within the limits of the accuracy of registration algorithm, which is usually the size of voxels.

Unlike classical shape analysis [44, 45, 111, 199, 331], DBM tries to avoid anatomical landmarks in characterizing morphological changes. Since it is hard to identify the multitude of anatomical landmarks in brain images systematically, DBM is advantageous over the traditional landmark-based techniques. Since the introduction of deformation-based morphometry by Ashburner and

Friston in 2000 [14], there has been an explosion of morphometric studies that utilize DBM. For an overview, see [81].

Via image registration, biologically homologous points in two different images are identified and the mathematical transformation between these two points, called *deformation*, can be computed. The deformation is given as a 3D vector $d(x)$ at each voxel position x for the whole brain volume or at each mesh vertex for the brain surface. Although the idea of deformation originates from the elastic theory and continuum mechanics [62, 243], perhaps the first scientist to apply this concept to deform one biological structure to another closely related structure is D'arcy Thompson in his book *On Growth and Form* [358], where he deformed the skulls of human and primates, and other biological structures using deformable grids.

Mathematically this deformation can be represented as a transformation from a point x to a homologous point $x + u(x)$ in the Lagrangian coordinate system:

$$d(x) = x + u(x).$$

The vector u is called the *displacement vector field* in elastic deformation theory and it measures a relative movement of the point x [243]. If u^j is the displacement obtained from the non-linear registration of the j-th subject brain to the atlas brain $\mathcal{M}$, the relative displacement Δu between the subject 1 and 2 is $\Delta u = u^1 - u^2$ (Figure 7.3). Optimistically assuming that the image registration algorithm is accurate to within one voxel distance (usually 1 or 2 mm), the registration error seems to be relatively large in brain development. So one may be skeptical about whether deformation-based morphometry can possibly detect such small changes. Nevertheless it is still possible to pick out the signal when there are enough data [81].

Once we obtain the deformation from nonlinear image registration techniques such as a popular diffeomorphic image registration [410], we build a statistical model. Instead of modeling on deformation, it is easier to model on displacement:

$$u(x) = \mu(x) + \Sigma^{1/2}(x)\epsilon(x), \tag{7.1}$$

where $\Sigma(x)$ is the 3×3 symmetric positive-definite covariance matrix, which allows for correlations between components of the deformation and depends on the spatial coordinates x only [394, 56, 81]. Since Σ is symmetric positive-definite, the square-root of Σ always exists. The components of the error vector ϵ are assumed to be independent and identically distributed as smooth stationary Gaussian random fields with zero mean and unit standard deviation. The model has been widely used in localizing the regions of abnormal displacement differences.

7.3 Hotelling's T^2 Statistic

The Hotelling's T^2-statistic has been first used in detecting morphological changes in deformation-based morphometry (DBM) [56, 81, 86, 131, 194, 359, 396, 394, 388]. In particular, [56] derived the excursion probability of the Hotelling's T^2-field and applied it to detect gender specific morphological differences.

We are interested in detecting local regions of statistically significant changes in displacement using the linear model (7.1). This is a standard multivariate statistical inference problem and can be solved using the Hotelling's T^2 statistic [359, 194, 56, 131, 81]. Under the assumption (7.1), we are interested in testing if the two groups have the same displacement with respect to the template:

$$H_0 : \mu_1(x) = \mu_2(x) \text{ for all } x \text{ vs. } H_1 : \mu_1(x) \neq \mu_2(x) \text{ for some } x \in \mathbb{R}^d, \quad (7.2)$$

where μ_i is the unknown mean vector field for the i-th group. The inference is based on the Hotelling's T^2-statistic. Let us rewrite (7.1) for an individual subject using the group index i and the subject index j:

$$u^{ij}(x) = \mu^i(x) + \Sigma^{1/2}(x)\epsilon^{ij}(x),$$

where μ^{ij} are the i-th group mean vector and ϵ^{ij} are independent and identically distributed Gaussian random vector field. Let n_i be the number of subjects in the i-th group. The unknown i-th group mean μ^i is estimated as

$$\overline{\mu}^i = \frac{1}{n_i}\sum_{j=1}^{n_i} u^{ij}.$$

Testing hypotheses (7.2) is done by checking the significance of the mean difference $\overline{\mu}^2 - \overline{\mu}^1$.

The Hotelling's T^2 field framework is fairly flexible and can be applicable to wide variety of situations. If we only have one group, we can simply assume μ_2 to be a known vector field and treat the problem as a one-sample problem. In the case of a longitudinal study, where two scans per subject are available, we can take μ_1 as the growth velocity by dividing the displacement difference by the scan interval. Then we are testing if there is any significant growth over time. For instance, arrows in Figure 7.4 indicate the principal direction of the brain growth in normally developing children.

The pooled sample covariance matrix is given by

$$\widehat{\Sigma} = \frac{1}{n_1 + n_2 - 2}\Big[\sum_{j=1}^{n_1}(u^{1j} - \overline{\mu}^1)(u^{1j} - \overline{\mu}^1)' + \sum_{j=1}^{n_2}(u^{2j} - \overline{\mu}^2)(u^{2j} - \overline{\mu}^2)'\Big].$$

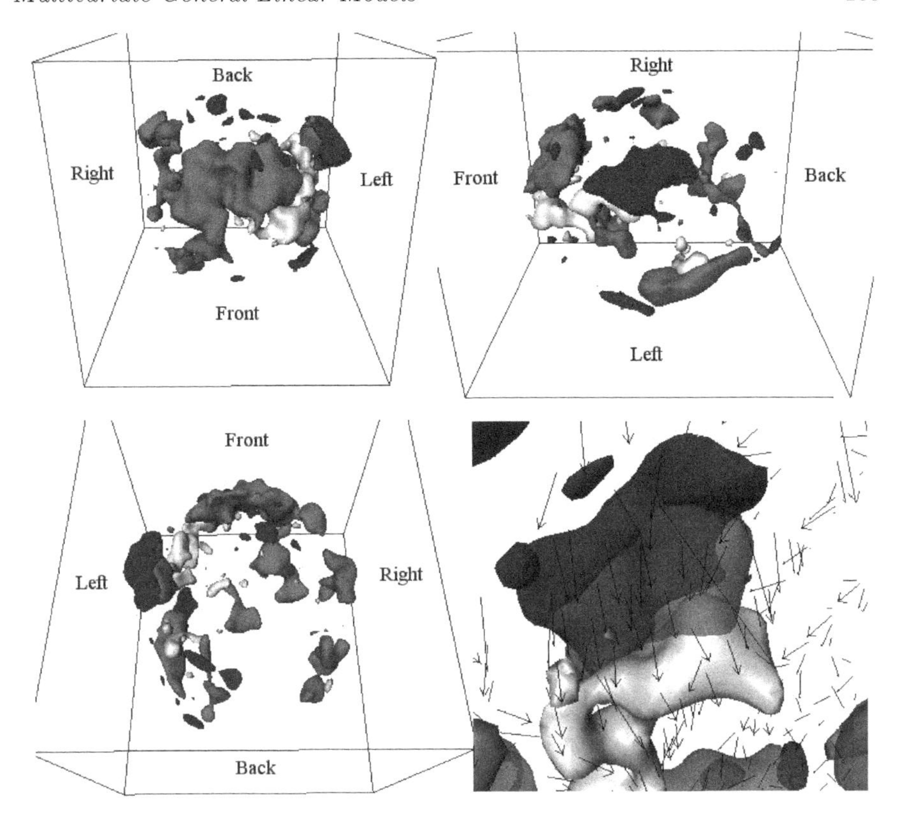

FIGURE 7.4
The p-value map of local volume increase (black), volume decrease (white), and structural displacement (gray) thresholded at 0.025, 0.025 and 0.05 (corrected) for normally developing children [81]. The p-value maps are superimposed on the template. The arrows represent the mean displacement difference over time (growth rate). The inference on the displacement difference is based on the Hotelling's T^2 statistic on the growth velocity field.

The significance of the group difference is then tested using the Hotelling's T^2-statistic

$$H(x) = \frac{n_1 n_2 (n_1 + n_2 - d - 1)}{d(n_1 + n_2)(n_1 + n_2 - 2)} (\overline{\mu}^2 - \overline{\mu}^1)' \widehat{\Sigma}^{-1} (\overline{\mu}^2 - \overline{\mu}^1).$$

At each x, under the null hypothesis of $\mu_1(x) = \mu_2(x)$, H is distributed as a F-statistic with d and $n_1 + n_2 - d - 1$ degrees of freedom. This is for two samples but a one-sample case is similar [81].

For the one-sample case, we assume μ_2 is a known constant vector field and the corresponding Hotelling's T^2 statistic is given by

$$H(x) = \frac{n_1(n_1 - d)}{d(n_1 - 1)}(\bar{\mu}^1 - \mu_2)'\widehat{\Sigma}^{-1}(\bar{\mu}^1 - \mu_2), \tag{7.3}$$

where the sample covariance is given by

$$\widehat{\Sigma} = \frac{1}{n_1 - 1}\sum_{j=1}^{n_1}(u^{1j} - \bar{\mu}^1)(u^{1j} - \bar{\mu}^1)'.$$

At each voxel x, under the hypothesis, $H(x)$ is distributed as a F-statistic with d and $n_1 - d$ degrees of freedom.

As pointed out in [14], the Hotelling's T^2 statistic does not directly localize regions within different structures, but rather identifies brain structures that have translated to different positions. It measures the relative positions of two particular voxels before and after the deformation. In the context of temporally varying brain morphology, where the brain tissue growth is an important concern, the statistic based on the displacement field should be taken as an indirect measure of brain growth [81]. A more direct morphological criterion that corresponds to the actual brain tissue growth is the Jacobian of the deformation field.

This simple 2D example illustrates how to write `MATLAB` routines for the Hotelling's T^2 statistic and MGLM.

Example 4 *We are interested in testing the equality of 2D vector measurements in two groups. The measurements for the first group is*

$$\begin{pmatrix} 0 \\ 1 \end{pmatrix}, \begin{pmatrix} 0 \\ 2 \end{pmatrix}, \begin{pmatrix} 0 \\ 3 \end{pmatrix}.$$

The measurements for the second group is

$$\begin{pmatrix} 2 \\ 0 \end{pmatrix}, \begin{pmatrix} 0 \\ 0 \end{pmatrix}.$$

The data is coded as

```
mu1= [0 1; 0 2; 0 3]
mu2=[2 0; 0 0]
n1=3; n2=2; d=2;
```

where `n1` and `n2` are the sample sizes and `d` is the dimension of the vector.

The sample means for the two groups and the mean difference are

$$\bar{\mu}^1 = \begin{pmatrix} 0 \\ 2 \end{pmatrix}, \bar{\mu}^2 = \begin{pmatrix} 1 \\ 0 \end{pmatrix}, \ \bar{\mu}^1 - \bar{\mu}^2 = \begin{pmatrix} -1 \\ 2 \end{pmatrix}.$$

The pooled variance is estimated as

$$\widehat{\Sigma} = \frac{1}{3}\begin{pmatrix} 2 & 0 \\ 0 & 2 \end{pmatrix}.$$

This is computed in `MATALB` as

```
m1 = mean(mu1,1)
m2=mean(mu2,1)

z1=mu1-kron(ones(n1,1),m1);
z2=mu2-kron(ones(n2,1),m2);

sigma= (z1'*z1 + z2'*z2)/(n1+n2-2)

sigma =

    0.6667         0
         0    0.6667
```

The inverse covariance matrix is then trivially

$$\widehat{\Sigma}^{-1} = \frac{3}{2}\begin{pmatrix} 1 & 0 \\ 0 & 1 \end{pmatrix}.$$

The Hotelling's T-square statistic value is then given as

$$h = \frac{3 \cdot 2(3+2-2-1)}{2(3+2)(3+2-2)}(\bar{\mu}^2 - \bar{\mu}^1)'\widehat{\Sigma}^{-1}(\bar{\mu}^2 - \bar{\mu}^1) = 3.$$

The Hotelling's T-square is distributed as a F-statistic with 2 and 2 degrees of freedom. The p-value is given by $P(H > h) = 0.25$, which is computed as `1 - fcdf(3,2,2)` in `MATLAB`. Since the p-value is large, the group difference is not significant. The whole procedure is implemented in `SurfStat2Hotel2.m` as

```
h=SurfStat2HotelT2(mu1,mu2)

h =

         t: 3
    pvalue: 0.2500
```

where `h.t` is the Hotelling's T^2 value and `h.pvalue` is the corresponding uncorrected p-value.

7.4 Multivariate General Linear Models

For a single contrast in the exploratory variables, Hotelling's T^2 can be used but for multiple contrasts, it is natural to set up a multivariate general linear model (MGLM) and construct a test statistic [396].

Consider the following MGLM at each fixed point along the template surface:

$$P_{n\times q} = X_{n\times p}B_{p\times q} + Z_{n\times r}G_{r\times q} + U_{n\times q}\Sigma_{q\times q}, \tag{7.4}$$

where P is the matrix of response vectors, X is the matrix of contrasted explanatory variables, and B is the matrix of unknown coefficients. Nuisance covariates are in the matrix Z and the corresponding coefficients are in the matrix G. The subscripts denote the dimension of matrices. The components of Gaussian random matrix U are zero mean and unit variance. Σ accounts for the covariance structure of between q variables. Then we are interested in testing the null hypothesis

$$H_0 : B = 0 \text{ vs. } H_1 : B \neq 0$$

For the reduced model corresponding to $B = 0$, the least squares estimator of G is given by solving $P = ZG$. i.e.

$$\widehat{G}_0 = (Z'Z)^{-1}Z'P.$$

We will assume that there is more sample size n than the number of parameters r to be estimated. The residual sum of squares of the reduced model is

$$E_0 = (P - Z\widehat{G}_0)'(P - Z\widehat{G}_0).$$

For the full model, the parameters are estimated by solving

$$P_{n\times q} = XB + ZG = [X, Z]_{n\times(p+r)} \begin{bmatrix} B \\ G \end{bmatrix}_{(p+r)\times q}.$$

The least squares estimation is given by

$$\begin{bmatrix} \widehat{B} \\ \widehat{G} \end{bmatrix} = ([X, Z]'[X, Z])^{-1}[X, Z]'P.$$

The corresponding residual sum of squared error is

$$E = (P - X\widehat{B} - Z\widehat{G})'(P - X\widehat{B} - Z\widehat{G}).$$

By comparing how large the residual E is against the residual E_0, we can determine the significance of coefficients B. However, since E and E_0 are matrices, we take a function of eigenvalues of E_0E^{-1} as a statistic. Since we expect the sample size n to be larger than q, there are q eigenvalues

$$\lambda_1, \lambda_2, \cdots, \lambda_q$$

satisfying

$$\det(E_0 - \lambda E) = 0.$$

This requires solving the generalized eigenvalue problem

$$E_0 v = \lambda E v$$

for eigenvectors v. The q eigenvectors give the orthogonal linear combinations of the responses that produces maximal univariate F statistics [126]. For instance, the Lawley-Hotelling trace is given by the sum of eigenvalues

$$\lambda_1 + \lambda_2 + \cdots + \lambda_q.$$

Wilks's Lambda is given by

$$(1 + \lambda_1)^{-1}(1 + \lambda_2)^{-1} \cdots (1 + \lambda_q)^{-1}.$$

Roy's maximum root statistic R is the largest eigenvalue $\max_j \lambda_j$. The distributions of these multivariate test statistics are approximately F. In the case there is only one eigenvalue, all these multivariate test statistics simplify to Hotelling's T-square statistic. The Hotelling's T-square statistic has been widely used in modeling 3D coordinates and deformations in brain imaging [56, 81, 131, 194, 359]. The random field theory for Hotelling's T-square statistic has been available for a while [56]. However, the random field theory for the Roy's maximum root was not developed until recently [355, 396].

The inference for Roy's maximum root is based on the Roy's union-intersection principle [306, 396], which simplifies the multivariate problem to a univariate linear model. Let us multiply an arbitrary constant vector $\nu_{3\times 1}$ on both sides of (7.4):

$$P\nu = XB\nu + ZG\nu + U\Sigma\nu. \tag{7.5}$$

Obviously (7.5) is a usual univariate linear model with a Gaussian noise. For the univariate testing on $B\nu = 0$, the inference is based on the usual F statistic with p and $n-p-r$ degrees of freedom, denoted as F_ν. Then Roy's maximum root statistic is given by

$$R = \max_\nu F_\nu.$$

Now it is obvious that the usual random field theory can be applied in correcting for multiple comparisons. The only trick is to increase the search space, in which we take the supreme of the F random field, from the template surface to a much higher dimension to account for maximizing over ν as well. Another way of defining RoyŠs maximum root is via maximal canonical correlations [396].

7.4.1 SurfStat

Keith Worsley's `SurfStat` package (`www.math.mcgill.ca/keith/surfstat`) has a built in `MATLAB` routine for determining the p-value for Roy's maximum root statistic so there is no need to worry about computational details for most users. `SurfStat` was developed to utilize a model formula and avoids the explicit use of design matrices and contrasts, which tend to be a hinderance to most end users not familiar with such concepts. `SurfStat` can import MNI [237], FreeSurfer (`surfer.nmr.mgh.harvard.edu`) based cortical mesh formats as well as other volumetric image data. A similar model formula approach is implemented in many statistics packages such as Splus (`www.insightful.com`) R (`www.r-project.org`) and SAS (`www.sas.com`). These statistics packages accept a linear model like

$$\texttt{P} = \texttt{Group} + \texttt{Age} + \texttt{Brain}$$

as the direct input for linear modeling avoiding the need to explicitly state the design matrix. `P` is a $n \times 3$ matrix of coordinates, `Age` is the age of subjects, `Brain` is the total brain volume of subject and `Group` is the categorical group variable. This type of model formula has yet to be implemented in widely used SPM or AFNI packages. Here we illustrate `SurfStat` package by showing the step-by-step command lines for multivariate linear models.

7.5 Case Study: Surface Deformation Analysis

The data published in [206, 205] is used throughout the section. We have 3 Tesla T1-weighted inverse recovery fast gradient echo MRI, collected in 124 contiguous 1.2-mm axial slices (TE=1.8 ms; TR=8.9 ms; flip angle = 10°; FOV = 240 mm; 256 × 256 data acquisition matrix) of 69 middle aged and elderly adults ranging from 38 to 79 years (mean age = 58.04 ± 11.34 years). The data were originally collected for a national study on the health and well-being in the aging population, called MIDUS (midlife in US). The website explaining additional details of the study is given in `midus.wisc.edu`.

There are 23 males and 46 females in the study. Trained raters manually segmented the amygdala and hippocampus structures from the T1-weighted images. Brain tissues in the MRI scans were automatically segmented using Brain Extraction Tool (BET) [332]. Then we performed a nonlinear image registration using the diffeomorphic shape and intensity averaging technique with cross-correlation as similarity metric through Advanced Normalization Tools (ANTS) [19]. Using the deformation field obtained from warping the individual image to the template, we aligned the amygdala and hippocampus binary masks to the template space. The normalized masks were then averaged to produce the subcortical structure template. The isosurfaces of the subcortical structure template were extracted using the marching cube algorithm [232]. The amygdala and hippocampus surface template is given in Figure 7.6.

The whole data set is saved in a single `MAT` file `midus.mat` that contains the surface coordinates of the template, displacement vector fields for all 69 subjects, age and gender of the subjects. The following codes for

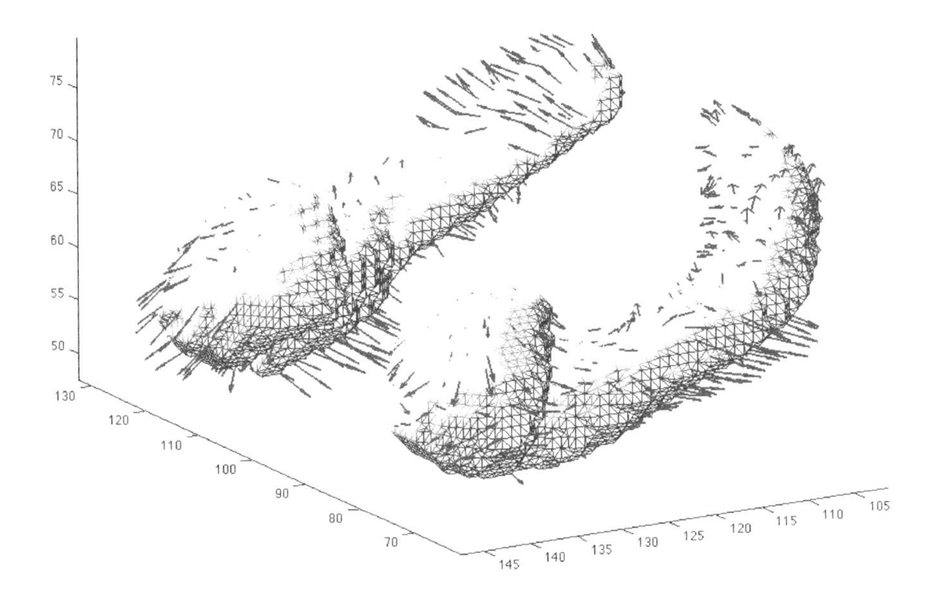

FIGURE 7.5
Displacement vector field of the 10th subject on top of the template surfaces. The end points of the arrows are where the surfaces of the 10th subject are.

performing univariate and multivariate general linear models are given in `http://brainimaging.waisman.wisc.edu/~chung/BIA`. Figure 7.6 is generated by running the codes:

```
load midus.mat

amygl=template{1,1}
amygr=template{1,2}
hippol=template{2,1}
hippor=template{2,2}

figure; figure_wire(amygl,'black','red')
hold on; figure_wire(amygr,'black','blue')
hold on; figure_wire(hippol,'black','yellow')
hold on; figure_wire(hippor,'black','green')
view([-120 15])
```

The left amygdala surface template is given as the structured array variable `amygl` and the right hippocampus template is given as `hippor`. Other surface variable names are similarly given. Each surface has a different number of mesh vertices and face numbers. For instance, `amygl` has the data structure of the form

```
amygl =

    vertices: [1296x3 double]
       faces: [2588x3 double]
```

The surface amygl has 1296 mesh vertices and 2588 triangles. At each mesh vertex, we have the displacement vector field. The displacement vector field is stored in the variable displacement, which is $69 \times 2 \times 2$ cells. Since there are 4 individual structures, there are 4 cell entries. The displacement vectors of the 10th subject left amygdala is accessed by displacement{10,1,1}. The cell entry is a usual matrix format, which is stored in the variable dispal. Other vector fields are stored similarly:

```
dispal = displacement{10,1,1};
dispar = displacement{10,1,2};
disphl = displacement{10,2,1};
disphr = displacement{10,2,2};
```

The variable dispal is a matrix of size 1296×3. The first few entries are

```
dispal =

   -2.4857    0.2103   -0.3367
   -2.3826    0.3100   -0.0352
   -2.9083    0.0469   -0.2653
   .
   .
   .
```

This is the response vector we will use in MGLM. Four vector fields dispal, dispar, disphl and disphr are displayed separately (Figure 7.5) as

```
figure;
hold on; figure_quiver2(amygl, dispal, 9)
hold on; figure_quiver2(amygr, dispar, 9)
hold on; figure_quiver2(hippol, disphl, 9)
hold on; figure_quiver2(hippor, disphr, 9)
view([-120 15])
```

Since the vector fields are dense, we only subsample 1 out of every 9 vectors and display. The subsampling rate is controlled in the last argument.

7.5.1 Univariate Tests in SurfStat

In the data set, we have brain, the total volume of brain excluding cerebellum in terms of mm^3, and age in terms of years. gender is the categorical variable consisting of 1 (female) and 0 (male).

```
gender =

     1     1     1     1     1     0 ...
```

SurfStat requires 3D array of vector fields to be arranged as 69 (number of subjects) $\times$ 2444 (number of vertices) $\times$ 3 (dimension). For example, the displacements for the left hippocampus surfaces `disphl` are given by

```
 disphl=[];
for i=1:69
    disphl(i,:,:)=displacement{i,2,1};
end

size(disphl)

ans =

          69          2444             3
```

From now on, we will only show statistical analysis for the left hippocampus since we are applying the same procedures to the other three surfaces as well. Let us test for the effect of gender on the length of displacement while accounting for brain size and age variations. This is done by the t-test via `SurfStatT` routine. The example below will be self-evident.

```
displ2 = sqrt(sum(disphl.^2,3));

Gender=term(gender);
Age=term(age);
Brain=term(brain);

slm1 = SurfStatLinMod( disphl2, 1+ Age +Brain + Gender);
slm = SurfStatT( slm1, gender);
hold on; figure_trimesh( hippol ,slm.t)
```

Equivalently we can test for the effect of gender using the F-test via `SurfStatF`. The problem with the F test is that it does not provide the directional information so the T-test is recommended for testing one contrast (Figure 7.7).

```
slm0 = SurfStatLinMod( disphl2, 1+Brain +Age);
slm1 = SurfStatLinMod( disphl2, 1+Brain + Gender+Age);
slm = SurfStatF( slm1, slm0);
hold on; figure_trimesh( hippol ,slm.t)
```

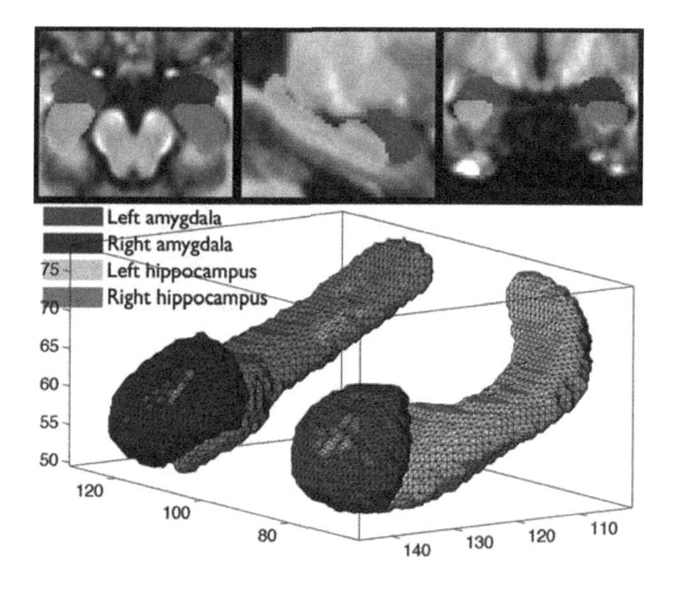

FIGURE 7.6
Subcortical structure template is superimposed on top of MRI [206, 205]. The figure is generated by Seung-Goo Kim of Seoul National University. (See color insert.)

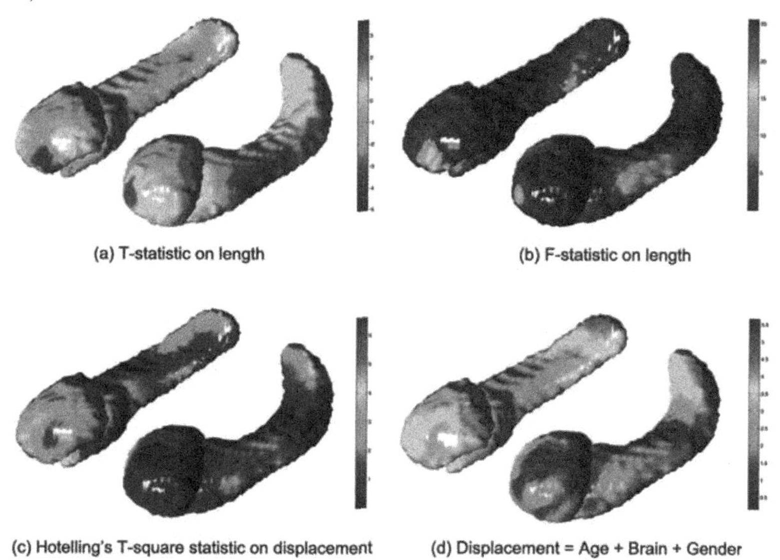

FIGURE 7.7
The gender effect on the length of displacement using (a) t-statistic and (b) F-statistic. The effect of gender on the displacement vector using (c) Hotelling's T^2 statistic and (d) Roy's maximum root. (See color insert.)

7.5.2 Multivariate Tests in `SurfStat`

The problem with the univariate tests is that they are based on data reduction of the displacement vector field into a scalar field. So we are not utilizing full information for the data analysis. In order to use the vector field without the data reduction, we need MGLM. Let us test the effect of `gender` on the displacement vectors using the Hotelling's T^2 statistic, which is implemented in `hotelT2`. For this, we need to separate the displacement into male and female groups using `find` function as follows.

```
disphlm = disphl(find(~gender),:,:);
disphlf = disphl(find(gender),:,:);
h =SurfStat2HotelT2(disphlm, disphlf);
figure; figure_trimesh( hippol ,h.t)
```

Equivalently, we can take the maximum of the t-statistics for all possible linear combinations of the x, y and z coordinates such as $x + 0.5y - 1.2z$. This maximum t gives the square root of Hotelling's T^2. In `SurfStat`, statistical analysis proceeds exactly as for univariate data and uses the same statistical routines. The resulting Hotelling's T^2 value is displayed in Figure 7.7.

The problem with the Hotelling's T^2 statistic method is the lack of controls for age and brain size differences. This gives a motivation for using MGLM. To test the effect of `Gender`, the model of the from `displacement = Age + Brain + Group` is fitted using

```
slm0 = SurfStatLinMod( disphl, Age + Brain + Gender);
slm = SurfStatT( slm0, gender);
figure; figure_trimesh( hippol ,slm.t)
```

The result is given in Figure 7.7. In this chapter, we discussed the multivariate general linear models without surface-based smoothing and multiple comparisons issues, which will be covered in later chapters.

8

Cortical Surface Analysis

In this chapter, we introduce a unified statistical approach to deformation-based morphometry applied to the cortical surface. The cerebral cortex has the topology of a 2D highly convoluted sheet. As the brain develops over time, the cortical surface area, thickness, curvature and total gray matter volume change (Figure 8.1). It is highly likely that such age-related surface changes are not uniform. By measuring how such surface metrics change over time, the regions of the most rapid structural changes can be localized. We avoided using surface flattening, which distorts the inherent geometry of the cortex in our analysis and it is only used in visualization. To increase the signal to noise ratio, diffusion smoothing, which generalizes Gaussian kernel smoothing to an arbitrary curved cortical surface, has been developed and applied to surface data. Afterwards, statistical inference on the cortical surface will be performed via random fields theory. As an illustration, we demonstrate how this new surface-based morphometry can be applied in localizing the cortical regions of the gray matter tissue growth and loss in the brain images longitudinally collected in the group of children and adolescents. This chapter follows closely to the methods and applications given in [83].

8.1 Introduction

The cerebral cortex has the topology of a 2-dimensional convoluted sheet. Most of the features that distinguish these cortical regions can only be measured relative to that local orientation of the cortical surface [97]. As the brain develops over time, cortical surface area as well as cortical thickness and the curvature of the cortical surface change. As shown in the previous normal brain development studies, the growth pattern in developing normal children is nonuniform over whole brain volume [81, 140, 273, 361]. Between ages 12 and 16, the corpus callosum and the temporal and parietal lobes show the most rapid brain tissue growth and some tissue degeneration in the subcortical regions of the left hemisphere [81, 361]. It is equally likely that such age-related changes with respect to the cortical surface are not uniform as well. By measuring how geometric metrics such as the cortical thickness, curvature

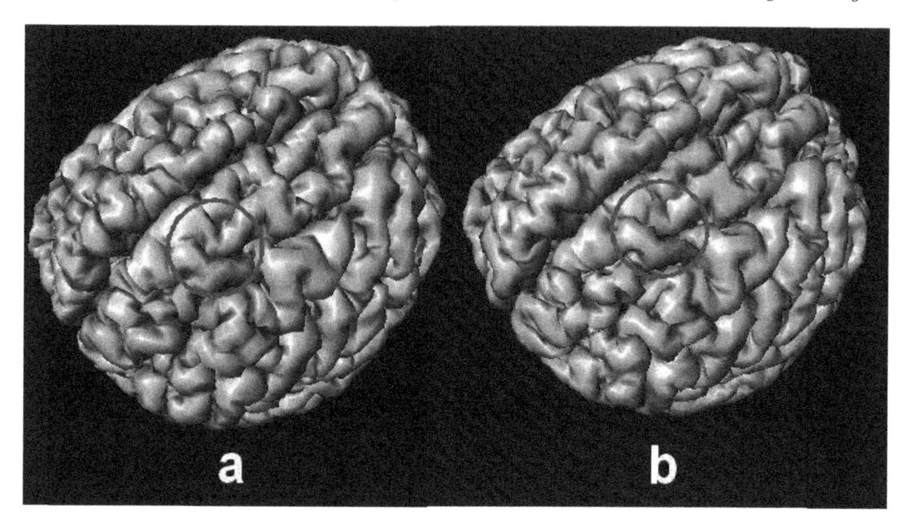

FIGURE 8.1
The outer cortical surfaces of a single subject at age 14 (a) and at age 19 (b) showing globally similar cortical patterns. If we observe the circled regions carefully, we can see slight shape variations.

and local surface area change over time, any statistically significant brain tissue growth or loss in the cortex can be detected.

The first obstacle in developing surfaced-based morphometry is the automatic segmentation or extraction of the cortical surfaces from MRI. It requires first correcting RF inhomogeneity artifacts. We have used nonparametric nonuniform intensity normalization method (N3), which eliminates the dependence of the field estimate on anatomy [330]. The next step is the tissue classification into three types: gray matter, white matter and cerebrospinal fluid (CSF). An artificial neural network classifier [211, 270] or a mixture model cluster analysis [147] can be used to segment the tissue types automatically. After the tissue classification, the cortical surface is usually generated as a smooth triangular mesh. The most widely used method for triangulating the surface is the marching cubes algorithm [232]. Level set method [320] or deformable surfaces method [100] is also available. In our study, we have used the anatomic segmentation using proximities (ASP) method [237], which is a variant of the deformable surfaces method, to generate cortical triangular meshes that have the topology of a sphere. Brain substructures such as the brain stem and the cerebellum were removed. Then an ellipsoidal mesh that already had the topology of a sphere was deformed to fit the shape of the cortex guaranteeing the same topology. The resulting triangular mesh will consist of 40,962 vertices and 81,920 triangles with the average internodal distance of 3 mm. Partial voluming is a problem with the tissue classifier but topology constraints used in the ASP method were shown to provide some correction by incorporating much neuroanatomical *a priori* information [237]. The triangular meshes are not constrained to lie on voxel boundaries. Instead the triangular meshes can cut through a voxel, which can be considered as correcting where the true boundary ought to be. Once we have a triangular mesh as the realization of the cortical surface, we can model how the cortical surface deforms over time.

In modeling the surface deformation, a proper mathematical framework might be found in both differential geometry and fluid dynamics. The concept of the *evolution of phase-boundary* in fluid dynamics [110, 155], which describes the geometric properties of the evolution of boundary layer between two different materials due to internal growth or external force, can be used to derive the mathematical formula for how the surface changes. It is natural to assume the cortical surfaces to be a smooth 2-dimensional *Riemannian manifold* parameterized by u^1 and u^2:

$$\mathbf{X}(u^1, u^2) = \{x_1(u^1, u^2), x_2(u^1, u^2), x_3(u^1, u^2) : (u^1, u^2) \in D \subset \mathbb{R}^2\}.$$

A more precise definition of a Riemannian manifold and a *parameterized surface* can be found in [46, 107, 213]. If D is a unit square in $\mathbb{R}^2$ and a surface is topologically equivalent to a sphere then at least two different global parameterizations are required. Surface parameterization of the cortical surface has been done previously in [363, 196]. From the surface parameterization, Gaussian and mean curvatures of the brain surface can be computed and used to characterize its shape [97, 150, 196]. In particular, S.C. Joshi *et al.* (1995) used the quadratic surface in estimating the Gaussian and mean curvature of the cortical surfaces [196].

By combining the mathematical framework of the evolution of phase-boundary with the statistical framework developed for 3D whole brain volume deformation [81], anatomical variations associated with the deformation of the cortical surface can be statistically quantified. Using the same stochastic assumption on the deformation field used in [81], we will localize the region of brain shape changes based on three surface metrics: area dilatation rate, cortical thickness and curvature changes and show how these metrics can be used to characterize the brain surface shape changes over time.

As an illustration of our unified approach to deformation-based surface morphometry, we will demonstrate how the surface-based statistical analysis can be applied in localizing the cortical regions of tissue growth and loss in brain images longitudinally collected in a group of children and adolescents.

8.2 Modeling Surface Deformation

Let $\mathbf{U}(\mathbf{x}, t) = (U_1, U_2, U_3)^t$ be the 3D displacement vector required to deform the structure at $\mathbf{x} = (x_1, x_2, x_3)$ in gray matter Ω_0 to the homologous structure

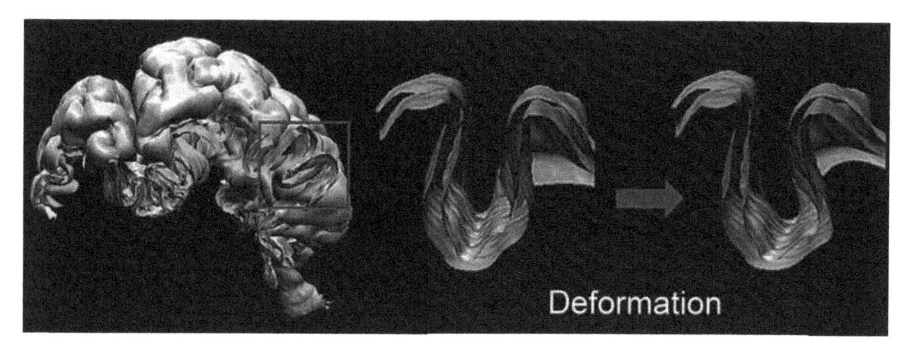

FIGURE 8.2
Gray matter deformation causes the geometry of the both outer and inner cortical surfaces to change. The deformation of the surfaces can be written as $\mathbf{x} \to \mathbf{x} + \mathbf{U}(\mathbf{x}, t)$, where $\mathbf{U}$ is the surface displacement vector field.

after time T. Whole gray matter volume Ω_0 will deform continuously and smoothly to Ω_t via the deformation $\mathbf{x} \to \mathbf{x}+\mathbf{U}$ while the cortical boundary $\partial\Omega_0$ will deform to $\partial\Omega_t$. The cortical surface $\partial\Omega_t$ may be considered as consisting of two parts: the outer cortical surface $\partial\Omega_t^{out}$ between the gray matter and CSF and the inner cortical surface $\partial\Omega_t^{in}$ between the gray and white matter (Figure 8.2), i.e.

$$\partial\Omega_t = \partial\Omega_t^{out} \cup \partial\Omega_t^{in}.$$

Although we will exclusively deal with the deformation of the cortical surfaces, our deformation-based surface morphometry can be equally applicable to the boundary of any brain substructure.

We propose the following stochastic model on the displacement velocity $\mathbf{V} = \partial\mathbf{U}/\partial t$, which has been used in the analysis of whole brain volume deformation [81]:

$$\mathbf{V}(\mathbf{x}) = \mu(\mathbf{x}) + \Sigma^{1/2}(\mathbf{x})\epsilon(\mathbf{x}), \mathbf{x} \in \Omega_0, \tag{8.1}$$

where μ is the mean displacement velocity and $\Sigma^{1/2}$ is the 3×3 symmetric positive definite covariance matrix, which allows for correlations between components of the displacement fields. The components of the error vector ϵ are are assumed to be independent and identically distributed as smooth stationary Gaussian random fields with zero mean and unit variance. It can be shown that the normal component of the displacement velocity $\mathbf{V} = \partial\mathbf{U}/\partial t$ restricted on the boundary $\partial\Omega_0$ uniquely determine the evolution of the cortical surface.

Estimating the surface displacement fields $\mathbf{U}$ and the surface extraction can be performed at the same time by the ASP algorithm. First, an ellipsoidal mesh placed outside the brain was shrunk down to the surface $\partial\Omega_0^{in}$. The vertices of the resulting inner mesh are indexed and the ASP algorithm will deform the inner mesh to fit the outer surface $\partial\Omega_0^{out}$. by minimizing a cost function that involves bending, stretch and other topological constraints [237].

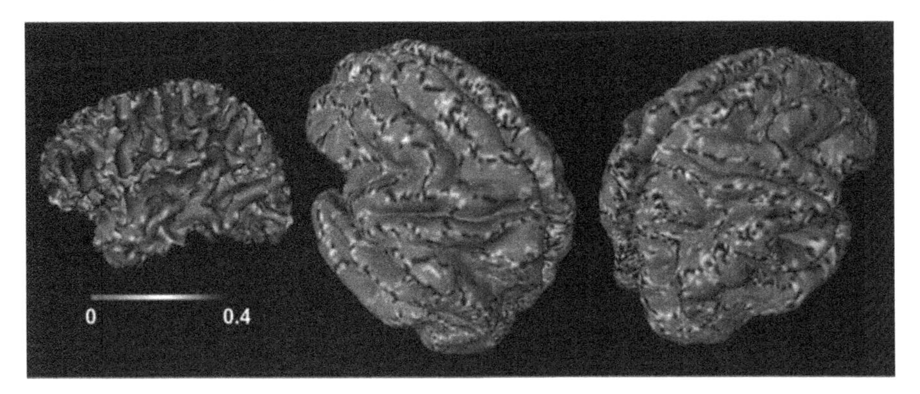

FIGURE 8.3
Individual gyral patterns mapped onto the surface atlas $\partial\Omega_{atlas}$. The gyral patterns are extracted by computing the bending metric on the inner cortical surface (left). The middle and right figures show the mapping of the gyral pattern of a single subject (left) onto the atlas surface. The gyri of the subject match the gyri of the atlas illustrating a close homology between the surface of an individual subject and surface atlas. If there is no homology between the corresponding vertices, we would have complete misalignment.

The vertices indexed identically on both meshes will lie within a very close proximity and these define the automatic linkage in the ASP algorithm. To generate the outer surface $\partial\Omega_t^{out}$ at later time T, we start with the inner surface $\partial\Omega_0^{in}$, and then deform it to match the outer surface $\partial\Omega_t^{out}$ by minimizing the same cost function. Starting with the same mesh in two outer surface extractions, each point on $\partial\Omega_0^{in}$ gets mapped to corresponding points on $\partial\Omega_0^{out}$ and $\partial\Omega_t^{out}$. This method assumes that the shape of the cortical surface does not appreciably change within the subject. This assumption is valid in the case of brain development for a short period of time as illustrated in Figure 8.1, where the global sulcal geometry remains stable for a five year interval, although local cortical geometry shows slight changes. As reported in [83], the displacement is less than 1 mm/year in average for the same data set while the average internodal distance in triangular meshes we are using is 3 mm. So the displacements are relatively small compared to the size of mesh itself.

Constructing surface atlas $\partial\Omega_{atlas}$, where the statistical parametric maps (SPM) of surface metrics will be formed, is done by averaging the coordinates of corresponding vertices that have the same indices. This atlas construction method has been first introduced in [237], where it is used to create the cortical thickness map for 150 normal subjects. The geometrical constraints such as stretch and bending terms in the ASP algorithm enforces a relatively consistent correspondence on the cortical surface. Figure 8.3 shows the mapping of the gyral pattern (red and yellow lines) of a single subject onto the atlas surface. The gyri of the subject match the gyri of the atlas. Note the full anatomical details still presented in $\partial\Omega_{atlas}$ even after the vertex averaging. Major sulci such as the central sulcus and superior temporal sulcus are clearly identifiable. If there is no homology between corresponding vertices, one would only expect to see featureless dispersion of points.

8.3 Surface Parameterization

The ASP method generates triangular meshes consisting of 81,920 triangles evenly distributed in size. In order to quantify the shape change of the cortical surface, *surface parameterization* is an essential part. We model the cortical surface as a smooth 2D Riemannian manifold parameterized by two parameters u^1 and u^2 such that any point $\mathbf{x} \in \partial\Omega_0$ can be uniquely represented as

$$\mathbf{x} = \mathbf{X}(u^1, u^2)$$

for some parameter space $\mathbf{u} = (u^1, u^2) \in D \subset \mathbb{R}^2$. We will try to avoid global parameterization such as tensor B-splines, which are computationally expensive compared to a local surface parameterization. Instead, a quadratic polynomial

$$z(u^1, u^2) = \beta_1 u^1 + \beta_2 u^2 + \beta_3 (u^1)^2 + \beta_4 u^1 u^2 + \beta_5 (u^2)^2 \tag{8.2}$$

will be used as a local parameterization fitted via least-squares estimation. Using the least-squares method, these coefficients β_i can be estimated. The numerical implementation can be found in [64]. Slightly different quadratic surface parameterizations are also used in estimating curvatures of a macaque monkey brain surface [196, 202]. Once β_i are estimated, $\mathbf{X}(u^1, u^2) = \big(u^1, u^2, z(u^1, u^2)\big)^t$ becomes a local surface parameterization of choice.

8.3.1 Quadratic Parameterization

Quadratic surface parameterization is a smoothing technique to fit the given points $\mathbf{p}_0, \ldots, \mathbf{p}_m$ to a polynomial function of the form

$$f(x, y) = \sum_{i+j \le p} \beta_{ij}\, x^i y^j. \tag{8.3}$$

For $\mathbf{p}_i = (x^i, y^i, z^i)$, $(p+1)(p+2)/2$ unknown coefficients β_{ij} are chosen to minimize the residual

$$\sum_{i=0}^{m} \left[z^i - f(x^i, y^i)\right]^2.$$

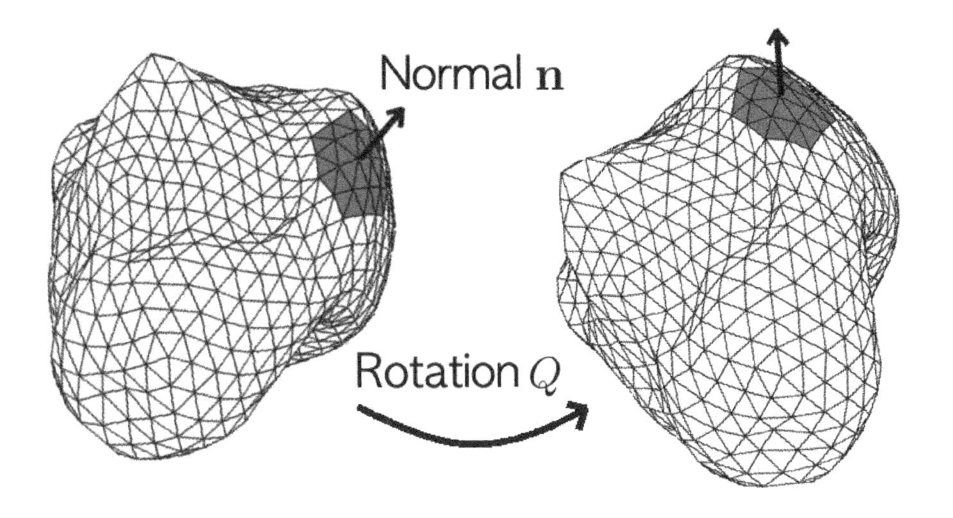

FIGURE 8.4
A surface has been rotated by an orthogonal matrix Q so that the normal vector $\mathbf{n}$ is aligning with the x_3-axis.

using the least squares method. The drawback of the polynomial regression is that there is a tendency to weave the outer most vertices to find vertices in the center. Therefore this is not advisable to directly fit (8.3) when the z-coordinate values rapidly change. Polynomial regression is not recommended for global surface parameterization.

In estimating various differential geometric measures such as the Laplace-Beltrami operator or curvatures, it is not necessary to find such global parameterization of the surface $\mathcal{M}$. A local surface parameterization in the neighborhood of $\mathbf{p}$ can be obtained via the projection of the local surface onto the tangent plane $T_{\mathbf{p}}(\mathcal{M})$.

Let Q be an orthogonal matrix which rotates the normal vector $\mathbf{n} = (n_1, n_2, n_3)$ to align with the x_3-axis, i.e. $Q\mathbf{n} = (0, 0, 1)'$. Algebraic manipulation shows that

$$Q = \begin{pmatrix} n_3 & 0 & -\sqrt{n_1^2 + n_2^2} \\ 0 & 1 & 0 \\ \sqrt{n_1^2 + n_2^2} & 0 & n_3 \end{pmatrix} \begin{pmatrix} \frac{n_1}{\sqrt{n_1^2+n_2^2}} & \frac{n_2}{\sqrt{n_1^2+n_2^2}} & 0 \\ -\frac{n_2}{\sqrt{n_1^2+n_2^2}} & \frac{n_1}{\sqrt{n_1^2+n_2^2}} & 0 \\ 0 & 0 & 1 \end{pmatrix}$$

is such an orthogonal matrix assuming $n_1, n_2 \neq 0$. If $n_1 = n_2 = 0$, we can take Q to be an identify matrix.

Let $x \in \mathcal{M}$ be a point in the neighborhood of p. Under the transformation

$(u^1, u^2, u^3)' : x \to Q(x - p)$, the local surface patch translates to the origin and then rotates by Q (Figure 8.4). With respect to the new coordinates (u^1, u^2, u^3), the local surface patch can be explicitly written as the standard explicit form $u_3 = f(u^1, u^2)$ for some function f assuming local smoothness of the surface. By identifying (u^1, u^2) as the parameter space $\mathcal{N}$, we have the following local parameterization in the neighborhood of p:

$$X(u^1, u^2) = p + Q'\Big(u^1, u^2, f(u^1, u^2)\Big)'. \tag{8.4}$$

Then the basis on the tangent plane $T_p(\mathcal{M})$ is

$$X_1 = Q'\Big(1, 0, \frac{\partial f}{\partial u^1}\Big|_{(0,0)}\Big)'$$

and

$$X_2 = Q'\Big(0, 1, \frac{\partial f}{\partial u^2}\Big|_{(0,0)}\Big)'.$$

In the neighborhood of $(0, 0)$, we have the Taylor approximation of the function f:

$$f(u^1, u^2) = \beta_1 u^1 + \beta_2 u^2 + \beta_3 (u^1)^2 + \beta_4 u^1 u^2 + \beta_5 (u^2)^2 + \cdots. \tag{8.5}$$

Since we are forcing the function f to pass through the origin, there is no constant term in the Taylor expansion. The problem of estimating the parameters β_j can be formulated in terms of least-squares estimation. For m neighboring points $p_1, \cdots, p_m$, let

$$u_i = (u_i^1, u_i^2, u_i^3)' = Q(p_i - p).$$

Then the unknown parameters β_i are chosen to be the least-squares estimates of linear equations $Y = \mathbf{X}\beta$, where

$$\beta = (\beta_1, \cdots, \beta_5)',$$

$$Y = (u_1^3, \cdots, u_m^3)'$$

and the $m \times 5$ matrix $\mathbf{X}$ is given by

$$\mathbf{X} = \begin{pmatrix} u_1^1 & u_1^2 & (u_1^1)^2 & u_1^1 u_1^2 & (u_1^2)^2 \\ u_2^1 & u_2^2 & (u_2^1)^2 & u_2^1 u_2^2 & (u_2^2)^2 \\ \cdots & \cdots & \cdots & \cdots & \cdots \\ u_m^1 & u_m^2 & (u_m^1)^2 & u_m^1 u_m^2 & (u_m^2)^2 \end{pmatrix}.$$

The least-squares estimation is

$$\widehat{\beta} = (\mathbf{X}'\mathbf{X})^- \mathbf{X}'Y, \tag{8.6}$$

where $^-$ denotes a generalized inverse, which can be obtained through the singular value decomposition [216].

8.3.2 Numerical Implementation

We will use the hippocampus surface data hippocampus.mat as an example.

```
load hippocampus.mat;

figure;figure_wire(hippoleft,'k','w')
view([140 10])
```

The MATLAB implementation requires Toolbox Graph written by Gabriel Peyré of Université Paris-Dauphine (www.mathworks.com/matlabcentral/fileexchange/5355-toolbox-graph) in computing the surface normal vectors.

```
coord=hippoleft.vertices';
tri=hippoleft.faces;

path(path, './toolbox_graph');
[n nf]=compute_normal(coord',tri);
```

The variable n contains the surface normal at mesh vertices while nf contains the surface normal at mesh faces. In order to compute the quadratic surface at the first mesh vertex, it is necessary to rotate the surface normal to the x_3 coordinate.

```
i_point=1;

[nbr, degree] = FINDnbr(tri);
nbr_list=nbr(i_point,1:degree(i_point));
coord_nbr = coord(:,nbr_list)- ...
            kron(ones(1,degree(i_point)),coord(:,i_point));

deno= sqrt(n(1)^2+n(2)^2);
if deno < 0.1
    Q=eye(3);
else
    Q1 = [[n(1)/deno,n(2)/deno,0];
        [-n(2)/deno,n(1)/deno,0];
        [0,      0,        1]];
    Q2 = [[n(3),0, -deno];
        [0,       1,       0];
        [deno,0,n(3)]];
    Q=Q2*Q1;
end;
```

This computes the orthonormal matrix Q. The above computation is exact if Q*n(:,1) gives the unit vector (0,0,1)'. Then we rotate the surface coordinates by the orthonormal matrix Q and fit the quadratic polynomial in the least squares fashion.

```
rot_coord_nbr = Q*coord_nbr;

z = [rot_coord_nbr(3,:)]';
x = [rot_coord_nbr(1,:)]';
y = [rot_coord_nbr(2,:)]';

X=[x, y, x.^2, 2*x.*y, y.^2];
beta = (X\z)'
```

For the first mesh vertex, the estimated coefficients are as follows.

```
>> beta

beta =

   -0.1198   -0.0335   -0.8999    0.0565   -0.0035
```

8.4 Surface-Based Morphological Measures

As in the case of local volume change in the whole brain volume [81], the rate of cortical surface area expansion or reduction may not be uniform across the cortical surface. Extending the idea of volume dilatation, we introduce new concepts of surface area, curvature, cortical thickness dilatation and their rate of change over time.

Suppose that the cortical surface $\partial\Omega_t$ at time T can be parameterized by the parameters $\mathbf{u} = (u^1, u^2)$ such that any point $\mathbf{x} \in \partial\Omega_t$ can be written as $\mathbf{x} = \mathbf{X}(\mathbf{u}, t)$. Let $\mathbf{X}_i = \partial\mathbf{X}/\partial u^i$ be a partial derivative vector. The *Riemannian metric tensor* g_{ij} is given by the inner product between two vectors $\mathbf{X}_i$ and $\mathbf{X}_j$, i.e.

$$g_{ij}(t) = \langle \mathbf{X}_i, \mathbf{X}_j \rangle.$$

The Riemannian metric tensor g_{ij} measures the amount of the deviation of the cortical surface from a flat Euclidean plane. The Riemannian metric tensor enables us to measure lengths, angles and areas in the cortical surface. Let $g = (g_{ij})$ be a 2×2 matrix of metric tensors. From [64], the rate of metric tensor change is approximately

$$\frac{\partial g}{\partial t} \approx 2(\nabla \mathbf{X})^t (\nabla \mathbf{V}) \nabla \mathbf{X}, \tag{8.7}$$

where $\mathbf{V} = \partial\mathbf{U}/\partial t$ and $\nabla\mathbf{X} = (\mathbf{X}_1, \mathbf{X}_2)$ is a 3×2 gradient matrix. We are not directly interested in the metric tensor change itself but rather functions of g and $\partial g/\partial t$, which will be used to measure surface area and curvature change.

8.4.1 Local Surface Area Change

The total surface area of the cortex $\partial\Omega_t$ is given by

$$\|\partial\Omega_t\| = \int_D \sqrt{\det g}\, d\mathbf{u},$$

where $D = X^{-1}(\partial\Omega_t)$ is the parameter space [213]. The term $\sqrt{\det g}$ is called the *infinitesimal surface area element* and it measures the transformed area of the unit square in the parameter space D via the transformation $X(\cdot, t) : D \to \partial\Omega_t$. The infinitesimal surface area element can be considered as a generalization of Jacobian, which has been used in measuring local volume in whole brain volume in tensor-based morphometry [81]. The *local surface area dilatation rate* Λ_{area} or the rate of local surface area change per unit surface area is then defined as

$$\Lambda_{area} = \frac{1}{\sqrt{\det g}} \frac{\partial \sqrt{\det g}}{\partial t}, \tag{8.8}$$

which can be further simplified as $\Lambda_{area} = \partial(\ln \sqrt{\det g})/\partial t$. If the whole gray matter Ω_t is parameterized by 3D curvilinear coordinates $\mathbf{u} = (u^1, u^2, u^3)$, then the dilatation rate $\partial(\ln \sqrt{\det g})/\partial t$ becomes the local volume dilatation rate Λ_{volume} first introduced in deformation-based morphometry [81]. Therefore, the concepts of local area dilatation and volume dilatation rates are essentially equivalent.

In our study, two MR scans were collected for each subject at different times. Let t_1^j and t_2^j to be the times scans were taken for subject j. Then the local surface area dilatation rate Λ_{area}^j for subject j is estimated as a finite difference:

$$\Lambda_{area}^j = \frac{\sqrt{\det g(t_2^j)} - \sqrt{\det g(t_1^j)}}{(t_2^j - t_1^j)\sqrt{\det g(t_1^j)}},$$

where $g(t)$ is the matrix evaluated at T. Other dilatation rates that will be introduced later can be estimated in a similar fashion.

Instead of using metric tensors g_{ij}, it is possible to formulate local surface area change in terms of the areas of the corresponding triangles. However, this formulation assigns surface area change values to each face instead of each vertex and this might cause a problem in both surface-based smoothing and statistical analysis, where values are defined on vertices. Defining scalar values on vertices from face values can be done by the weighted average of face values. It is not hard to develop surface-based smoothing and statistical analysis on values defined on faces but the cortical thickness and the curvature metric will be defined on vertices so we will end up with two separate approaches: one for metrics defined on vertices and the other for metrics defined on faces. Therefore, our metric tensor approach seems to provide a basis of unifying surface metric computations, surface-based smoothing and statistical analysis together.

Under the assumption of stochastic model (8.1), the area dilatation rate can be approximately distributed as Gaussian:

$$\Lambda_{area}(\mathbf{x}) = \lambda_{area}(\mathbf{x}) + \epsilon_{area}(\mathbf{x}), \tag{8.9}$$

where

$$\lambda_{area} = tr[g^{-1}(\nabla X)^t(\nabla\mu)\nabla X]$$

is the mean area dilatation rate and ϵ_{area} is a mean zero Gaussian random field defined on the cortical surface [64]. The area dilatation rate is invariant under parameterization, i.e. the area dilatation rate will always be the same no matter which parameterization is chosen. λ_{area} can be estimated by the sample mean

$$\bar{\Lambda}_{area} = \frac{1}{n}\sum_{j=1}^{n}\Lambda_{area}^{j}$$

and the significance of the mean will be tested via T statistic.

So far our statistical modeling is centered on localizing regions of rapid morphological changes on the cortical surface but both local and global morphological measures are important in the characterization of brain deformation. Global morphometry is relatively easy compared to local morphometry with respect to modeling and computation. The total surface area $\|\partial\Omega_t\|$ can be estimated by the sum of the areas of 81,920 triangles generated by the ASP algorithm. Then we define the total surface area dilatation rate as

$$\Lambda_{total\ area} = \frac{\partial \ln \|\partial\Omega_t\|}{\partial t}.$$

It can be shown that, under assumption (8.1), the total surface area dilatation rate $\Lambda_{total\ area}$ is distributed as a Gaussian random variable and hence a statistical inference on total surface area change will be based on a simple T test [64]. This measure will be used in determining the rate of the total surface area decreases in both outer and inner cortical surfaces between ages 12 and 16.

8.4.2 Local Gray Matter Volume Change

Local volume dilatation rate Λ_{volume} for whole brain volume is defined in [81] using the Jacobian of deformation $\mathbf{x} \to \mathbf{x} + \mathbf{U}(\mathbf{x})$ as

$$\Lambda_{volume} = \mathrm{tr}(\nabla\mathbf{V}) = \frac{\partial}{\partial t}\mathrm{tr}(\nabla\mathbf{U})$$

and used successfully in detecting the regions of brain tissue growth and loss in the whole brain volume. Compared to the local surface area change metric, the local volume change measurement is more sensitive to small deformation.

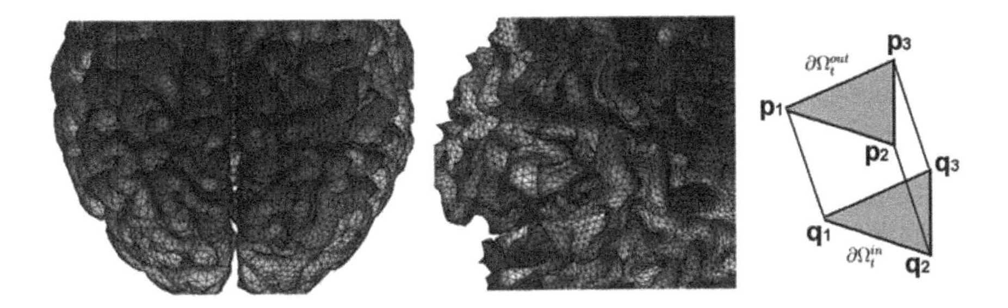

FIGURE 8.5
Outer (left) and inner (middle) triangular meshes. Triangle $(\mathbf{p}_1, \mathbf{p}_2, \mathbf{p}_3) \in \partial\Omega_t^{out}$ on the outer surface will have corresponding triangle $(\mathbf{q}_1, \mathbf{q}_2, \mathbf{q}_3) \in \partial\Omega_t^{in}$ on the inner surface. A convex-hull from 6 points $\{\mathbf{p}_1, \mathbf{p}_2, \mathbf{p}_3, \mathbf{q}_1, \mathbf{q}_2, \mathbf{q}_3\}$ will then form a triangular prism and the collection of 81,920 triangular prisms becomes the whole gray matter.

If a unit cube increases its sides by one, the surface area will increase by $2^2 - 1 = 3$ while the volume will increase by $2^3 - 1 = 7$. Therefore, the statistical analysis based on the local volume change will be at least twice more sensitive compared to that of the local surface area change. So the local volume change should be able to pick out gray matter tissue growth pattern even when the local surface area change may not. In the result section, the highly sensitive aspect of local volume change in relation to local surface area change will be demonstrated.

The gray matter Ω_t can be considered as a thin shell bounded by two surfaces $\partial\Omega_t^{out}$ and $\partial\Omega_t^{in}$ with varying cortical thickness. In triangular meshes generated by the ASP algorithm, each of 81,920 triangles on the outer surface has a corresponding triangle on the inner surface (Figure 8.5). Let $\mathbf{p}_1, \mathbf{p}_2, \mathbf{p}_3$ be the three vertices of a triangle on the outer surface and $\mathbf{q}_1, \mathbf{q}_2, \mathbf{q}_3$ be the corresponding three vertices on the inner surface such that $\mathbf{p}_i$ is linked to $\mathbf{q}_i$ by the ASP algorithm. The triangular prism consists of three tetrahedra with vertices $\{\mathbf{p}_1, \mathbf{p}_2, \mathbf{p}_3, \mathbf{q}_1\}$, $\{\mathbf{p}_2, \mathbf{p}_3, \mathbf{q}_1, \mathbf{q}_2\}$ and $\{\mathbf{p}_3, \mathbf{q}_1, \mathbf{q}_2, \mathbf{q}_3\}$. Then the volume of the triangular prism is given by the sum of the determinants

$$D(\mathbf{p}_1, \mathbf{p}_2, \mathbf{p}_3, \mathbf{q}_1) + D(\mathbf{p}_2, \mathbf{p}_3, \mathbf{q}_1, \mathbf{q}_2) + D(\mathbf{p}_3, \mathbf{q}_1, \mathbf{q}_2, \mathbf{q}_3),$$

where

$$D(\mathbf{a}, \mathbf{b}, \mathbf{c}, \mathbf{d}) = |\det(\mathbf{a} - \mathbf{d}, \mathbf{b} - \mathbf{d}, \mathbf{c} - \mathbf{d})|/6$$

is the volume of a tetrahedron whose vertices are $\{\mathbf{a}, \mathbf{b}, \mathbf{c}, \mathbf{d}\}$. Afterwards, the total gray matter volume $\|\Omega_t\|$ can be estimated by summing the volumes of all 81,920 triangular prisms. Similar to the total surface area dilatation rate, we define the total gray matter volume dilatation rate as

$$\Lambda_{total\ volume} = \frac{\partial \ln \|\Omega_t\|}{\partial t}.$$

8.4.3 Cortical Thickness Change

The average cortical thickness for each individual is about 3mm [171]. Cortical thickness usually varies from 1mm to 4mm depending on the location of the cortex. In normal brain development, it is highly likely that the change of cortical thickness may not be uniform across the cortex. We will show how to localize the cortical regions of statistically significant thickness change in brain development. Our approach introduced here can also be applied to measuring the rate of cortical thinning, possibly associated with Alzheimer's disease. As in the case of the surface area dilatation, we introduce the concept of the cortical thickness dilatation, which measures cortical thickness change per unit thickness and unit time. There are many different computational approaches to measuring cortical thickness but we will use the Euclidean distance $d(\mathbf{x})$ from a point $\mathbf{x}$ on the outer surface $\partial\Omega_t^{out}$ to the corresponding point $\mathbf{y}$ on the inner surface $\partial\Omega_t^{in}$, as defined by the automatic linkages used in the ASP algorithm [237]. The vertices on the inner triangular mesh are indexed and the ASP algorithm can deform the inner mesh to fit the outer surface by minimizing a cost function that involves bending, stretch and other topological constraints. Therefore, both the outer and the inner surfaces should match sulci to sulci and gyri to gyri, and the vertices indexed identically on both surfaces would lie within a very close proximity. One advantage of the cortical thickness metric based on this automatic linkage is that it is less sensitive to fluctuations in surface normals and regions of high curvature [237]. A validation study for the assessment of the accuracy of the cortical thickness measure based on the ASP algorithm has been performed and found to be valid for most of the cortex [197]. There is also an alternate method for automatically measuring cortical thickness based on the Laplace equation [190].

Let $d(\mathbf{x}) = \|\mathbf{x} - \mathbf{y}\|$ be the cortical thickness computed as the usual Euclidean distance between $\mathbf{x} \in \partial\Omega^{out}$ and $\mathbf{y} \in \partial\Omega^{in}$. We define the *cortical thickness dilatation rate* as the rate of the change of the thickness per unit thickness and unit time, i.e.

$$\Lambda_{thickness} = \frac{\partial \ln d(\mathbf{x})}{\partial t}.$$

Under the assumption of stochastic model (8.1), the thickness dilatation rate can be approximately distributed as Gaussian:

$$\Lambda_{thickness}(\mathbf{x}) = \lambda_{thickness}(\mathbf{x}) + \epsilon_{thickness}(\mathbf{x}),$$

where $\lambda_{thickness}$ is the mean cortical thickness dilatation rate and $\epsilon_{thickness}$ is a mean zero Gaussian random field. Unlike the surface area change metric, cortical thickness can only be defined locally but we can compute the within average thickness dilatation rate for subject j:

$$\Lambda^j_{avg\ thickness} = \frac{1}{\|\partial\Omega_0\|} \int_{\mathbf{x}\in\partial\Omega_0} \Lambda^j_{thickness}(\mathbf{x})\, d\mathbf{x}. \tag{8.10}$$

Then $\bar{\Lambda}_{avg\ thickness}$ will measure the between and within average cortical thickness dilatation rate.

8.4.4 Curvature Change

When the surface $\partial\Omega_0$ deforms to $\partial\Omega_t$, curvatures of the surface change as well. The *principal curvatures* can characterize the shape and location of the sulci and gyri, which are the valleys and crests of the cortical surfaces [27, 196, 202]. By measuring the curvature changes, rapidly folding and cortical regions can be localized. Let κ_1 and κ_2 be the two principal curvatures as defined in [46, 213]. The principal curvatures can be represented as functions of β_is in quadratic surface (8.2). To measure the amount of folding, we define bending metric K as a function of the principal curvatures:

$$K = \frac{\kappa_1^2 + \kappa_2^2}{2} + \alpha.$$

We may arbitrarily set $\alpha = 0.001$. If the cortical surface is flat, bending metric K obtains the minimum 0.001. The larger the bending metric, the more surface will be crested as shown in Figure 8.12.

We define the *local curvature dilatation rate* as

$$\Lambda_{curvature} = \frac{\partial \ln K}{\partial t}. \tag{8.11}$$

Under the linear model (8.1), it can be similarly shown that the curvature dilatation is approximately distributed as a Gaussian random field [64]:

$$\Lambda_{curvature}(\mathbf{x}) = \lambda_{curvature}(\mathbf{x}) + \epsilon_{curvature}(\mathbf{x}),$$

where $\lambda_{curvature}$ is the mean curvature dilatation rate and $\epsilon_{curvature}$ is a mean zero Gaussian random field. As in the case of the average cortical thickness dilatation rate (8.10), the average curvature dilatation rate can be computed and used as a global measure:

$$\Lambda_{avg\ curvature} = \frac{1}{\|\partial\Omega_0\|} \int_{\partial\Omega_0} \Lambda_{curvature}(\mathbf{x})\, d\mathbf{x}.$$

A similar integral approach has been taken to measure the amount of bending in the 2D contour of the corpus callosum [277].

8.5 Surface-Based Diffusion Smoothing

In order to increase the signal-to-noise ratio (SNR) [109, 305], Gaussian kernel smoothing is desirable in many statistical analyses . For example, Figure 8.10

shows fairly large variations in cortical thickness of a single subject displayed on the average brain atlas Ω_{atlas}. By smoothing the data on the cortical surface, the SNR will increase if the signal itself is smooth and in turn, it will be easier to localize the morphological changes. However, due to the convoluted nature of the cortex whose geometry is non-Euclidean, we can not directly apply Gaussian kernel smoothing on the cortical surface. Gaussian kernel smoothing of functional data $f(\mathbf{x}), \mathbf{x} = (x_1, \ldots, x_n) \in \mathbb{R}^n$ with FWHM (*full width at half maximum*) $= 4(\ln 2)^{1/2}\sqrt{t}$ is defined as the convolution of the Gaussian kernel with f:

$$F(\mathbf{x}, t) = \frac{1}{(4\pi t)^{n/2}} \int_{\mathbb{R}^n} e^{-(x-y)^2/4t} f(y) dy. \quad (8.12)$$

Formulation (8.12) can not be directly applied to the cortical surfaces. However, by reformulating Gaussian kernel smoothing as a solution of a diffusion equation on a Riemannian manifold, the Gaussian kernel smoothing approach can be generalized to an arbitrary curved surface. This generalization is called *diffusion smoothing* and has been used in the analysis of fMRI data on the cortical surface [10]. It can be shown that (8.12) is the integral solution of the n-dimensional diffusion equation

$$\frac{\partial F}{\partial t} = \Delta F \quad (8.13)$$

with the initial condition $F(\mathbf{x}, 0) = f(\mathbf{x})$, where

$$\Delta = \frac{\partial^2}{\partial x_1^2} + \cdots + \frac{\partial^2}{\partial x_n^2}$$

is the Laplacian in n-dimensional Euclidean space. Hence the Gaussian kernel smoothing is equivalent to the diffusion of the initial data $f(\mathbf{x})$ after time T. When applying diffusion smoothing on curved surfaces, the smoothing somehow has to incorporate the geometrical features of the curved surface and the Laplacian Δ should change accordingly. The extension of the Euclidean Laplacian to an arbitrary Riemannian manifold is called the *Laplace-Beltrami operator* [13, 213]. The approach taken in [10] is based on a local flattening of the cortical surface and estimating the planar Laplacian, which may not be as accurate as our estimation based on the finite element method (FEM). Further, our direct FEM approach completely avoids any local or global surface flattening. For the given Riemannian metric tensor g_{ij}, the Laplace-Beltrami operator Δ is defined as

$$\Delta F = \sum_{i,j} \frac{1}{|g|^{1/2}} \frac{\partial}{\partial u^i} \Big(|g|^{1/2} g^{ij} \frac{\partial F}{\partial u^j} \Big), \quad (8.14)$$

where $(g^{ij}) = g^{-1}$ [13]. Note that when g becomes a 2 $\times$2 identity matrix, the Laplace-Beltrami operator in (8.14), simplifies to a standard 2D Laplacian:

$$\Delta F = \frac{\partial^2 F}{\partial (u^1)^2} + \frac{\partial^2 F}{\partial (u^2)^2}.$$

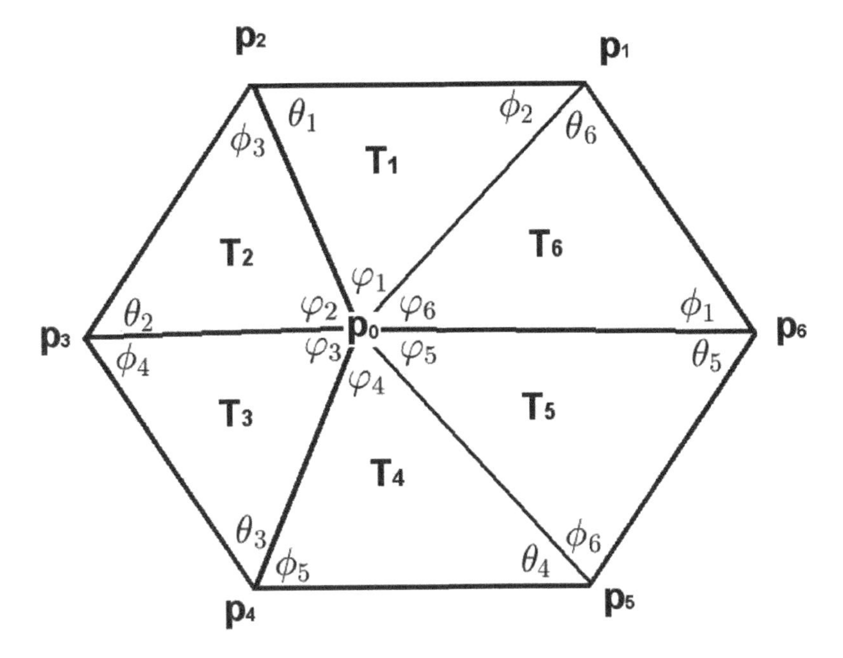

FIGURE 8.6
A typical triangulation in the neighborhood of $\mathbf{p} = \mathbf{p}_0$. When ASP algorithm is used, the triangular mesh is constructed in such a way that it is always pentagonal or hexagonal.

Using the FEM on the triangular cortical mesh generated by the ASP algorithm, it is possible to estimate the Laplace-Beltrami operator as the linear weights of neighboring vertices.

Let $\mathbf{p}_1, \cdots, \mathbf{p}_m$ be m neighboring vertices around the central vertex $\mathbf{p} = \mathbf{p}_0$. Then the estimated Laplace-Beltrami operator is given by

$$\widehat{\Delta F}(\mathbf{p}) = \sum_{i=1}^{m} w_i \big(F(\mathbf{p}_i) - F(\mathbf{p})\big)$$

with the weights

$$w_i = \frac{\cot\theta_i + \cot\phi_i}{\sum_{i=1}^{m} \|T_i\|},$$

where θ_i and ϕ_i are the two angles opposite to the edge connecting $\mathbf{p}_i$ and $\mathbf{p}$, and $\|T_i\|$ is the area of the i-th triangle (Figure 8.6).

This is an improved formulation from the previous attempt in diffusion smoothing on the cortical surface [10], where the Laplacian is simply estimated as the planar Laplacian after locally flattening the triangular mesh consisting of nodes $\mathbf{p}_0, \cdots, \mathbf{p}_m$ onto a flat plane. In the numerical implementation, we

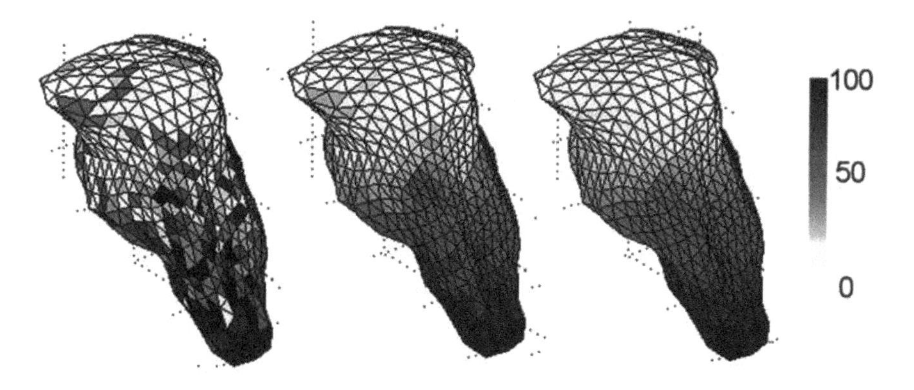

FIGURE 8.7
Diffusion smoothing simulation on a triangular mesh consisting of 1280 triangles. This smaller mesh is the surface of the brain stem. The artificial signal was generated with Gaussian noise to illustrate the smoothing process. From the left to right, initial signal, after 10 iterations with $\delta t = 0.5$ and after 20 iterations with $\delta t = 0.5$.

have used formula

$$\begin{aligned}\cot\theta_i &= \frac{\langle \mathbf{p}_{i+1} - \mathbf{p}, \mathbf{p}_{i+1} - \mathbf{p}_i \rangle}{2\|T_i\|}\\ &= \mathrm{c} \quad \boldsymbol{\phi}_i \frac{\langle \mathbf{p}_{i-1} - \mathbf{p}, \mathbf{p}_{i-1} - \mathbf{p}_i \rangle}{2\|T_i\|}\end{aligned}$$

and

$$\|T_i\| = \frac{1}{2}\|(\mathbf{p}_{i+1} - \mathbf{p}) \times (\mathbf{p}_i - \mathbf{p})\|$$

Afterwards, the finite difference scheme is used to iteratively solve the diffusion equation at each vertex $\mathbf{p}$:

$$\frac{F(\mathbf{p}, t_{n+1}) - F(\mathbf{p}, t_n)}{t_{n+1} - t_n} = \widehat{\Delta} F(\mathbf{p}, t_n),$$

with the initial condition $F(\mathbf{p}, 0) = f(\mathbf{p})$. After N-iterations, the finite difference scheme gives the diffusion of the initial data f after duration $N\delta t$. If the diffusion were applied to Euclidean space, it would be equivalent to Gaussian kernel smoothing with

$$\text{FWHM} = 4(\ln 2)^{1/2}\sqrt{N\delta t}.$$

Computing the linear weights for the Laplace-Beltrami operator takes a fair amount of time (about 4 minutes in `MATLAB` running on a Pentium III machine), but once the weights are computed, it is applied throughout the whole iteration repeatedly and the actual finite difference scheme takes only two minutes for 100 iterations. Figure 8.7 illustrates the process of diffusion smoothing. Unlike Gaussian kernel smoothing, smoothing is an iterative procedure. However, it should be emphasized that Gaussian kernel smoothing is a special case of diffusion smoothing restricted to Euclidean space.

8.6 Statistical Inference on the Cortical Surface

All of our morphological measures such as surface area, cortical thickness and curvature dilatation rates are modeled as Gaussian random fields on the cortical surface, i.e.

$$\Lambda(\mathbf{x}) = \lambda(\mathbf{x}) + \epsilon(\mathbf{x}), \mathbf{x} \in \partial\Omega_{atlas}, \tag{8.15}$$

where the deterministic part λ is the mean of the metric Λ and ϵ is a mean zero Gaussian random field. This theoretical model assumption has been checked using both Lilliefors test [87] and quantile-quantile plots (qq-plots) [160]. The qq-plot displays quantiles from an empirical distribution on the vertical axis versus theoretical quantiles from a Gaussian distribution on the horizontal axis. It is used to check graphically if the empirical distribution follows the theoretical Gaussian distribution. If the data comes from a Gaussian field, then the qq-plot should be close to a straight line (Figure 8.8 (a-b)). If the data comes from a lognormal distribution, it may not form a straight line (Figure 8.8 (c)). Because it is not possible to view qq-plots for every vertex on the cortex, we measured the correlation coefficients γ of the vertical and horizontal coordinates in the qq-plots. If the empirical distribution comes from Gaussian, γ should asymptotically converge to 1. For Gaussian simulation, $\gamma = 0.98 \pm 0.01$ and for lognormal simulation $\gamma = 0.84 \pm 0.08$ on the cortex. For the cortical thickness data, which has been filtered with the diffusion smoothing, $\gamma = 0.96 \pm 0.03$. So it does seems that the smoothed cortical thickness metric can be modeled as a Gaussian random field. Using Lilliefors statistic, we statistically tested the Gaussian assumption. The Lilliefors test, which is a special case of the Komogorov-Smirnov test, looks at the maximum difference between the empirical and a theoretical Gaussian distribution when the mean and the variance of the distribution are not known. Since the Lilliefors statistics of the cortical thickness metric are mostly smaller than the cutoff value of 0.19 at 1% level (0.16 at 5% level), there is no reason to reject model (8.15). (Figure 8.9).

Gaussian kernel smoothed images tend to reasonably follow random field assumptions when a fairly large FWHM is used. Diffusion smoothing is equivalent to Gaussian kernel smoothing locally in conformal coordinates on the cortex [64]. Also P-value for local maxima formula is quite stable even if a slightly different assumption such as non-isotropy is used (Worsley *et al.*,

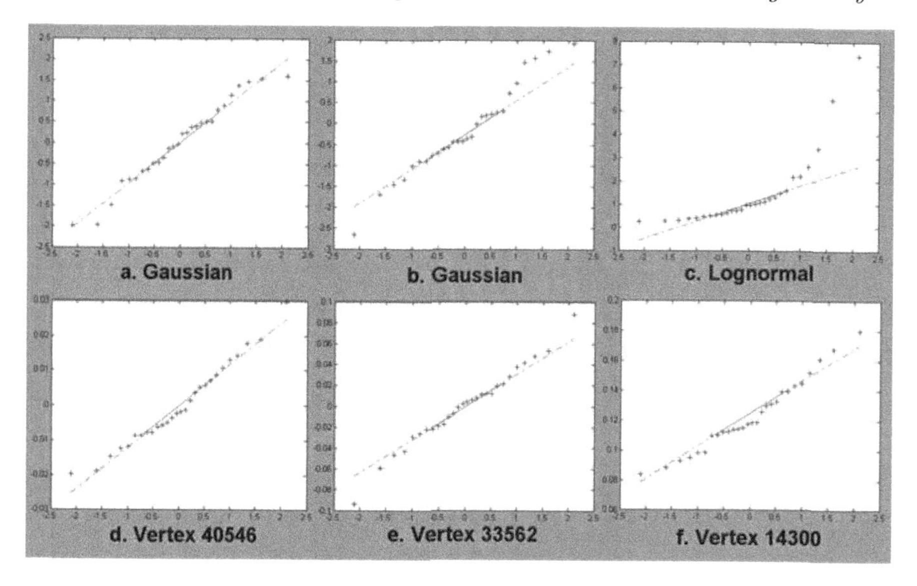

FIGURE 8.8
Checking Gaussian assumption for the cortical thickness metric. The horizontal axis displays the quantiles of a Gaussian distribution while the vertical axis displays the quantiles of an empirical distribution. How closely the scatter points lie along the straight line gives an idea if the underlying empirical distribution follows the theoretical Gaussian distribution. (a-b) Computer simulation of Gaussian distributions. (c) Computer simulation of a lognormal distribution (d-f) qq-plots of three randomly selected vertices.

1999b). Therefore, detecting the region of statistically significant $\lambda(\mathbf{x})$ for some $\mathbf{x}$ can be done via thresholding the maximum of the T random field defined on the cortical surface [394, 390].

The T random field on the manifold $\partial\Omega_{atlas}$ is defined as

$$T(\mathbf{x}) = \sqrt{n}\frac{M(\mathbf{x})}{S(\mathbf{x})}, \ \mathbf{x} \in \partial\Omega_{atlas}$$

where M and S are the sample mean and standard deviation of metric Λ over the n subjects. Under the null hypothesis

$$H_0 : \lambda(\mathbf{x}) = 0 \text{ for all } \mathbf{x} \in \partial\Omega_{atlas},$$

i.e. no structural change, $T(\mathbf{x})$ is distributed as a student's T with $n-1$ degrees of freedom at each voxel $\mathbf{x}$. The P value of the local maxima of the T field will give a conservative threshold, which has been used in brain imaging for a quite some time [394].

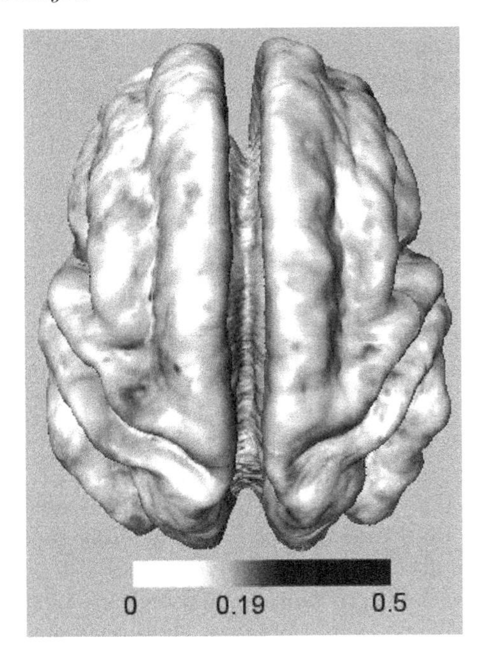

FIGURE 8.9
Lilliefors statistic measures the maximum difference between the empirical and a theoretical Gaussian distribution. Most of the cortex shows a value less than the cutoff value 0.19 indicating that the Gaussian random field assumption is valid.

For very high threshold y, we can show that

$$P\Big(\max_{\mathbf{x}\in\partial\Omega_{atlas}} T(\mathbf{x}) \geq y\Big) \approx \sum_{i=0}^{3} \phi_i(\partial\Omega_{atlas})\rho_i(y), \tag{8.16}$$

where ρ_i is the i-dimensional EC-density and the Minkowski functional ϕ_i are

$$\phi_0(\partial\Omega_{atlas}) = 2,\ \phi_1(\partial\Omega_{atlas}) = 0,\ \phi_2(\partial\Omega_{atlas}) = \|\partial\Omega_{atlas}\|,\ \phi_3(\partial\Omega_{atlas}) = 0$$

and $\|\partial\Omega_{atlas}\|$ is the total surface area of $\partial\Omega_{atlas}$ (Worsley, 1996a). When diffusion smoothing with given FWHM is applied to metric Λ on surface $\partial\Omega_{atlas}$, the 0-dimensional and 2-dimensional EC-density becomes

$$\rho_0(y) = \int_y^\infty \frac{\Gamma(\frac{n}{2})}{((n-1)\pi)^{1/2}\Gamma(\frac{n-1}{2})}\Big(1 + \frac{y^2}{n-1}\Big)^{-n/2} dy,$$

$$\rho_2(y) = \frac{1}{\text{FWHM}^2}\frac{4\ln 2}{(2\pi)^{3/2}}\frac{\Gamma(\frac{n}{2})}{(\frac{n-1}{2})^{1/2}\Gamma(\frac{n-1}{2})}y\Big(1 + \frac{y^2}{n-1}\Big)^{-(n-2)/2}.$$

Therefore, the excursion probability on the cortical surface can be approximated by the following formula:

$$P\Big(\max_{\mathbf{x}\in\partial\Omega_{atlas}} T(\mathbf{x}) \geq y\Big) \approx 2\rho_0(y) + \|\partial\Omega_{atlas}\|\rho_2(y).$$

We compute the total surface area $\|\partial\Omega_{atlas}\|$ by summing the area of each triangle in a triangulated surface. The total surface area of the average atlas brain is 275,800 mm^2, which is roughly the area of a 53×53 cm^2 sheet. We want to point out that the surface area of the average atlas brain is not the average surface area of 28 subjects. When 20mm FWHM diffusion smoothing is used on the template surface $\partial\Omega_{atlas}$, 2.5% thresholding gives

$$P\Big(\max_{\mathbf{x}\in\partial\Omega_{atlas}} T(\mathbf{x}) \geq 5.1\Big) = P\Big(\max_{\mathbf{x}\in\partial\Omega_{atlas}} T(\mathbf{x}) \leq -5.1\Big) \approx 0.025.$$

Our surface-based smoothing and analysis are checked on null data. The null data is created by reversing time for a randomly chosen half of the subjects. In the null data, the mean time difference $t_2 - t_1$ is -0.24 year so the statistical analysis presented here should not detect any morphological changes. For the cortical thickness dilatation rate, the maximum and the minimum t-values are 3.1808 and -3.9570 respectively, well below the threshold 5.1 and -5.1 indicating that the analysis obviously did not detect any statistically significant morphological changes.

8.7 Results

Twenty-eight normal subjects were selected based on the same physical, neurological and psychological criteria described in [141]. This is the same data set reported in [81], where the Jacobian of the 3D deformation was used to detect statistically significant brain tissue growth or loss in 3D whole brain via deformation-based morphometry. 3D Gaussian kernel smoothing used in this study is not sensitive to the interfaces between the gray, white matter and CSF. Gaussian kernel smoothing tends to blur gray matter volume increase data across the cortical boundaries. So in some cases, statistically significant brain tissue growth could be found in CSF. Our deformation-based surface morphometry can overcome this inherent shortcoming associated with the previous morphometric analysis.

Two T1-weighted MR scans were acquired for each subject at different times on the same GE Sigma 1.5T superconducting magnet system. The first scan was obtained at the age $t_1 = 11.5 \pm 3.1$ years (min 7.0 years, max 17.8 years) and the second scan was obtained at the age $t_2 = 16.1 \pm 3.2$ years (min 10.6 years, max 21.8 years). The time difference between the first and the second scan was 4.6 ± 0.9 years (min time difference 2.2 years, max time difference 6.4 years). Using the automatic image processing pipeline [412], MR images were spatially normalized into standardized stereotactic space via a global affine transformation [85, 350]. Subsequently, an automatic tissue-segmentation algorithm based on a supervised artificial neural network classifier was used to classify each voxel as CSF, gray matter and white matter. Afterwards, a triangular mesh for each cortical surface was generated by deforming a mesh to fit the proper boundary in a segmented volume using the ASP algorithm. As described in the previous section, the ASP algorithm is used to extract the surface and compute the displacement field on the outer cortical surface. Then we computed the local area dilatation, the cortical thickness and the curvature dilatation rates. Such surface metrics are then filtered with 20 mm FWHM diffusion smoothing. Image smoothing should improves the power of detection and compensates for some of registration errors [81].

8.7.1 Gray Matter Volume Change

The total gray matter volume dilatation rate for each subject was computed by computing the volume of triangular prisms that forms gray matter. The mean total gray matter volume dilatation rate $\bar{\Lambda}_{total\ volume} = -0.0050$. This 0.5% annual decrease in the total gray matter volume is statistically significant (T value of -4.45). There has been substantial developmental studies on gray matter volume reduction for children and adolescents [94, 140, 278, 293, 300, 340]. Our result confirms these studies. However, the ROI-based volumetry used in the previous studies did not allow investigators to detect local volume change within the ROIs. Our local volumetry based on the deformation field can overcome the limitations of the ROI-based volumetry.

Brain tissue growth and loss based on the local volume dilatation rate was detected in the whole brain volume that includes both gray and white matter [81]. The morphometric analysis used in [81] generates 3D statistical parametric map (SPM) of brain tissue growth. By superimposing the 3D SPM with the triangular mesh of the cortical surface of the atlas brain, we get gray matter volume change SPM restricted onto the cortical surface (Figure 8.11). Although it is an ad hoc approach, the resulting SPM projected on to the atlas brain seems to confirm some of the results in [140, 361]. In particular, [140] reported that frontal and parietal gray matters decrease but temporal and occipital gray matters increase even after age 12. In our analysis, we found local gray matter volume growth in the parts of temporal, occipital, somatosensory and motor regions but did not detect any volume loss in the frontal lobe. Instead we found statistically significant structural movements without accompanying volume decreases [81].

8.7.2 Surface Area Change

We measured the total surface area dilatation rate for each subject by computing the total area of triangular meshes on both outer and inner cortical surfaces. The mean surface areas for all 28 subjects area are

$$\|\partial\Omega_{t_1}^{out}\| = 302180 \text{ mm}^2, \ \|\partial\Omega_{t_1}^{in}\| = 289380 \text{ mm}^2,$$

$$\|\partial\Omega_{t_2}^{out}\| = 229940 \text{ mm}^2, \ \|\partial\Omega_{t_2}^{in}\| = 221520 \text{ mm}^2.$$

So we can see that both the outer and inner surface area tend to decrease between ages 12 and 16. The mean total area dilatation rate for 28 subjects was found to be

$$\bar{\Lambda}_{total\ area}^{out} = -0.0093, \ \bar{\Lambda}_{total\ area}^{in} = -0.0081.$$

0.8 to 0.9% decrease of both surface area change per year is statistically significant (T value of -9.2 and -7.5 respectively). So it does seem that the outer surface shrinks faster than the inner surface.

After knowing that the total surface areas shrink, we need to identify the regions of local surface area growth or reduction. The surface area dilatation rates were computed for all subjects, then smoothed with 20mm FWHM diffusion smoothing to increase the signal-to-noise ratio. Averaging over 28 subjects, local surface area change was found to be between -15.79% and 13.78% per year. In one particular subject, we observed between -106.5% and 120.3% of the local surface area change over a 4 year time span. Figure 8.11 is the T map of the cortical surface area dilatation showing cortical tissue growth pattern. Surface area growth and decrease were detected by $T > 5.1$ and $T < -5.1$ ($P < 0.05$, corrected) respectively, showing statistically significant local surface expansion in Broca's area in the left hemisphere and local surface shrinkage in the left superior frontal sulcus. Most of surface reduction seems to be concentrated near the frontal region.

8.7.3 Cortical Thickness Change

The growth pattern of cortical thickness change is found to be very different but closely related to that of local surface area change. The average cortical thickness at t_1 is 2.55 mm while the average cortical thickness at t_2 is 2.60 mm. The average cortical thickness dilatation rate across all within and between subjects was found to be $\bar{\Lambda}_{avg\ thickness} = 0.012$. This 1.2% annual increase in the cortical thickness is statistically significant (T value of 4.86). However, three out of 28 subjects showed from 0.6 up to 2% of cortical thinning. Although there are regions of cortical thinning present in all individual subjects, our statistical analysis indicates that the overall global pattern of thickness increase is a more predominant feature between ages 12 and 16. Also we localized the region of statistically significant cortical thickness increase by thresholding the t map of the cortical thickness dilatation rate by 5.1 (Figure 8.10). Cortical thickening is widespread on the cortex. The most predominant thickness increase was detected in the left superior frontal sulcus, which is the same location we detected local surface area reduction while

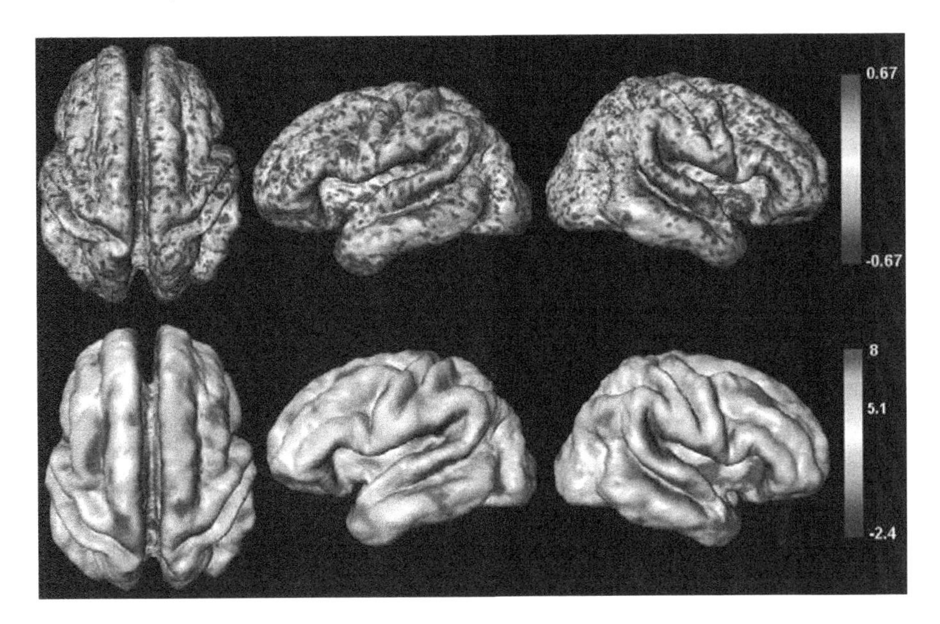

FIGURE 8.10
Top: Cortical thickness dilatation rate for a single subject. Due to such large measurement variations, surface-based smoothing is required. Bottom: T statistic map thresholded at the corrected p-value of 0.05. (See color insert.)

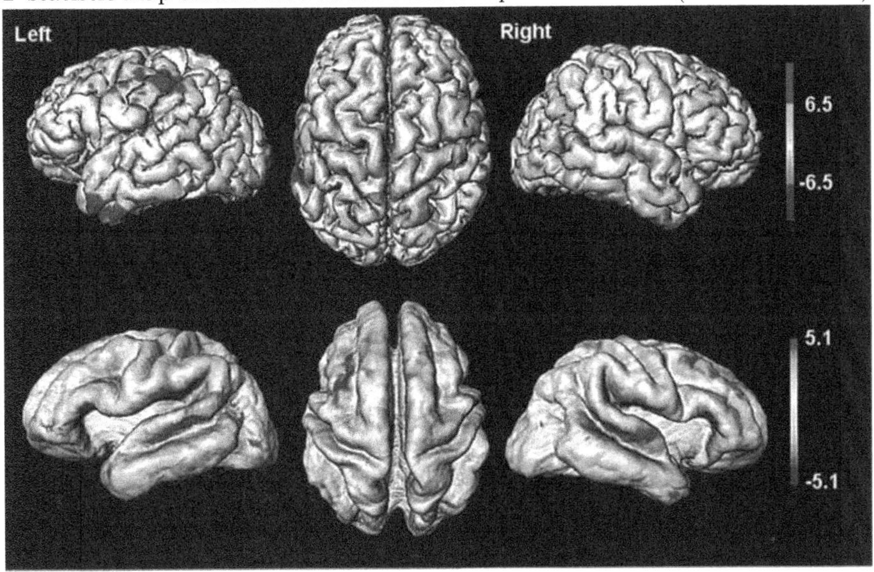

FIGURE 8.11
T-statistic map of local gray matter volume change (top) and cortical surface area dilatation rate (bottom) between ages 12 and 16. (See color insert.)

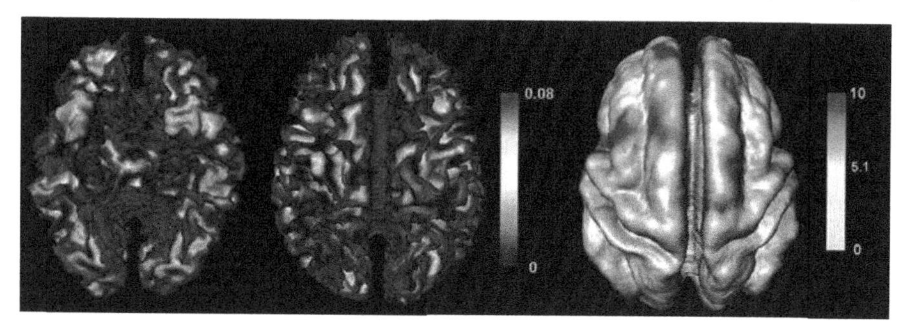

FIGURE 8.12
Left and middle: Cortical curvature measuring surface bending of a 14 year old subject. Right: T-statistic map thresholded at 5.1 (corrected) showing statistically significant region of curvature increase.

local gray matter volume remains the same. So it seems that while there is no gray matter volume change, the left superior frontal sulcus undergoes cortical thickening and surface area shrinking and perhaps this is why we did not detect any local volume change in this region.

The most interesting result found so far is that there is almost *no* statistically significant local cortical thinning detected on the whole cortical surface between ages 12 and 16. As we have shown, the total inner and outer surface areas as well as the total volume of gray matter decrease. So it seems that all these results are in contradiction. However, if the rate of the total surface area decrease is faster than the rate of the total volume reduction, it is possible to have cortical thickening. To see this, suppose we have a shallow solid shell with constant thickness h, total volume V and total surface area A. Then $V = hA$. It can be shown that the rate of volume change per unit volume can be written as

$$\frac{dV}{V} = \frac{dh}{h} + \frac{dA}{A}.$$

Using our dilatation notation,

$$\bar{\Lambda}_{total\ volume} = \bar{\Lambda}_{avg\ thickness} + \bar{\Lambda}^{out}_{total\ area}.$$

In our data,

$$\bar{\Lambda}_{total\ volume} = -0.0050 > \bar{\Lambda}^{out}_{total\ area} = -0.0093,$$

so we should have increase in the cortical thickness. However, we want to point out that this argument is heuristic because the cortical thickness is not uniform across the cortex. Sowell *et al.* (2001) reported cortical thinning or gray matter density decrease in the frontal and parietal lobes in a similar age group [334]. The thickness measure they used is based on gray matter density, which measures the proportion of gray matter within a sphere with

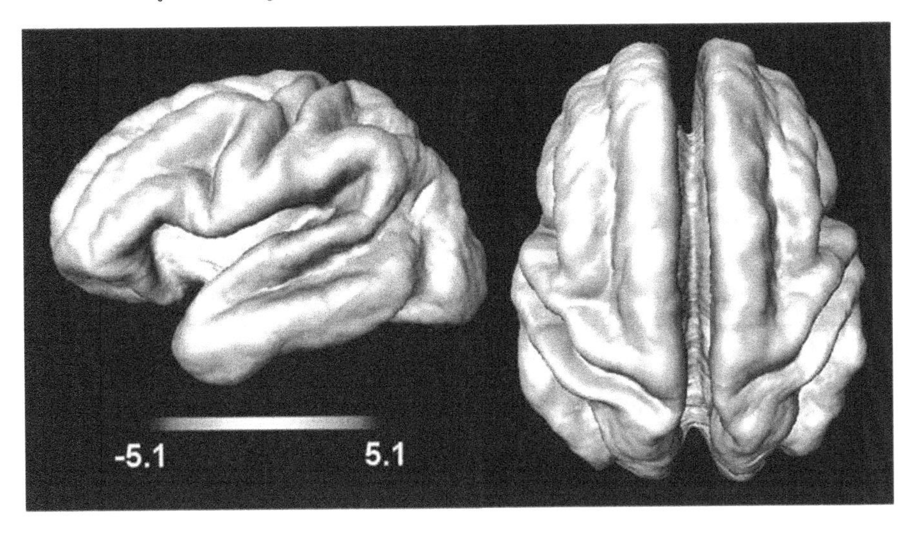

FIGURE 8.13
T-statistic map of the cortical dilatation rate for null data. The null data was created by reversing time for half of the subjects chosen randomly. In the null data, the mean time difference is -0.24 year, so our statistical analysis should not detect any morphological changes.

fixed radius between 5 to 15mm around a point on the outer cortical surface [334, 360]. However, the gray matter density not only measures the cortical thickness but also the amount of bending. If a point is chosen on a gyrus, the increase in the bending energy will correspond to the increase in gray matter density. So the region of gray matter density decrease reported in [334] more closely resembles the region of curvature increase (Figure 8.12) than the region of cortical thickness change (Figure 8.10). Because they measure different anatomical quantities, it is hard to directly compare the result reported in [334] to our result.

8.7.4 Curvature Change

Our study might be the first to use the curvature as the direct measure of anatomical changes in normal brain development. We measured curvature dilatation rate for each subject. The average curvature dilatation rate was found to be $\bar{\Lambda}_{avg\ curvature} = 2.50$. A 250% increase is statistically significant (T value of 19.42). Local curvature change was detected by thresholding the T statistic of the curvature dilatation rate at 5.1 (corrected). The superior frontal and middle frontal gyri show curvature increase. It is interesting to note that between these two gyri we have detected cortical thickness increase and local surface area decrease. It might be possible that cortical thickness increase and local surface area shrinking in the superior frontal sulcus causes the bending in the neighboring middle and superior frontal gyri. Such an interacting dynamic pattern has been also detected in [81], where gray matter tissue growth causes the inner surface to translate toward the region of white matter tissue reduction.

We also found *no* statistically significant local curvature decrease over the whole cortex. While the gray matter is shrinking in both total surface area and volume, the cortex itself seems to get folded to give increasing curvature.

8.8 Discussion

The surface-based morphometry presented here can statistically localize the regions of cortical thickness, area and curvature change at a local level without specifying the regions of interest (ROI). This ROI-free approach might be best suitable for exploratory whole brain morphometric studies. Our analysis successfully avoids artificial surface flattening [10, 12], which can destroy the inherent geometrical structure of the cortical surface. It seems that any structural or functional analysis associated with the cortex can be performed without surface flattening if appropriate mathematics is used.

Our metric tensor formulation gives us an added advantage that not only it can be used to measure local surface area and curvature change of the cortex but also it is used for generalizing Gaussian kernel smoothing on the cortex via diffusion smoothing. Since it is a direct generalization of Gaussian kernel smoothing, the diffusion smoothing should locally inherit many mathematical and statistical properties of Gaussian kernel smoothing applied to standard 3D whole brain volume. The modification for any other triangular mesh can be easily done. We tried to combine and unify morphometric measurement, image smoothing and statistical inference in the same mathematical and statistical framework.

As an illustration of this powerful unified approach, we applied it to a group of normal children and adolescents to see if we can detect the region of anatomical changes in gray matter. It is found that the cortical surface area and gray matter volume shrink, while the cortical thickness and curvature tend to increase between ages 12 and 16 with a highly localized area of cortical thickening and surface area shrinking found in the superior frontal sulcus at the same time. It seems that the increase in thickness and the decrease in the superior frontal sulcus might cause increased folding in the middle and superior frontal gyri.

Our unified deformation-based surface morphometry can be also used as a tool for future investigations of neurodevelopmental disorders where surface analysis of either the cortex or brain substructures would be relevant.

9

Heat Kernel Smoothing on Surfaces

The emphasis of this chapter is placed on the development of kernel-based surface data smoothing and analysis framework. The human brain cortex is a highly convoluted surface. Due to the convoluted non-Euclidean surface geometry, data smoothing and analysis on the cortex are inherently difficult. When measurements lie on a curved surface, it is natural to assign kernel smoothing weights based on the geodesic distance along the surface rather than the Euclidean distance. We present a new data smoothing framework that addresses this problem implicitly without actually computing the geodesic distance and presents its statistical properties. Afterwards, the statistical inference is based on the random field theory based multiple comparison correction. As an illustration, we have applied the method in detecting the regions of abnormal cortical thickness in 16 high functioning autistic children. This chapter follows closely to the methods first presented in [76, 77].

9.1 Introduction

The human cerebral cortex has the topology of a 2D highly convoluted gray matter shell of average thickness of 3mm. The thickness of the gray matter shell is usually referred to as the *cortical thickness* and can be obtained from magnetic resonance images (MRI). The cortical thickness can be used as an anatomical index for quantifying cortical shape variations. The thickness measures are obtained after a sequence of image processing steps which are described briefly here. The first step is to classify each voxel into three different tissue types: cerebrospinal fluid (CSF), gray matter, and white matter. The CSF/gray matter interface is called the *outer cortical surface* while the gray/white matter interface is called the *inner cortical surface.* These two surfaces bound the gray matter. The mainstream approach in representing the cortical surface has been to use a fine triangular mesh that is constructed from deformable surface algorithms [100, 237]. Cortical thickness is estimated by computing the distance between the two triangular meshes [190, 237, 249, 120, 97]. In order to compare cortical thickness measures across subjects, it is necessary to align the cortical surfaces via surface registration algorithms [301, 363, 97, 99, 121]. For cross-comparison between subjects, surfaces are registered into the *template surface* which serves as reference coordinates.

The image segmentation, thickness computation and surface registration procedures are expected to introduce noise into the thickness measure. In order to increase the signal-to-noise ratio (SNR) and smoothness of data for the random field theory, some type of data smoothing is necessary [203]. For 3D whole brain MRIs, Gaussian kernel smoothing is widely used to smooth data, in part, due to its simplicity in numerical implementation. The Gaussian kernel weights an observation according to its Euclidean distance. However, data residing on the convoluted brain surface fails to be isotropic in the Euclidean sense. On the curved surface, a straight line between two points is not the shortest distance so one may incorrectly assign less weights to closer observations. So when the observations lie on the cortical surface, it is more natural to assign the weights based on the geodesic distance along the surface. Previously *diffusion smoothing* has been developed for smoothing data along the cortex before the random field based multiple comparison correction [10, 82, 79]. By solving a diffusion equation on a manifold, Gaussian kernel smoothing can be indirectly generalized. Although diffusion smoothing has been used widely in image analysis starting with [275], most of previous work is about surface fairing [239, 351, 353]. There are a very few publications that smooth out observations defined on surface for data analysis [10, 79, 82, 247]. The drawback of the previous diffusion smoothing approach is the need for setting up a finite element method (FEM) to solve the diffusion equation numerically and making the algorithm converges [79]. To address this problem, we have developed a simpler and more efficient method based on the heat kernel convolution on a manifold.

As an illustration, the method was applied to groups of autistic and normal subjects, and we were able to detect the regions of statistically significant cortical thickness difference between the groups.

9.2 Heat Kernel Smoothing

The cortical surface $\partial\Omega$ can be assumed to be a C^2 Riemannian manifold [196]. Let $p = X(u^1, u^2) \in \partial\Omega$ be the parametric representation of $\partial\Omega$. We assume the following model on thickness measure Y:

$$Y(p) = \theta(p) + \epsilon(p),$$

where $\theta(p)$ is a mean thickness function and $\epsilon(p)$ is a zero-mean random field, possibly a Gaussian white noise process, with covariance function $R_\epsilon(p, q)$. The Laplace-Beltrami operator Δ corresponding to the surface parameterization

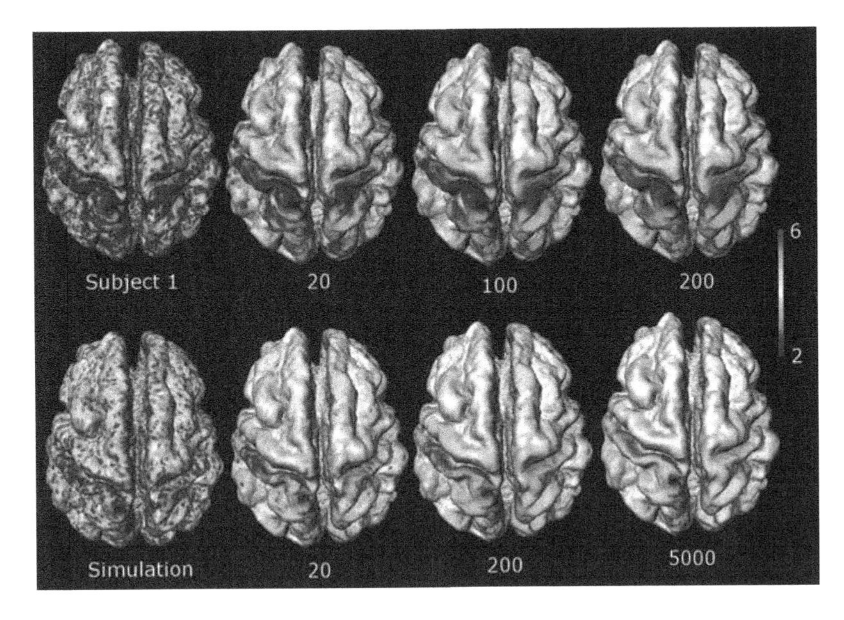

FIGURE 9.1
Heat kernel smoothing of cortical thickness (top) and simulated data (bottom) with $\sigma = 1$ and different number of iterations. (See color insert.)

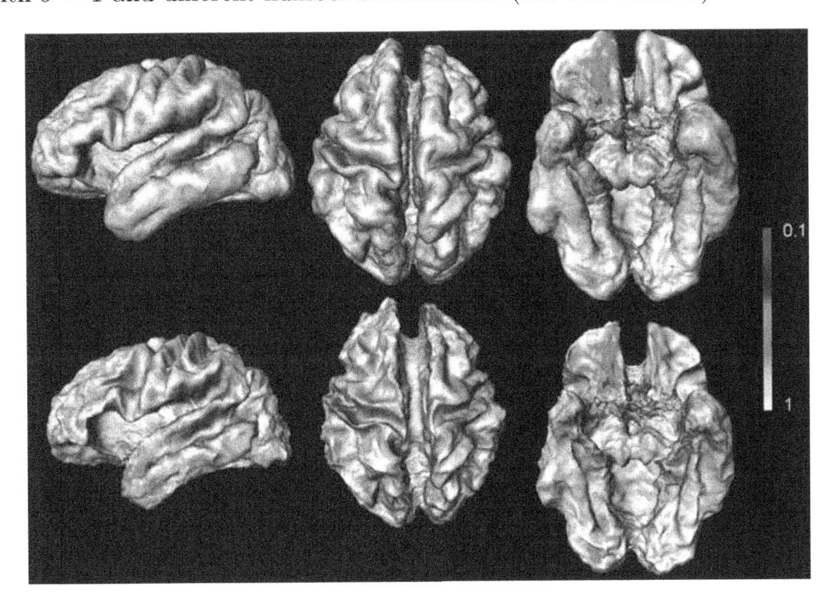

FIGURE 9.2
Corrected p-value maps of F-statistic removing the effect of age and relative gray matter volume difference projected onto the outer and inner surfaces. (See color insert.)

$p = X(u^1, u^2) \in \partial\Omega$ is given by

$$\Delta = \frac{1}{\det g^{1/2}} \sum_{i,j=1}^{2} \frac{\partial}{\partial u^i}\Big(\det g^{1/2} g^{ij} \frac{\partial}{\partial u^j}\Big),$$

where $g = (g_{ij})$ is the Riemannian metric tensor. Solving the equation

$$\Delta\psi = \lambda\psi,$$

we order eigenvalues

$$0 = \lambda_0 \leq \lambda_1 \leq \lambda_2 \leq \cdots$$

and corresponding eigenfunctions $\psi_0, \psi_1, \cdots$. The eigenfunctions ψ_j form orthonormal basis of $L^2(\partial\Omega)$, the L^2 space of functions defined on $\partial\Omega$. On the unit sphere, the eigenvalues are $m(m+n-1)$ and the corresponding eigenfunctions are spherical harmonics Y_{lm} ($|m| \leq l, 0 \leq l$) [372]. On an arbitrary surface, the explicit representation of eigenvalues and eigenfunction are only obtained through numerical methods.

Definition 11 *The* heat kernel $K_\sigma(p,q)$ *is analytically defined as*

$$K_\sigma(p,q) = \sum_{j=0}^{\infty} e^{-\lambda_j \sigma}\psi_j(p)\psi_j(q), \tag{9.1}$$

where σ is the bandwidth of the kernel [37, 304].

When $g_{ij} = \delta_{ij}$, the heat kernel becomes the Gaussian kernel, which is the probability density of $N(0, \sigma^2)$, i.e.

$$K_\sigma(p,q) = \frac{1}{(2\pi\sigma)^{1/2}} \exp\Big[-\frac{\|p-q\|^2}{2\sigma^2}\Big].$$

Hence the heat kernel is a natural extension of the Gaussian kernel. This can be interpreted as the transition probability density for an isotropic diffusion process with respect to the surface area element [374]. The kernel is symmetric, i.e. $K_\sigma(p,q) = K_\sigma(q,p)$ and isotropic with respect to the geodesic distance $d(p,q)$. The property of a kernel being isotropic needs some explanation.

Definition 12 *Let $d(p,q)$ be the geodesic distance between p and q. Then function f is* isotropic *on surface $\partial\Omega$ if $f(p) =$ constant for all point p on geodesic circle $d(0,p) =$ constant.*

Since $K_\sigma(p,q)$ has two arguments while symmetric, the isotropic property holds for either one of the arguments.

Definition 13 *Heat kernel smoothing [77] of functional measurement Y is the convolution:*

$$K_\sigma * Y(p) = \int_{\partial\Omega} K_\sigma(p,q) Y(q)\, dq. \tag{9.2}$$

The heat kernel can be written in terms of basis function expansion:

$$K_\sigma * Y(p) = \sum_{j=0}^{\infty} \alpha_j \phi_j(p),$$

where

$$\alpha_j = e^{-\lambda_j \sigma} \int_{\partial\Omega} \phi_j(q) Y(q)\, dq$$

are the weighted Fourier coefficients [71].

Definition 14 *The* heat kernel estimator *of unknown signal $\theta(p)$ is*

$$\widehat{\theta}_\sigma(p) = K_\sigma * Y(p).$$

As $\sigma \to 0$, $K_\sigma(p,q)$ becomes the Dirac delta function $\delta(p,q)$ so the heat kernel estimator becomes unbiased as $\sigma \to 0$, i.e.

$$\lim_{\sigma \to 0} \mathbb{E}\widehat{\theta}_\sigma(p) = \theta(p).$$

As σ gets larger, the bias increases. However the total bias over all the cortex is always zero. This is stated as following.

Theorem 2

$$\int_{\partial\Omega} [\theta(p) - \mathbb{E}\widehat{\theta}_\sigma(p)]\, dp = 0.$$

Theorem 2 can be seen from

$$\int_{\partial\Omega} \mathbb{E}\widehat{\theta}_\sigma(p)\, dp = \int_{\partial\Omega}\int_{\partial\Omega} K_\sigma(p,q)\theta(p)\, dq\, dp = \int_{\partial\Omega} \theta(p)\, dp.$$

We used the fact that K is a probability density so it integrates over $\partial\Omega$. Let us list important nontrivial properties of heat kernel smoothing.

Theorem 3 *$K_\sigma * Y$ is the unique solution of the following isotropic diffusion equation at time $t = \sigma^2/2$ [77, 304]:*

$$\frac{\partial f}{\partial t} = \Delta f,\ f(p,0) = Y(p), p \in \partial\Omega \tag{9.3}$$

This is a well known result [304]. Theorem 3 implies that the heat kernel smoothing isotropically assigns weights on $\partial\Omega$.

Theorem 4

$$\underbrace{K_\sigma * \cdots * K_\sigma}_{k \ times} * Y = K_{\sqrt{k}\sigma} * Y.$$

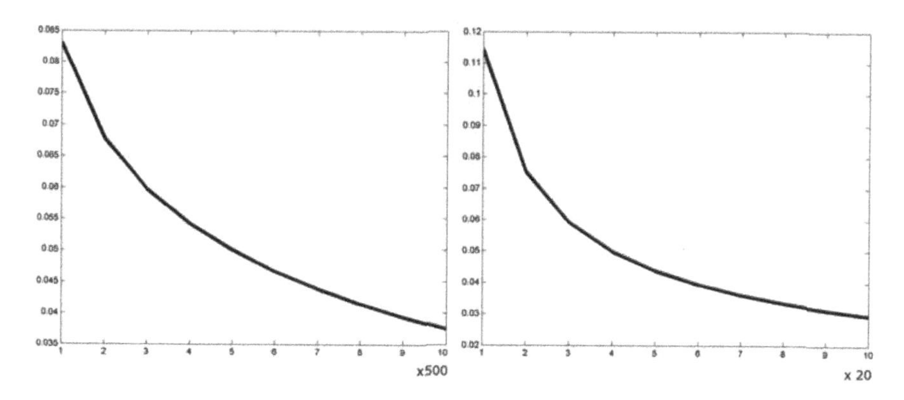

FIGURE 9.3
Left: Within-subject variance plotted over the number of iterations of heat kernel smoothing with $\sigma = 1$. Decreasing variance implies the convergence of the heat kernel smoothing to the mean thickness (Theorem 5). Right: Between-subject variance plotted over the number of iterations illustrating Theorem 6.

This can be seen as a scale space property of diffusion. From Theorem 3, $K_\sigma * (K_\sigma * Y)$ can be taken as the diffusion of signal $K_\sigma * Y$ after time $\sigma^2/2$ so that $K_\sigma * (K_\sigma * Y)$ is the diffusion of signal Y after time σ^2. Hence

$$K_\sigma * K_\sigma * Y = K_{\sqrt{2}\sigma} * Y.$$

Arguing inductively we see that the general statement holds. We will denote the k-fold iterated kernel as $K_\sigma^{(k)} = \underbrace{K_\sigma * \cdots * K_\sigma}_{k \text{ times}}$. This is the basis of our iterated heat kernel smoothing. Heat kernel with a large bandwidth will be performed by iteratively applying heat kernel smoothing with a smaller bandwidth. For instance iterated heat kernel smoothing with $\sigma = 1$ and $k = 200$ will generate heat kernel smoothing with the effective bandwidth of $\sqrt{200} = 14.14$mm. Figure 9.1 shows the process of iterated heat kernel smoothing.

Theorem 5

$$\lim_{\sigma \to \infty} K_\sigma * Y = \frac{\int_{\partial\Omega} Y(q)\, dq}{\mu(\partial\Omega)}.$$

Here $\mu(\partial\Omega)$ is the total surface area of $\partial\Omega$. This theorem shows that when we choose large bandwidth, heat kernel smoothing converges to the sample mean of data on $\partial\Omega$.

Figure 9.1 (bottom) shows the convergence of heat kernel smoothing to the within-subject mean cortex 4mm as the bandwidth increases. Figure 9.3

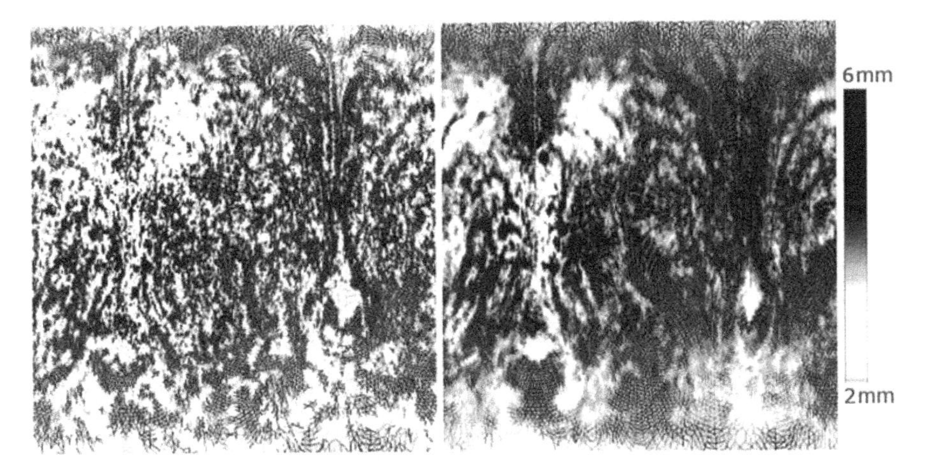

FIGURE 9.4
Thickness maps are projected onto a unit square. Left: original noisy thickness map. Right: Heat kernel smoothing with $\sigma = 1$ and $k = 200$ iterations.

(left) shows the convergence of the within-subject variance indirectly implying $K_\sigma * Y$ converges to a constant, which is the average thickness over the cortex.

It is natural to assume the measurements $Y(p)$ and $Y(q)$ to have less correlation when p and q are away so we assume the covariance function to be $R_\epsilon(p,q) = \rho(d(p,q))$ for some nondecreasing function ρ. Then we can show the variance reduction property of heat kernel smoothing.

Theorem 6 $\mathbf{Var}[K_\sigma * Y(p)] \leq \mathbf{Var}Y(p)$ *for each* $p \in \partial\Omega$.

Figure 9.3 (right) shows the between-subject variance decreases as σ increases.

The problem with the heat kernel smoothing on an arbitrary surface is that the explicit analytic form of the heat kernel is unknown. To address this problem we use the *parametrix expansion* of the heat kernel [304, 374]:

$$K_\sigma(p,q) = \frac{1}{(2\pi\sigma)^{1/2}} \exp\Big[-\frac{d^2(p,q)}{2\sigma^2}\Big]\big[1 + O(\sigma^2)\big] \tag{9.4}$$

for small $d(p,q)$. This expansion spells out the exact form of the kernel for small bandwidth. When the metric is flat, the heat kernel becomes a Gaussian kernel, reconfirming that heat convolution is a generalization of Gaussian kernel. The expansion is the basis of our heat kernel smoothing formulation. Heat kernel smoothing with a large bandwidth will be decomposed into iterated kernel smoothing. We will truncate and normalize the heat kernel using the first order term. For each $p \in \partial\Omega$, we define

$$\widetilde{K}_\sigma(p,q) = \frac{\exp\Big[-\frac{d^2(p,q)}{2\sigma^2}\Big]\mathbf{1}_{B_p}(q)}{\int_{B_p}\exp\Big[-\frac{d^2(p,q)}{2\sigma^2}\Big]\,dq}, \tag{9.5}$$

where $\mathbf{1}_{B_p}$ is an indicator function defined on a small compact domain containing B such that $\mathbf{1}_{B_p}(q) = 1$ if $q \in B_p$ and $\mathbf{1}_{B_p}(q) = 0$ otherwise. Note that for each fixed p, $\widetilde{K}_\sigma(p,q)$ defines a probability distribution in B_p and it converges to $K_\sigma(p,q)$ as $\sigma \to 0$ in B_p This implies

$$\widetilde{K}_\sigma^{(k)} * Y(p) \to K_\sigma^{(k)} * Y(p) \text{ as } \sigma \to 0.$$

For a discrete triangular mesh, we can take B_p to be a set of points containing p and its neighboring nodes $q_1, \cdots, q_m$, and take a discrete measure on B_p, which still make (9.5) a probability distribution. This can be viewed as a *Gaussian kernel Nadaraya-Watson* type smoothing extended to manifolds [63]. Figure 9.4 shows a flattened thickness map illustrating how heat kernel smoothing can enhance the thickness pattern by increasing the signal-to-noise ratio.

Figure 9.1 shows the heat kernel smoothing on both a single subject data and simulated data. The cortical thickness measures are projected onto a template that has less surface folding to show the progress of smoothing. The bottom figure shows the progress of heat kernel smoothing with $\sigma = 1$ and up to $k = 5000$ iterations applied to a simulated data. From 12 normal subjects, the mean thickness $\theta(p)$ and variance $R_\epsilon(p,p)$ functions are estimated and used to generate random fields with a Gaussian white noise. It begin to show the convergence to the sample mean thickness over all cortex.

9.3 Numerical Implementation

The heat kernel smoothing is implemented in MATLAB. For demonstration, we use the first autistic subject in AUTISM.coordinates.mat.

```
load AUTISM.coordinates.mat

surf.vertices=squeeze(autism_coord(1,:,:))';
surf.faces=tri;
thick_au=squeeze(sqrt(sum((autism_coord-autism_coordw).^2,2)));
figure;figure_trimesh(surf,thick_au(1,:)')
view([0 90]); caxis([2 6]); colormap('hot')
```

Heat kernel smoothing routine inputs surface signal, surface, bandwidth and the number iterations as the input arguments:

```
smoothed=hk_smooth(thick_au(1,:),surf,1,50);
figure ;figure_trimesh(surf,smoothed);
view([0 90]); caxis([2 6]); colormap('hot')
```

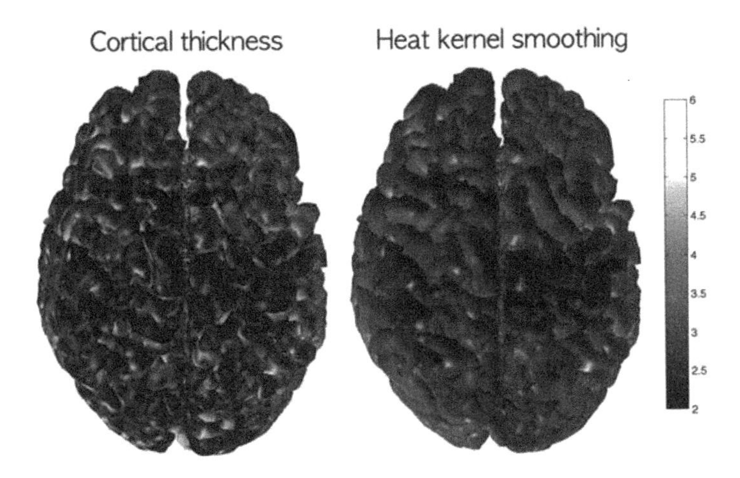

FIGURE 9.5
Cortical thickness of the first autism subject and its heat kernel smoothing with bandwidth $\sigma = 1$ and $k = 50$ iterations.

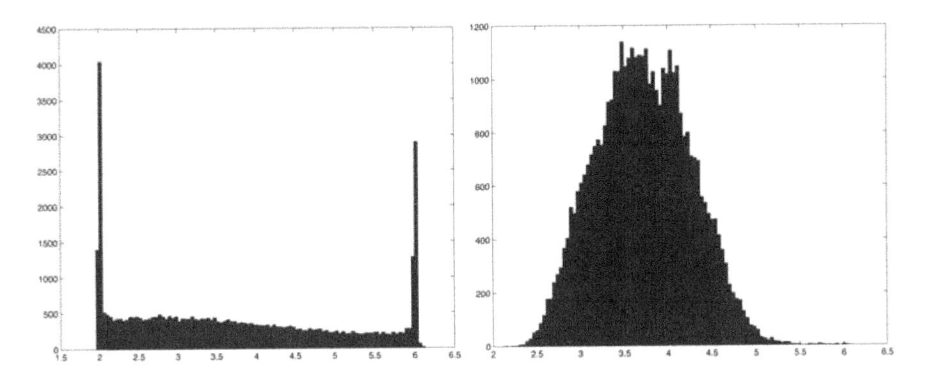

FIGURE 9.6
Cortical thickness of the first autism subject and its heat kernel smoothing with bandwidth $\sigma = 1$ and $k = 50$ iterations.

The resulting smoothed cortical surface is shown in Figure 9.5. The thickness histogram behaves better for the random field based statistical inference. The histogram change is shown in Figure 9.6.

```
figure;  hist(thick_au(1,:), 100)
figure ; hist(smoothed, 100)
```

We will also demonstrate heat kernel smoothing with the mandible surface data `load mandible77subjects.mat`, which has 77 mandible surfaces. We average 77 surfaces to obtain the population mean surface `ss`. Then with

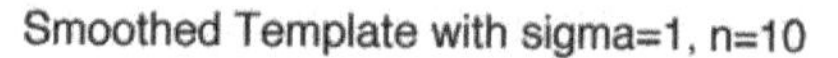

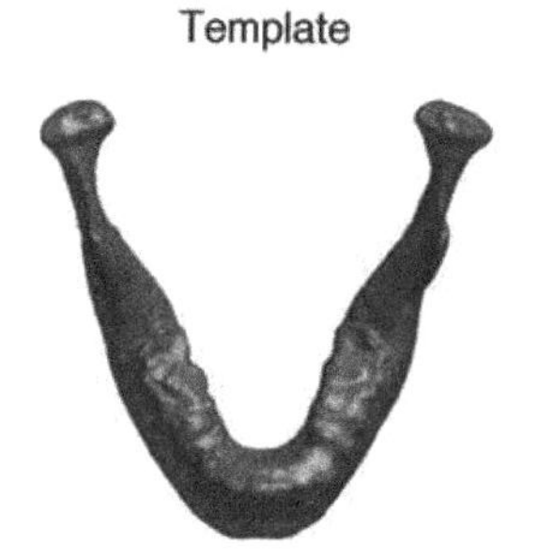

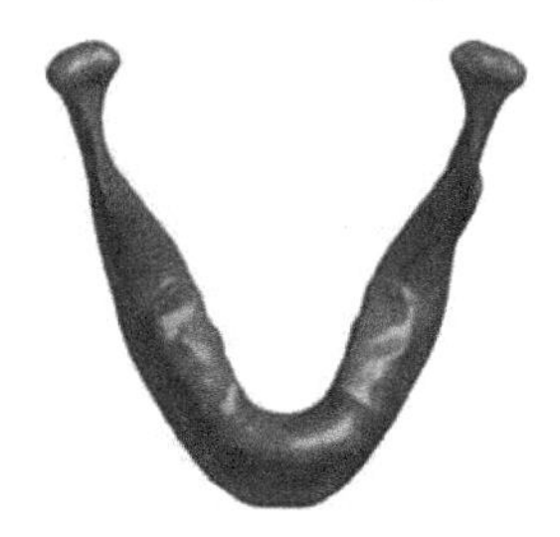

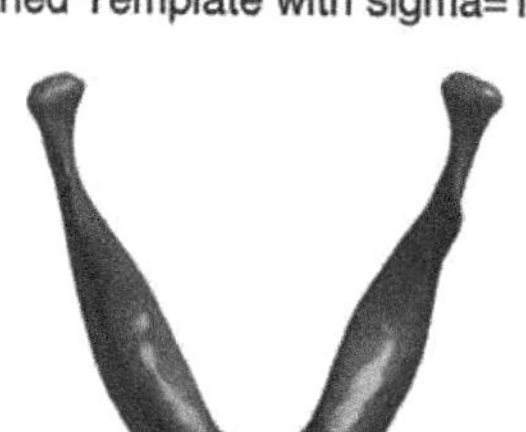

FIGURE 9.7
Heat kernel smoothing applied to mandible surface with $\sigma = 1$ and $k = 10, 40, 150$ iterations.

`sigma = 1`, we perform heat kernel smoothing with 1, 40 and 150 iterations on the surface coordinates. The results are shown in Figure 9.7.

```
load mandible77subjects.mat

ss.faces=faces;
ss.vertices=voxelsize(44)*mean(vertices,3);
figure; subplot(2,2,1); figure_patch(ss,[0.74 0.71 0.61],0.7);
view([90 60]); camlight;
title('Template')

sigma=1;
n_smooth=10;
ss.vertices=hk_smooth2(ss.vertices,ss, sigma,n_smooth);
subplot(2,2,2); figure_patch( ss,[0.74 0.71 0.61],0.7);
view([90 60]); camlight;
title('Smoothed Template with sigma=1, n=10')

n_smooth=40;
ss.vertices=hk_smooth2(ss.vertices,ss,sigma,n_smooth);
```

```
subplot(2,2,3); figure_patch( ss,[0.74 0.71 0.61],0.7);
view([90 60]); camlight;
title('Smoothed Template with sigma=1, n=50')

n_smooth=150;
ss.vertices=hk_smooth2(ss.vertices,ss,sigma,n_smooth);
subplot(2,2,4); figure_patch( ss,[0.74 0.71 0.61],0.7);
view([90 60]); camlight;
title('Smoothed Template with sigma=1, n=200')
```

9.4 Random Field Theory on Cortical Manifold

Here we will describe how to perform multiple comparisons on $\partial\Omega$ using the random field theory. The random field theory based approach is widely used for correcting multiple comparisons in 3D whole brain volume but rarely used on 2D cortical manifolds [10, 82, 83]. First we combine both the autistic and the control subjects in a single indexing j and set up a general linear model (GLM) on cortical thickness Y_j for subject j:

$$K_\sigma * Y_j(p) = \lambda_1(p) + \lambda_2(p) \cdot \texttt{age}_j + \lambda_3(p) \cdot \texttt{volume}_j + \beta(p) \cdot \texttt{group}_j + \epsilon_j \quad (9.6)$$

is used. Here the dummy variable `group` is 1 for the autistic subjects and 0 for the normal subjects. `volume` is the total gray matter volume for subject j. The total gray matter volume is estimated by computing the volume bounded by the both outer and inner surfaces [83]. The error is modeled as a smooth Gaussian random field which is viewed as the heat kernel convolution with Gaussian white noise, i.e. $\epsilon_j = K_\sigma * W$. Then we test the group difference by performing a hypothesis testing:

$$H_0 : \beta(p) = 0 \text{ for all } p \in \partial\Omega$$

v.s.

$$H_1 : \beta(p) \neq 0 \text{ for some } p \in \partial\Omega.$$

The test statistic is the ratio of the sum of the squared residual errors under the null and alternate models. Under H_0, the test statistic is a F random random field with 1 and $n = n_1 + n_2 - 4$ degrees of freedom [388]. The null hypothesis is the intersection of collection of the hypothesis

$$H_0 = \bigcap_{p \in \partial\Omega} H_0(p),$$

where $H_0(p) : \beta(p) = 0$ for each fixed p. The type I error for the multiple comparisons is then given by

$$
\begin{aligned}
\alpha &= P\Big(\bigcup_{p \in \partial\Omega} \{F(p) > h\} \Big) = 1 - P\Big(\bigcap_{p \in \partial\Omega} F(p) \leq h\} \Big) \\
&= 1 - P(\sup_{p \in \partial\Omega} F(p) \leq h) = P(\sup_{p \in \partial\Omega} F(p) > h)
\end{aligned}
$$

for some h. The resulting p-value is usually called the *corrected p-value*. The distribution of $\sup_{p \in \partial\Omega} F(p)$ is asymptotically given as

$$
P(\sup_{p \in \partial\Omega} F(p) > h) \approx \sum_{d=0}^{2} \phi_d(\partial\Omega)\rho_d(h) \tag{9.7}
$$

where ϕ_d are the d-dimensional Minkowski functionals of $\partial\Omega$ and ρ_d are the d-dimensional Euler characteristic (EC) density of F-field with $\alpha = 1$ and $\beta = n$ degrees of freedom [388]. The Minkowski functionals are $\phi_0 = 2, \phi_1 = 0, \phi_2 = \text{area}(\partial\Omega)/2 = 49,616\text{mm}^2$, the half area of the template cortex $\partial\Omega$. The EC density is given by

$$
\begin{aligned}
\rho_0(h) &= \int_h^\infty \frac{\Gamma(\frac{\alpha+\beta}{2})}{\Gamma(\frac{\alpha}{2})\Gamma(\frac{\beta}{2})} \frac{\alpha}{\beta} \Big(\frac{\alpha x}{\beta}\Big)^{\frac{(\alpha-2)}{2}} \Big(1 + \frac{\alpha x}{\beta}\Big)^{-\frac{(\alpha+\beta)}{2}} dx, \\
\rho_2(h) &= \frac{\lambda}{2\pi} \frac{\Gamma(\frac{\alpha+\beta-2}{2})}{\Gamma(\frac{\alpha}{2})\Gamma(\frac{\beta}{2})} \Big(\frac{\alpha h}{\beta}\Big)^{\frac{(\alpha-2)}{2}} \Big(1 + \frac{\alpha h}{\beta}\Big)^{-\frac{(\alpha+\beta-2)}{2}} \\
&\times \Big[(\beta - 1)\frac{\alpha h}{\beta} - (\alpha - 1)\Big]
\end{aligned}
$$

where λ measures the smoothness of fields ϵ and given as $\lambda = 1/(2\sigma^2)$. The resulting corrected p-values maps for F field is shown in Figure 9.2. The main use of the corrected p-value maps are the localization and visualization of thickness difference.

9.5 Case Study: Cortical Thickness Analysis

T1-weighted MR scans were acquired for 16 autistic and 12 control subjects on a 3-Tesla GE SIGNA scanner. They are all right-handed males. Sixteen autistic subjects were diagnosed with high functioning autism. The average age is 17.1 ± 2.8 is for the control subjects and 16.1 ± 4.5 for the autistic subjects. The complete description of the data set, image acquisition parameters, the subsequent image processing routines, and the interpretation of the resulting statistical parametric maps is provided in [76]. Each image underwent several image preprocessing steps. Image intensity nonuniformity was corrected using the nonparametric nonuniform intensity normalization method [330]. Then using the automatic image processing pipeline, [412] the image was spatially normalized into the Montreal neurological institute (MNI) stereotaxic space using a global affine transformation [85]. Subsequently, an automatic tissue-segmentation algorithm based on a supervised artificial neural network classifier was used to classify each voxel as cerebrospinal fluid (CSF), gray matter, or white matter [211]. Brain substructures such as the brain stem and the cerebellum were removed automatically. Triangular meshes for inner and outer cortical surfaces were obtained by a deformable surface algorithm [237]. Such a *deformable surface* approach has the advantage that the surface topology can be fixed to be spherical and the deformation process can maintain a non-intersecting surface at all times, obviating the need for topology correction [97]. The mesh starts as an ellipsoid located outside the brain and is shrunk to obtain the inner cortical surface. Then the inner surface is expanded, with constraints, to obtain the outer cortical surface. The triangular meshes are not constrained to lie on voxel boundaries. Instead, the triangular meshes can cut through a voxel, which serves to reduce discretization error and partial volume effect. Thickness is measured using the natural anatomical homology between vertices on the inner and outer cortical surface meshes, since the outer surface is obtained by deforming the inner surface.

Afterwards, thickness measures are smoothed with heat kernel smoothing with parameters $\sigma = 1$ and $k = 200$ giving the effective smoothness of $\sqrt{200} = 14.14$ mm. A surface-to-surface registration to a template surface was performed to facilitate vertex-by-vertex inter-subject thickness comparison. [301]. We have formulated it as a registration problem of two functional data on a unit sphere [301, 363]. First a mapping from a cortical surface onto the sphere is established while recording the mapping. Then cortical curvatures are mapped onto the sphere. The two curvature functions on the sphere are aligned by solving a regularization problem that tries to minimize the discrepancy between two functions while maximizing the smoothness of the alignment in such a way that the pattern of gyral ridges are matched smoothly [301]. This alignment is projected back to the original surface using the recorded mapping. This regularization mechanism produces a smooth deformation field, with very little folding. The deformation field is parameterized using a triangulated mesh and the algorithm proceeds in a coarse-to-fine manner, with four levels of mesh resolution. For details on surface registration, see [76, 77].

After smoothing out thickness measurements, statistical analysis is performed following the procedures described in the previous section. The resulting corrected p-value map (< 0.1) for the F statistic is projected onto the template surface for visualization. Figure 9.2 shows statistically significant regions of cortical thickness between two groups. After removing the effect of age and total gray matter volume difference, the statistically significant regions of thickness decreases are highly localized at the right inferior orbital prefrontal cortex, the left superior temporal sulcus and the left occipito-temporal gyrus in autistic subjects.

9.6 Discussion

This chapter introduced heat kernel smoothing and its statistical properties for data analysis on the cortical manifolds. The technique can be used in smooth out data that is necessary in the random field theory based multiple comparison correction. We have applied the methodology in detecting the regions of abnormal cortical thickness in a group of autistic subjects; however, the approach is not limited to a particular clinical population. The algorithm is implemented in `MATLAB` and freely available to download on the web `http://www.stat.wisc.edu/~mchung/softwares/hk`. The sample cortical meshes and their thickness measures can be also downloaded from the above website for researchers.

10

Cosine Series Representation of 3D Curves

In this chapter, we present a novel cosine series representation for encoding fiber bundles consisting of multiple 3D curves. The coordinates of curves are parameterized as coefficients of cosine series expansion. We address the issue of registration, averaging and statistical inference on curves in a unified Hilbert space framework. Unlike traditional splines, the proposed method does not have internal knots and explicitly represents curves as a linear combination of cosine basis. This simplicity in the representation enables us to design statistical models, register curves and perform subsequent analysis in a more unified statistical framework than splines.

The proposed representation is applied in characterizing the abnormal shape of white matter fiber tracts passing through the splenium of the corpus callosum in autistic subjects. For an arbitrary tract, a 19 degree expansion is usually found to be sufficient to reconstruct the tract with 60 parameters. This chapter is based on the publication [67].

10.1 Introduction

Diffusion tensor imaging (DTI) has been used to characterize the microstructure of biological tissues using magnitude, anisotropy and anisotropic orientation associated with diffusion[29]. It is assumed that the direction of greatest diffusivity is most likely aligned to the local orientation of the white matter fibers. White matter tractography offers the unique opportunity to characterize the trajectories of white matter fiber bundles noninvasively in the brain. Whole brain tractography studies routinely generate up to half million tracts per brain. Various deterministic tractography have been used to visualize and map out major white matter pathways in individuals and brain atlases [30, 60, 88, 218, 254, 255, 364, 406]; however, tractography data can be challenging to interpret and quantify. Recent efforts have attempted to cluster [266] and automatically segment white matter tracts [267] as well as characterize tract shape parameters [32]. Many of these techniques can be quite computationally demanding. Clearly efficient methods for representing tract shape, regional tract segmentation and clustering, tract registration and quantification would be of tremendous value to researchers.

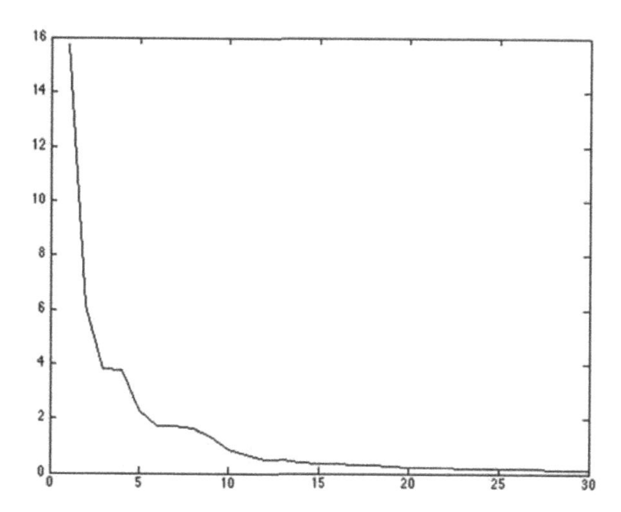

FIGURE 10.1
The error plot displays the average reconstruction error in millimeter (vertical) vs. degree (horizontal).

In this chapter, we present a novel approach for parameterizing white matter fiber tract shapes using a new Fourier descriptor. Fourier descriptors have been around for many decades for modeling mainly planar curves [276, 338]. They have been previously used to classify tracts [32]. The Fourier coefficients are computed by the Fourier transform that involves both the sine and cosine series expansion. Then the sum of the squared coefficients are obtained up to degree 30 for each tract and the k-means clustering is used to classify the fibers globally. Our approach differs from [32] in that we obtain local shape information employing cosine series only, without using both the cosine and sine series making our representation more compact. Using our new compact representation, we demonstrate how to quantify abnormal patterns of white matter fibers passing through the splenium of the corpus callosum for autistic subjects. The authors conclude that a downside of using Fourier descriptors is that they are not local and it is not possible to make a statement about a specific portion of the curve. Although the Fourier coefficients are global and mainly used for globally classifying shapes [323], it is still possible to obtain local shape information and make a statement about local shape characteristics [71]. In this chapter, we propose to use the Fourier descriptor as a parameterization for local shape representation.

Splines have also been widely used for modeling and matching 3D curves [84, 151, 209]. Unfortunately, splines are not easy to model and to manipulate explicitly compared to Fourier descriptors, due to the introduction of internal

FIGURE 10.2
Left: control points (dots) are obtained from the second order Runge-Kutta streamline algorithm. Subsampled 500 tracts with length larger than 50mm are only shown here. Right: nineteen degree cosine series representation of tracts.

knots. In Clayden et al. [84], the cubic-B spline is used to parameterize the median of a set of tracts for tract dispersion modeling. Matching two splines with different numbers of knots is not computationally trivial and has been solved using a sequence of ad-hoc approaches. In Gruen et al. [151], the optimal displacement of two cubic spline curves is obtained by minimizing the sum of squared Euclidean distances. The minimization is nonlinear so an iterative updating scheme is used. On the other hand, there is no need for any numerical optimization in obtaining the matching in our method due to the very nature of the Hilbert space framework. We will show that the optimal matching is embedded in the representation itself. Instead of using the squared distance of coordinates, others have used the curvature and torsion as features to be minimized to match curves [153, 209, 223]. In particular, Corouge et al. used cubic B-splines for representation fiber tracts and then curvature and torsion were used as [91].

In this chapter, we (i) introduce a more compact Fourier descriptor that uses only half the number of bases; (ii) show that curve matching can be done without any numerical optimization; and (iii) show how to perform a statistical inference on fiber bundles consisting of multiple 3D curves. The `MATLAB` implementation for the cosine series representation can be found in `brainimaging.waisman.wisc.edu/~chung/tracts`.

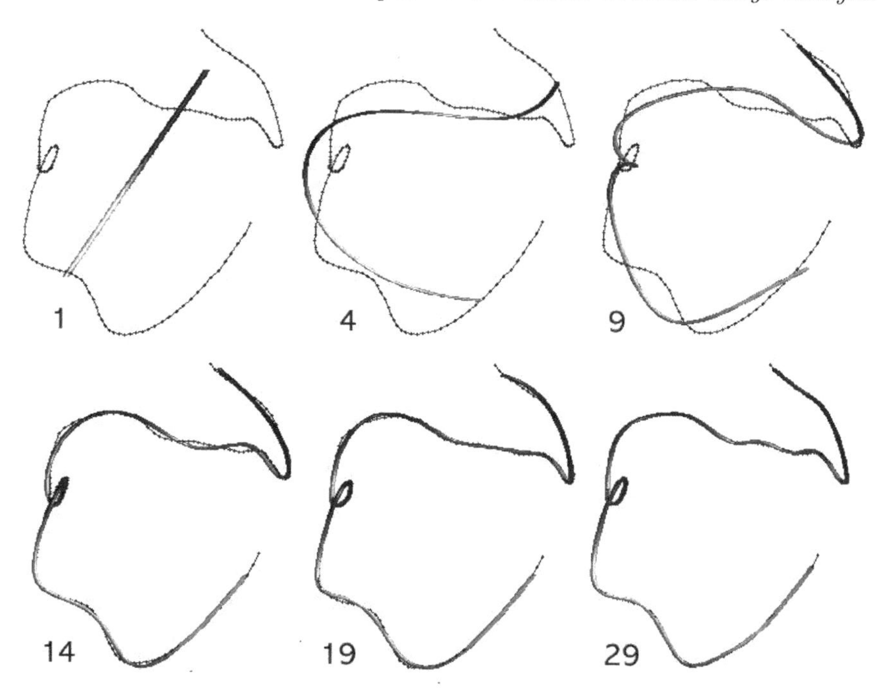

FIGURE 10.3
Cosine representation of a tract at various degrees. Dots are control points obtained from a streamline based tractography. The degree 1 representation is a straight line that fits all the control points in a least squares fashion. The degree 19 representation is used throughout the book.

10.2 Parameterization of 3D Curves

We are interested in encoding a smooth curve $\mathcal{M}$ consisting of n noisy ordered control points $p_1, \cdots, p_n$. Consider a mapping $\boldsymbol{\zeta}^{-1}$ that maps the control point p_j onto the unit interval $[0,1]$ as

$$\boldsymbol{\zeta}^{-1} : p_j \rightarrow \frac{\sum_{i=1}^{j} \|p_i - p_{i-1}\|}{\sum_{i=1}^{n} \|p_i - p_{i-1}\|} = t_j. \tag{10.1}$$

This is the ratio of the arc-length from the point p_1 to p_j, to p_1 to p_n. We let this ratio be t_j. We assume $\boldsymbol{\zeta}^{-1}(p_1) = 0$. The ordering of the control points is also required in obtaining smooth one-to-one mapping. Then we parameterize the smooth inverse map

$$\boldsymbol{\zeta} : [0,1] \rightarrow \mathcal{M}$$

as a linear combination of smooth basis functions.

10.2.1 Eigenfunctions of 1D Laplacian

Consider the space of square integrable functions in $[0,1]$ denoted by $\mathcal{L}^2[0,1]$. Let us solve the eigenequation

$$\Delta\psi + \lambda\psi = 0 \tag{10.2}$$

in $\mathcal{L}^2[0,1]$ with 1D Laplacian $\Delta = \frac{d^2}{dt^2}$. The eigenfunctions $\psi_0, \psi_1, \cdots$ form an orthonormal basis in $\mathcal{L}^2[0,1]$. Instead of solving (10.2) in the domain $[0,1]$, we solve it in the larger domain $\mathbb{R}$ with the periodic constraint

$$\psi(t+2) = \psi(t). \tag{10.3}$$

The eigenfunctions are then Fourier sine and cosine basis

$$\psi_l = \sin(l\pi t), \cos(l\pi t)$$

with the corresponding eigenvalues $\lambda_l = l^2\pi^2$. The period 2 constraint forces the basis function expansion to be only valid in the intervals $\cdots, [-2,-1], [0,1], [2,3], \cdots$ while there are gaps in $\cdots, (-1,0), (1,2), (3,4), \cdots$. We can fill the gap by padding with zeros but this will result in the Gibbs phenomenon (ringing artifacts) [71] at the points of jump discontinuities.

One way of filling the gap automatically while making the function continuous across the whole intervals is by putting the constraint of evenness, i.e.

$$\psi(t) = \psi(-t) \tag{10.4}$$

Then the only eigenfunctions satisfying two constraints (10.3) and (10.4) are the cosine basis of the form

$$\psi_0(t) = 1, \psi_l(t) = \sqrt{2}\cos(l\pi t) \tag{10.5}$$

with the corresponding eigenvalues $\lambda_l = l^2\pi^2$ for integers $l > 0$. The constant $\sqrt{2}$ is introduced to make the eigenfunctions orthonormal in $[0,1]$ with respect to the inner product

$$\langle \psi_l, \psi_m \rangle = \int_0^1 \psi_l(t)\psi_m(t)\,dt = \delta_{lm}, \tag{10.6}$$

where δ_{lm} is the Dirac-delta function. With respect to the inner product, the norm $\|\cdot\|$ is then defined as

$$\|\psi\| = \langle \psi, \psi \rangle^{1/2}.$$

10.2.2 Cosine Representation

Denote the coordinates of $\boldsymbol{\zeta}$ as $(\zeta_1, \zeta_2, \zeta_3)$. Then each coordinate is modeled as

$$\zeta_i(t) = \mu_i(t) + \epsilon_i(t), \tag{10.7}$$

where μ_i is an unknown smooth function to be estimated and ϵ_i is a zero mean random field, possibly Gaussian. Instead of estimating μ_i in $\mathcal{L}^2[0,1]$, we estimate in a smaller subspace $\mathcal{H}_k$, which is spanned by up to the k-th degree eigenfunctions:

$$\mathcal{H}_k = \{\sum_{l=0}^{k} c_l \psi_l(t) : c_l \in \mathbb{R}\} \subset \mathcal{L}^2[0,1].$$

Then the least squares estimation of μ_i in $\mathcal{H}_k$ is given by

$$\widehat{\mu_i} = \arg\min_{f \in \mathcal{H}_k} \|f - \zeta_i(t)\|^2.$$

Obviously, the minimization is simply given as the k-th degree expansion:

$$\widehat{\mu_i} = \sum_{l=0}^{k} \langle \zeta_i, \psi_l \rangle \psi_l. \tag{10.8}$$

With this motivation in mind, we have the following k-th degree *cosine series representation* for a 3D curve:

$$\zeta_i(t) = \sum_{l=0}^{k} c_{li} \psi_l + \epsilon_i(t), \tag{10.9}$$

where ϵ_i is a zero mean random field. Figure 10.3 shows the cosine series representation up to degree 19. It is also possible to have a slightly different but equivalent model that will be used for statistical inference. Assuming Gaussian random field, ϵ_i can be expanded using the given basis ψ_l as follows.

$$\epsilon_i(t) = \sum_{l=0}^{k} Z_l \psi_l(t) + e_i(t),$$

where $Z_l \sim N(0, \tau_l^2)$ are possibly *correlated* Gaussian and e_i is the residual error field that can be neglected in practice. This is the direct consequence of the Karhunen-Loeve expansion [3, 109, 214, 399]. Therefore we can equivalently model (10.9) as

$$\zeta_i(t) = \sum_{l=0}^{k} X_l \psi_l(t) + e_i(t), \tag{10.10}$$

where $X_l \sim N(c_{li}, \tau_l^2)$.

10.2.3 Parameter Estimation

We only observe the curve $\mathcal{M}$ in a finite number of control points $\zeta_j(t_1), \cdots, \zeta_j(t_n)$ so we further need to estimate the Fourier coefficient $c_{li} = \langle \zeta_i, \psi_l \rangle$ as follows. At control points we have normal equations

$$Y_{n\times 3} = \Psi_{n\times k} C_{k\times 3},$$

where

$$Y_{n\times 3} = \begin{pmatrix} \zeta_1(t_1) & \zeta_2(t_1) & \zeta_3(t_1) \\ \zeta_1(t_2) & \zeta_2(t_2) & \zeta_3(t_2) \\ \vdots & \vdots & \vdots \\ \zeta_1(t_n) & \zeta_2(t_n) & \zeta_3(t_n) \end{pmatrix},$$

$$\Psi_{n\times k} = \begin{pmatrix} \psi_0(t_1) & \psi_1(t_1) & \cdots & \psi_k(t_1) \\ \psi_0(t_2) & \psi_1(t_2) & \cdots & \psi_k(t_2) \\ \vdots & \vdots & \ddots & \vdots \\ \psi_0(t_n) & \psi_1(t_n) & \cdots & \psi_k(t_n) \end{pmatrix},$$

$$C_{k\times 3} = \begin{pmatrix} c_{01} & c_{02} & c_{03} \\ c_{11} & c_{12} & c_{13} \\ \vdots & \vdots & \vdots \\ c_{k1} & c_{k2} & c_{k3} \end{pmatrix}.$$

The coefficients are simultaneously estimated in the least squares fashion as

$$\widehat{C} = (\Psi'\Psi)^{-1}\Psi'Y.$$

The proposed least squares estimation technique avoids using the Fourier transform (FT) [32, 52, 152]. The drawback of the FT is the need for a predefined regular grid system so some sort of interpolation is needed. The advantage of the cosine representation is that, instead of recording the coordinates of all control points, we only need to record $3\cdot(k+1)$ number of parameters for all possible tract shape. This is a substantial data reduction considering that the average number of control points is 105 (315 parameters). We recommend readers to use $10 \le k \le 30$ degrees for most applications. In our application, we have used degree $k = 19$ through out the paper (Figure 10.1). This gives the average absolute error of 0.26mm along the tract.

10.2.4 Optimal Representation

We have explored the possibly of choosing the optimal number of bases using a stepwise model selection framework. This model selection framework for Fourier descriptors was first presented in [71, 73]. Although increasing the degree of the representation increases the goodness-of-fit, it also increases the number of estimated coefficients linearly. So it is necessary to stop the series expansion at the degree where the goodness-of-fit and the number of coefficients balance out.

Assuming up to the $(k-1)$-degree representation is proper in (10.9), we determine if adding the k-degree term is statistically significant by testing

$$H_0 : c_{ki} = 0.$$

Let the k-th degree *sum of squared errors* (SSE) for the i-th coordinate be

$$\text{SSE}_k = \sum_{j=1}^{n} \Big[\zeta_i(t_j) - \sum_{l=0}^{k} \widehat{c_{li}} \psi_l(t_j) \Big]^2,$$

where $\widehat{c_{li}}$ are the least squares estimation. The plot of SSE for varying degree $1 \leq k \leq 50$ a particular tract is shown in Figure 10.4. As the degree k increases, SSE decreases until it flattens out. So it is reasonable to stop the series expansion when the decrease in SSE is no longer significant. Under H_0, the test statistic F follows

$$F = \frac{\text{SSE}_{k-1} - \text{SSE}_k}{\text{SSE}_{k-1}/(n-k-2)} \sim F_{1,n-k-2},$$

the F-distribution with 1 and $n-k-2$ degrees of freedom. We compute the F statistic at each degree and stop increasing the degree of expansion if the corresponding p-value first becomes bigger than the pre-specified significance $\alpha = 0.01$. The forward model selection framework hierarchically builds the cosine series representation from lower to higher degree.

In Fourier descriptor and spherical harmonic representation literature, the issue of the optimal degree has not been addressed properly and the degree is simply selected based on a pre-specified error bound [52, 135, 135, 152, 322, 323]. Since the stepwise model selection framework chooses the optimal degree for each coordinate separately, we have chosen the maximum of optimal degrees for all coordinates. The optimal degree changes if a different tract is chosen. For instance, the optimal degrees for 4987 randomly chosen whole brain white matter tracts longer than 30mm are 13.94 ± 7.02 and the upper 80 percentile is approximately 19. For simplicity in numerical implementation and inference, it is crucial to choose the same fixed degree for all tracts. We do not want to choose the degree 14 as optimal since then about 50% of tracts will not be represented optimally. Therefore, we have chosen the degree corresponding to the upper 80 percentile to be used throughout the book.

We have also checked if the optimal degree is related to the length but found no relation. The correlation between the length of tracts and the optimal degree is 0.06, which is statistically insignificant. The increased degree should correspond to the increased curvature and bending rather than the the length of tracts. This issue is left to future research and we did not pursue it any further.

10.3 Numerical Implementation

The numerical implementation for the cosine series representation is based on `MATLAB` 7.5. Using the second order Runge-Kutta streamline algorithm with

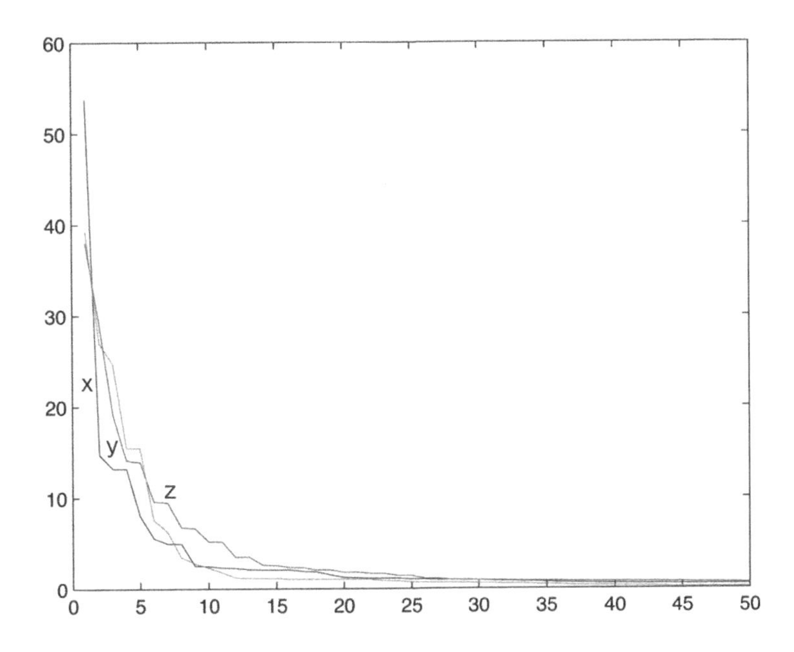

FIGURE 10.4
The plot of the sum of squired errors (SSE) for degree between 1 and 50. SSE rapidly flattens out around degree 15-20. The plots are for x, y and z coordinates from the left to right.

tensor deflection (TEND), we saved the whole brain tracts as `*.Bflot` files. The file is about 500-1000GB and depends on the number of fiber tracts.

```
SL=get_streamlines('001_spd_tracts.Bfloat',[1.5 1.75 2.25]);
tract=SL{1000}';
figure; plot3(tract(1,:),tract(2,:),tract(3,:),'.b');
```

In this example, the variable `SL` consists of 10000 tracts and 1000-th tract is extracted and stored as `tract`. To parameterize `tract`, we run

```
[arc_length  para]=parameterize_arclength(tract);
```

Note that this is not the arclength parameterization. Instead we parameterize the arclength as a unit interval. The i-th control point in `tract(:,i)` is mapped to `para(i)`, where `para` is a number between 0 and 1:

```
figure;
subplot(3,1,1); plot(para, tract(1,:));
subplot(3,1,2); plot(para, tract(2,:));
subplot(3,1,3); plot(para, tract(3,:));
```

The 19-th degree cosine series representation is then given by

```
[wfs beta]=WFS_tracts(tract,para,19);
figure; plot3(wfs(1,:),wfs(2,:),wfs(3,:),'k')
```

The parameters of the representation `beta` are estimated using the least squares method and `wfs` is the coordinates of the representation. To visualize the fiber tract, we used the streamtube representation:

```
figure_streamtube(tract, 1, [1 .3 .3]);
```

10.4 Modeling a Family of Curves

Using the cosine series representation, we show how to analyze a collection of fiber bundles consisting of similarly shaped curves. The ability to register one tract to another tract is necessary to establish anatomical correspondence for a subsequent population study. Since curves are represented as combinations of cosine functions, the registration will be formulated as a minimization problem in the subspace $\mathcal{H}_k$ which avoids brute-force style numerical optimization schemes given in [151, 153, 209, 223, 294]. This simplicity makes the cosine series representation more well suited than the usual spline representation of curves [151] in subsequent statistical analysis.

10.4.1 Registering 3D Curves

With the abuse of notations, we will interchangeably use curves to be estimated and their estimation with the same notations when the meaning is clear. Let the cosine series representation of two curves $\boldsymbol{\eta}$ and $\boldsymbol{\zeta}$ be

$$\boldsymbol{\eta}(t) = \sum_{l=0}^{k} \boldsymbol{\eta}_l \psi_l(t), \tag{10.11}$$

$$\boldsymbol{\zeta}(t) = \sum_{l=0}^{k} \boldsymbol{\zeta}_l \psi_l(t) \tag{10.12}$$

where $\boldsymbol{\eta}_l$ and $\boldsymbol{\zeta}$ are the Fourier coefficient vectors.

Consider the displacement vector field $\mathbf{u} = (u_1, u_2, u_3)$ that is required to register $\boldsymbol{\zeta}$ to $\boldsymbol{\eta}$. We will determine an optimal displacement $\mathbf{u}$ such that the discrepancy between the deformed curve $\boldsymbol{\zeta} + \mathbf{u}$ and $\boldsymbol{\eta}$ is minimized with respect to a certain discrepancy measure ρ. The discrepancy ρ between $\boldsymbol{\eta}$ and $\boldsymbol{\zeta}$ is defined as the integral of the sum of squared distance:

$$\rho(\boldsymbol{\zeta}, \boldsymbol{\eta}) = \int_0^1 \|\boldsymbol{\zeta}(t) - \boldsymbol{\eta}(t)\|^2 \, dt. \tag{10.13}$$

The discrepancy ρ can be simplified as

$$\begin{aligned}\rho(\boldsymbol{\zeta}, \boldsymbol{\eta}) &= \int_0^1 \sum_{j=1}^3 \Big[\sum_{l=0}^k (\zeta_{lj} - \eta_{lj})\psi_l(t) \Big]^2 dt \\ &= \sum_{j=1}^3 \sum_{l=0}^k (\zeta_{lj} - \eta_{lj})^2.\end{aligned}$$

We have used the orthogonality condition (10.6) to simplify the expression. The algebraic manipulation will show that the optimal displacement $\mathbf{u}^*$, which minimizes the discrepancy between $\boldsymbol{\zeta} + \mathbf{u}$ and $\boldsymbol{\eta}$, is given by

$$\mathbf{u}^*(t) = \arg \min_{u_1, u_2, u_3 \in \mathcal{H}_k} \rho(\boldsymbol{\zeta} + \mathbf{u}, \boldsymbol{\eta}) \tag{10.14}$$

$$= \sum_{l=0}^k (\boldsymbol{\eta}_l - \boldsymbol{\zeta}_l)\psi_l(t). \tag{10.15}$$

The proof requires substituting

$$\mathbf{u}(t) = \sum_{l=0}^k \mathbf{u}_l \psi_l(t)$$

in the expression (10.14), which becomes the unconstrained positive definite quadratic program with respect to variables $\mathbf{u}_l = (u_{l1}, u_{l2}, u_{l3})$. So the global minimum always exists and is obtained when $\rho(\boldsymbol{\zeta} + \mathbf{u}^*, \boldsymbol{\eta}) = 0$. Figure 10.5 shows the schematic view of registration.

The simplicity of our approach is that curve registration is done by simply matching the corresponding Fourier coefficients without any sort of numerical optimization as in spline curve matching.

10.4.2 Inference on a Collection of Curves

Based on the idea of registering tracts by matching coefficients, we construct the average of a white fiber bundle consisting of m curves $\boldsymbol{\zeta}^1, \cdots, \boldsymbol{\zeta}^m$ by finding the optimal curve that minimizes the sum of all discrepancies in $\mathcal{H}_k$:

$$\overline{\boldsymbol{\zeta}}(t) = \arg \min_{\zeta_1, \zeta_2, \zeta_3 \in \mathcal{H}_k} \sum_{j=1}^m \rho(\boldsymbol{\zeta}^j, \boldsymbol{\zeta}).$$

Again the algebraic manipulation will show that the optimum curve is obtained by the average of representation:

$$\overline{\boldsymbol{\zeta}}(t) = \frac{1}{m} \sum_{j=1}^m \sum_{l=0}^k \boldsymbol{\zeta}_l^j \psi_l(t) = \sum_{l=0}^k \overline{\boldsymbol{\zeta}}_l \psi_l(t), \tag{10.16}$$

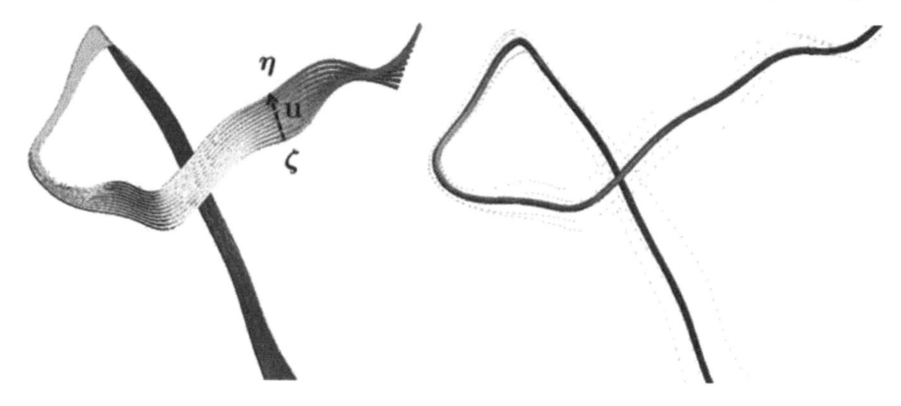

FIGURE 10.5
Left: the curve $\boldsymbol{\zeta}$ is registered to $\boldsymbol{\eta}$ by the displacement vector field $\mathbf{u}$. The other intermediate curves are generated by plotting $\boldsymbol{\zeta} + \alpha\mathbf{u}^*$ with $\alpha \in [0, 1]$ to show how the different amount of displacement deforms the curve $\boldsymbol{\zeta}$. Right: the average of a fiber bundle consisting of 5 tracts.

where $\overline{\boldsymbol{\zeta}}_l$ is the average coefficient vector

$$\overline{\boldsymbol{\zeta}}_l = \frac{1}{m}\sum_{j=1}^{m} \zeta_l^j.$$

Again, this simplicity is the consequence of the Fourier series having the best representation in the Hilbert space. So any optimization involving our quadratic discrepancy will simplify the expression as the sum of squared Fourier coefficients making the problem a fairly simple quadratic problem. As an illustration, we show how to average five tracts in Figure 10.5.

Similarly we can define the sample variance of m curves and it will turn out to be the cosine representation with the coefficient vector consisting of the sample variance of m coefficients. The construction of the sample variance of m curves should be fairly straightforward and we will not go into detail.

The next question we investigate is that given another population of curves $\boldsymbol{\eta}^1, \cdots, \boldsymbol{\eta}^n$, how to perform statistical inference on the equality of curve shape in the two populations. The null hypothesis of interest is then

$$H_0 : \overline{\boldsymbol{\zeta}} = \overline{\boldsymbol{\eta}}. \tag{10.17}$$

Here we again abused the notation so we are testing the equality of mean representations of populations. From the very property of the Fourier series in Hilbert space, the uniqueness of the cosine series representation is guaranteed so the two representations are equal if and only if the coefficients vectors match. Therefore, the equivalent hypothesis to (10.17) is given by

$$H_0' : \overline{\boldsymbol{\zeta}}_1 = \overline{\boldsymbol{\eta}}_1, \cdots, \overline{\boldsymbol{\zeta}}_k = \overline{\boldsymbol{\eta}}_k.$$

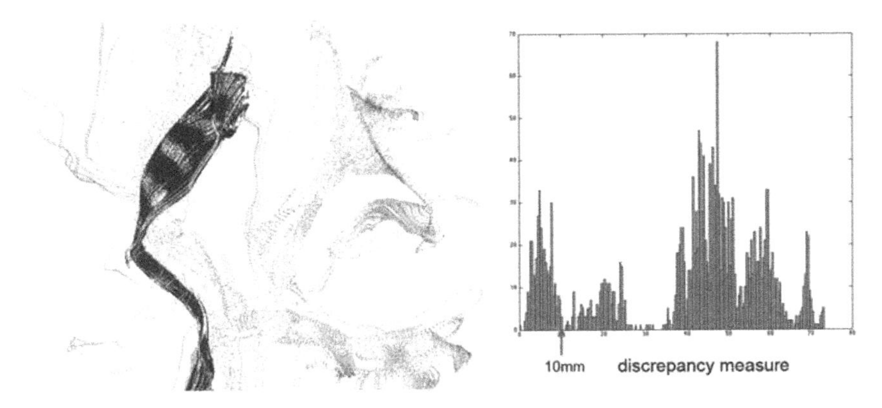

FIGURE 10.6
Histogram of discrepancy measure from a reference tract. Thresholding at 10mm gives the clustered black tracts.

Obviously this is a multiple comparisons problem. Under the Gaussian assumption in (10.10), testing the equality of the mean coefficient vector can be done using the Hotelling's T-square statistic. For correcting for the multiple comparisons, the Bonferroni correction can be used.

10.5 Case Study: White Matter Fiber Tracts

We have applied the cosine series representation to white matter fibers passing through the splenium of the corpus callosum. We have mainly chosen these fibers since the splenium is known to exhibit structural abnormality in autism [69, 222]

10.5.1 Image Acquisition

DTI data were acquired on a Siemens Trio 3.0 Tesla Scanner with an 8-channel, receive-only head coil. DTI was performed using a single-shot, spin-echo, EPI pulse sequence and SENSE parallel imaging (undersampling factor of 2). Diffusion-weighted images were acquired in 12 non-collinear diffusion encoding directions with diffusion weighting factor 1000 s/mm^2 in addition to a single reference image. Data acquisition parameters included the following: contiguous (no-gap) fifty 2.5mm thick axial slices with an acquisition matrix of 128x128 over a field of view (FOV) of 256mm, 4 averages, repetition time (TR) = 7000 ms, and echo time (TE) = 84 ms. Two-dimensional gradient echo images with two different echo times of 7 ms and 10 ms were obtained prior to the DTI acquisition for correcting distortions related to magnetic field

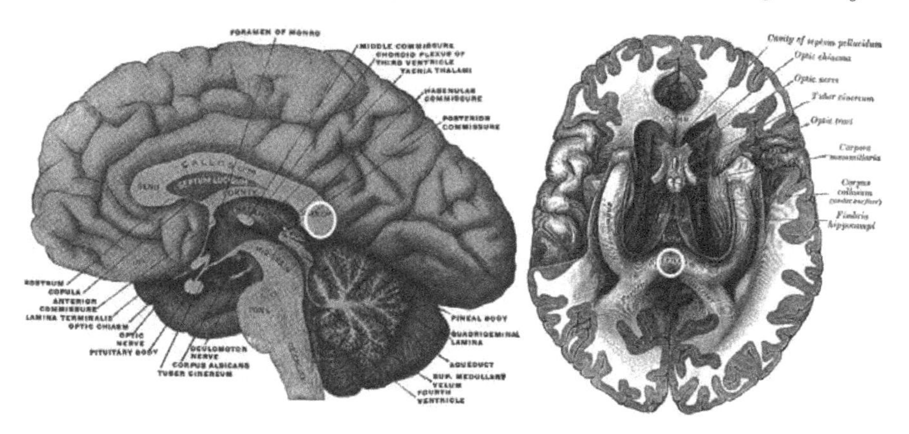

FIGURE 10.7
The splenium of the corpus callosum (marked in white circle) was manually masked and a streamline based tractography algorithm was applied to obtain white matter tracts passing through the splenium. The anatomic drawings are from the Wikipedia version of Gray's Anatomy [149].

inhomogeneities.

10.5.2 Image Processing

Eddy current related distortion and head motion of each data set were corrected using the Automated Image Registration (AIR) software [386] and distortions from field inhomogeneities were corrected using custom software algorithms based on [185]. Distortion-corrected diffusion weighted (DW) images were interpolated to $2 \times 2 \times 2$mm voxels and the six tensor elements were calculated using a multivariate log-linear regression method [29].

The images were isotropically resampled at 1 mm^3 resolution before applying the white matter tractography algorithm. The second order Runge-Kutta streamline algorithm with tensor deflection [218] was used. The trajectories were initiated at the center of the seed voxels and were terminated if they either reached regions with the factional anisotropy (FA) value smaller then 0.15 or if the angle between two consecutive steps along the trajectory was larger than $\pi/4$. Each tract consists of 105 ± 54 control points as shown in Figure 10.2. The distance between control points is 1mm. Whole brain tracts are stored as a binary file of about 600MB in size. Whole brain white matter tracts for 74 subjects are further aligned using the affine registration [183] of FA-maps to the average FA-map.

FIGURE 10.8
The average tract (thick tube) of 2149 fibers in a single subject. 2149 fiber tracts are subsampled to show few selective tracts. The average tract is obtained by averaging the coefficients of all 2149 cosine representations. The glass brain is obtained from the average fractional anisotropy map.

10.5.3 Cosine Series Representation

The splenium of the corpus callosum was manually masked by J.E. Lee [222]. See Figure 10.7 for the location of the splenium in the brain. Then the white matter tracts passing through a ball of radius 5mm at the splenium are identified. Each subject has 1943 ± 1148 number of tracts passing through the ball. The cosine series representation was constructed for each tracts and resulted in 60 coefficients for characterizing the single tract. The within-subject tract averaging was easily done within our representations by averaging the coefficients of the same degree (Figure 10.8). We are interested in testing the fiber shape difference between the groups.

10.5.4 Two Sample T-test

We have investigated the utility of the proposed parametric representation in discriminating the different populations (42 autistic vs. 32 control subjects) using two different tests.

The average tracts for all 74 subjects were obtained using the cosine series representation. The coefficients of the representation are used to discriminate the groups. The bar plots of all 20 coefficients for 3 coordinates are given in Figure 10.9. The significance of the mean coefficient difference for each degree is determined using the two sample T-test with unequal variance assumption. The corresponding p-value in $-\log_{10}$ scale is also given. The first three bars (black to gray) in each degree correspond to the p-values for three coordinates. The minimum p-values are 0.0362 (x coordinate, degree 15), 0.0093 (y

coordinate, degree 6) and 0.0023 (z coordinate, degree 8). Note that at least 4 coefficients (degree 0, 2, 6, 8) for the z coordinate show p-value smaller than 0.01. The Bonferroni correction was used to determine the overall significance across different degrees, we have used The T-statistics across different degrees. The Bonferroni corrected p-value for the 8-th degree coefficient of the z coordinate (by multiplying 20 to 0.0023) is 0.0456 indicating that there is significant group difference at the particular spatial frequency. Note that from (10.5), the 8-th degree corresponds to the spatial frequency of 4.

10.5.5 Hotelling's T-square Test

The problem of using T-test is that the inference has to be done for each coordinates separately. Although T-test gives additional localized information (about z coordinate values being responsible for shape difference), it is not really a clear cut conclusion so we require an overall measure of significance across different coordinates. Therefore, to avoid using T-test separately for each coordinate, we use the Hotelling's T-square statistic on the vector of 3 coefficients at each degree. The last bar (yellow) in the $-\log_{10} p$ plot shows the resulting p-values. These p-value should be interpreted as the measure of overall significance of three p-values obtained from the T-tests. The minimum p-value is 0.0047 at degree 6. After the Bonferroni correction by multiplying 20, we obtain the corrected p-value of 0.0939, which would be still considered as significant at $\alpha = 0.1$ level test.

10.5.6 Simulating Curves

We have performed a simulation study to determine if the proposed framework can detect small tract shape differences between two simulated samples of curves. Our simulations demonstrate the proposed cosine series representation works as expected.

Taking the parametric curve

$$(x, y, z) = (s \sin s, s \cos s, s), s \in [0, 10] \tag{10.18}$$

as a base for simulation, we have generated two groups of random curves. This gives a shape of a spiral with increasing radius along the z-axis. In `MATLAB`, the base curve is generated by

```
t=0:0.1:10
tract=[t.*sin(t); t.*cos(t); t];
figure; plot3(tract(1,:),tract(2,:),tract(3,:),'.b')

[arc_length  para]=parameterize_arclength(tract);
[wfs beta]=WFS_tracts(tract,para,19);
hold on; plot3(wfs(1,:),wfs(2,:),wfs(3,:),'r')
```

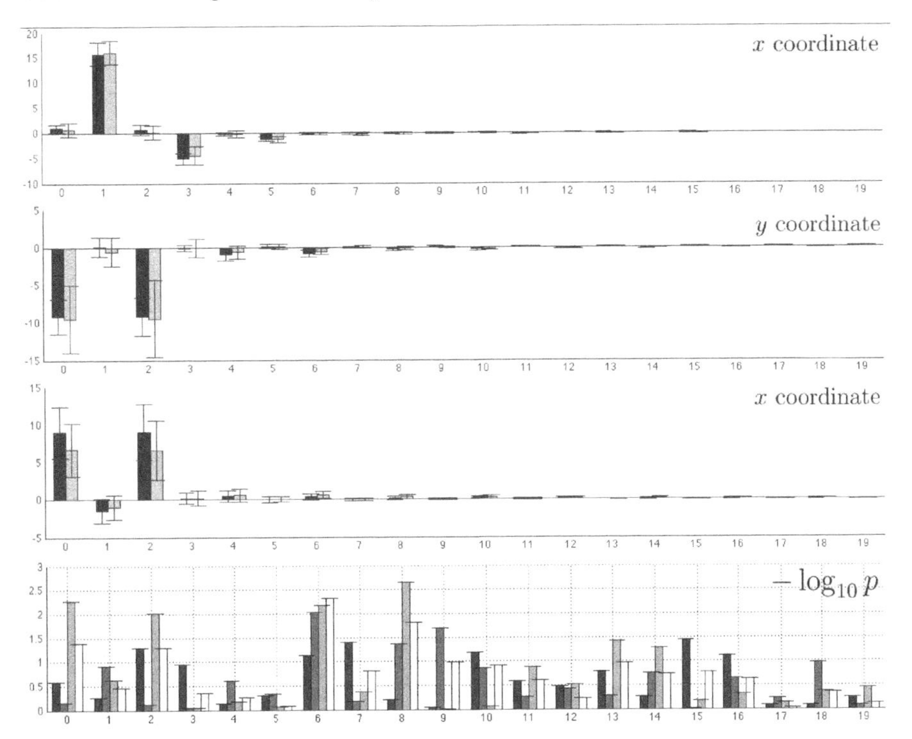

FIGURE 10.9
Bar plots of coefficients for autistic (gray) and control (black) groups. Each row corresponds to the x, y and z coordinates. The error bars are one standard deviation in each direction. The autistic subjects show larger variability compared to controls, which is consistent with the literature [76]. The last row is the p-value plots in $-\log_{10}$ scale. The p-values of the two sample T-test corresponding to x, y and z coordinates are given in three bars sequentially. The last bar (white) shows the p-values of the Hotelling's T-square test on coefficient vectors.

Taking this as the underlying signal, noises are added nonlinearly to obtain two groups of random curves. The first group (red curves in Figure 10.11) consists of 20 curves generated by

$$(x, y, z) = (s \sin(s + e_1), s \cos(s + e_2), s + e_3).$$

The second group (blue curves in Figure 10.11) consists of 20 curves generated by

$$(x, y, z) = ((s + e_4) \sin(s + 0.1), (s + e_5) \cos(s - 0.1), s - 0.1).$$

The non-additive noise is given to perturb (10.18) a little bit while to make our procedure blind to the underlying additive noise assumption (10.7). We have performed two different simulations with a different amount of noise variability.

Example 5 *Let* $e_1, e_2, e_3 \sim N(0, 0.1^2)$, $e_4, e_5 \sim N(0, 0.2^2)$. *The p-values are all less than 0.00000243 for Hotelling's T-square test. The corrected p-value after the Bonfrerroni correction is less than 0.00005 indicating very strong discrimination between the groups for every degree used. This is evident from the first figure in Figure 10.11, where we clearly see group separation.*

Example 6 *In the second simulation, we have increased the noise variability such that* $e_1, e_2, e_3 \sim N(0, 0.2^2)$ $e_4, e_5 \sim N(0, 0.5^2)$. *The smallest p-value is 0.0147. After correcting for multiple comparisons, we obtained the corrected p-value of 0.294 indicating weak group separation in almost all degrees used. The second figure in Figure 10.11 does not show any clear group separation between the groups.*

10.6 Discussion

We have presented a unified parametric model building technique for a bundle of 3D curves, and applied the method in discriminating the shape of white matter fibers passing through the splenium in autistic subjects. In this section, we discuss two major limitations of the cosine series representation.

10.6.1 Similarly Shaped Tracts

The cosine series representation assumes the correspondence of two end points of all fiber tracts. In practice, such assumption is not realistic. A short fiber may correspond to a segment of a longer fiber. In this case, the proposed method does not work properly. However, this is not the major limitation in our seed-based fiber tract modeling since it is guaranteed that two end points of all fiber tracts have to match. Also the tracts that pass through the splenium of the corpus callosum are somewhat similar in shape and length (Figure 10.10). Therefore, we do not need to worry about the case of matching a short tract to a longer tract. We are basically matching tracts of similar shape and length in this study.

10.6.2 Gibbs Phenomenon

Gibbs phenomenon (ringing artifacts) often arises in Fourier series expansion of discontinuous data. It is named after American physicist Josiah Willard Gibbs. In representing piecewise continuously differentiable data using the

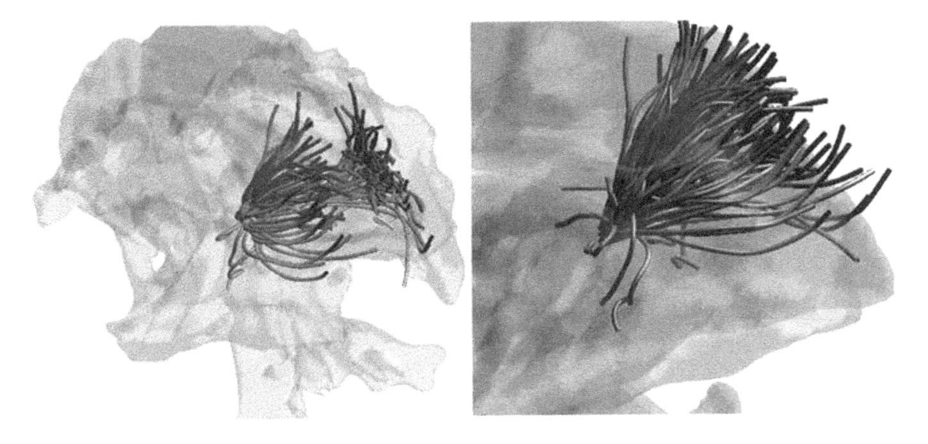

FIGURE 10.10
Each streamtube is the average tract in a subject. White matter fibers in controls (blue) are more clustered together with smaller spreading compared to autism (red). (See color insert.)

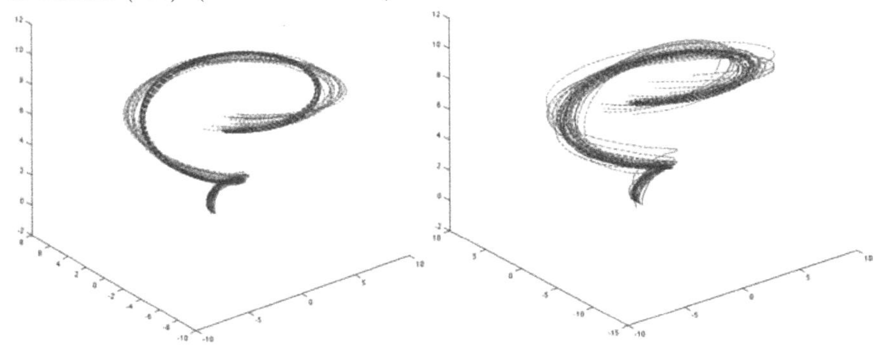

FIGURE 10.11
Simulated curves obtained from perturbing the basic curve shape $(x, y, z) = (s \sin s, s \cos s, s), s \in [0, 10]$. The first figure shows clear group separation while the second figure has too much overlap. We expect the cosine series representation to work extremely well for the first simulation while it may not work for the second simulation. (See color insert.)

Fourier series, the overshoot of the series happens at a jump discontinuity (Figure 10.12). The overshoot does not decease as the number of terms increases in the series expansion, and it converges to a finite limit called the Gibbs constant. The Gibbs phenomenon was first observed by Henry Willbraham in 1848 [380] but it did not attract any attention at that time. Then a Nobel prize laureate Albert Michelson constructed an harmonic analyzer, one of the first mechanical analogue computers, that was used to plot Fourier series and

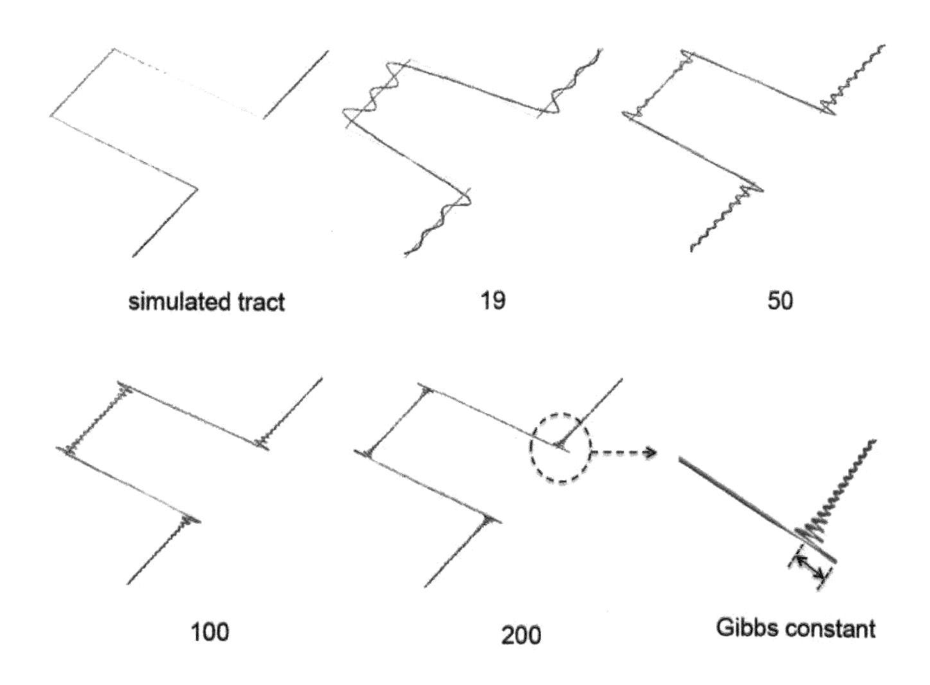

FIGURE 10.12
The Gibbs phenomenon is shown in the simulated tract. Overshoot near the sharp corners will not disappear even if the degree of expansion increases.

observed the phenomenon. He thought the phenomenon was caused by mechanical error but Josiah Willard Gibbs correctly explained the phenomenon as mathematical in 1899. Josiah Willard Gibbs rediscovered the phenomenon in 1898 [139]. Later mathematician Maxime Bocher named it the Gibbs phenomenon and gave a precise mathematical analysis in 1906 [43]. The Gibbs phenomenon associated with spherical harmonics were first observed by Herman Weyl in 1968. The history and the overview of Gibbs phenomenon can be found in the literature [124, 184].

The Gibbs phenomenon will likely arise in modeling arbitrary shaped curves with possible sharp corners. We have demonstrated the Gibbs phenomenon for a simulated tract with jump discontinuities in the following simulated example.

Example 7 *We have simulated 300 uniformly sampled control points along the parameterized curve* $(x, y, z) = (t, 0, t)$ *for* $t \in [1, 100) \cup [200, 300)$ *and* $(x, y, z) = (t, 1, t)$ *for* $t \in [200, 300)$. *Figure 10.12 only shows the part of the curve with jump discontinuities. The control points are fitted with the cosine representation with various degrees. As the degree increases to 200, the* *representation suffers from the severe ringing artifacts. The overshoot shown in Figure 10.12 will not disappear even if the degree of expansion goes to infinity. Note that white matter fibers are assumed to be smooth so we will not encounter the Gibbs phenomenon in modeling fibers.*

There are few available techniques for reducing Gibbs phenomenon [50] [148]. Most techniques are a variation on some sort of kernel methods. For instance, for the Fejer kernel K_n, it can be shown that

$$K_n * f \to f$$

for any, even discontinuous, $f \in \mathcal{L}^2[-\pi, \pi]$ as $n \to \infty$. It has the effect of smoothing the discontinuous signal f and in turn the convolution will not exhibit the ringing artifacts for sufficiently large n. Particularly related to Fourier and spherical harmonic descriptors, we have introduced an exponential weighting scheme [71, 73]. By weighting Fourier coefficients with exponentially decaying weights, the series expansion can converge faster and reduce the Gibbs phenomenon significantly.

Instead of the k-th degree expansion (10.8), we define the weighted Fourier expansion as

$$\sum_{l=0}^{k} e^{-\lambda_l \sigma} \langle f, \psi_l \rangle \psi_l \tag{10.19}$$

for some smoothing parameter σ. Then it can be shown that (10.19) is the finite series expansion of heat kernel smoothing $K_\sigma * f$, where the heat kernel is defined as

$$K_\sigma(t, s) = \sum_{l=0}^{\infty} e^{-\lambda_l \sigma} \psi_l(t) \psi_l(s).$$

The expansion (10.19) can be further shown to be the finite approximation to the solution of heat diffusion

$$\frac{\partial}{\partial \sigma} g = \Delta g, \; g(t, \sigma = 0) = f(t).$$

Since the weighting scheme makes the expansion converges to heat diffusion, the estimation at the jump discontinuity is smoothed out reducing the Gibbs phenomenon.

11

Weighted Spherical Harmonic Representation

There is a lack of a unified statistical modeling framework for cerebral shape asymmetry analysis in the literature. Most previous approaches start with flipping the 3D magnetic resonance images (MRI). The anatomical correspondence across the hemispheres is then established by registering the original image to the flipped image. A difference of an anatomical index between these two images is used as a measure of cerebral asymmetry. We present a radically different asymmetry analysis that utilizes a novel weighted spherical harmonic representation of cortical surfaces. The weighted spherical harmonic representation is a surface smoothing technique given explicitly as a weighted linear combination of spherical harmonics. This new representation is used to parameterize cortical surfaces, establish the hemispheric correspondence, and normalize cortical surfaces in a unified mathematical framework. The methodology has been applied in characterizing the cortical asymmetry of a group of autistic subjects. This chapter is mainly based on [73].

11.1 Introduction

Previous neuroanatomical studies have shown left occipital and right frontal lobe asymmetry, and left planum temporal asymmetry in normal controls [25, 200]. These studies mainly flip the whole brain 3D MRI to obtain the mirror reflected MRI with respect to the mid-sagittal cross-section. Then the anatomical correspondence across the hemispheres is established and a subsequent statistical analysis is performed at each voxel in the 3D MRI. Although this approach is sufficient for the voxel-based morphometry [14, 15], where we only need an approximate alignment of corresponding brain substructures, it may fail to properly align highly convoluted sulcal and gyral foldings of gray matter. In order to address this shortcoming inherent in 3D whole brain volume asymmetry analysis, we need a new 2D cortical surface based framework.

The human cerebral cortex has the topology of a 2D highly convoluted grey matter shell with an average thickness of 3mm. The outer boundary of the shell is called the *outer cortical surface* while the inner boundary is called the *inner cortical surface*. Cortical surfaces are segmented from magnetic resonance images (MRI) using a deformable surface algorithm and represented as a triangle mesh consisting of more than 40,000 vertices and 80,000 triangle elements [83, 237]. Once we obtain both the outer and inner cortical surfaces of a subject, *cortical thickness*, which is the distance between the outer and inner surfaces, is computed at each vertex of the outer surface [237]. Since different clinical populations are expected to show different patterns of cortical thickness variations, cortical thickness has been used as a quantitative index for characterizing a clinical population [76]. Cortical thickness varies locally by region and is likely to be influenced by aging, development and disease [26]. By analyzing how cortical thickness differs locally in a clinical population with respect to a normal population, neuroscientists can locate the regions of abnormal anatomical differences in the clinical population. Cortical thickness serves as a metric of interest in performing 2D cortical asymmetry analysis. However, there are various methodological issues associated with using triangle mesh data. Our novel 2D surface modeling framework called the *weighted spherical harmonic representation* [71] can address these issues in a unified mathematical framework.

Cortical surface mesh construction and cortical thickness computation are expected to introduce noise. To counteract this, surface-based data smoothing is necessary. For 3D whole brain volume-based method, Gaussian kernel smoothing, which weights neighboring observations according to their 3D Euclidean distance, has been used. However, for data that lie on a 2D surface, smoothing must be weighted according to the geodesic distance along the surface [10, 83]. It will be shown that the weighted spherical harmonic representation is a 2D surface-based smoothing technique, where the explicit basis function expansion is used to smooth out noisy cortical surface data. The basis function expansion corresponds to the solution of isotropic heat diffusion. Unlike the previous surface based smoothing that solves the heat equation nonparametrically [10, 83, 76], the result of the weighted spherical harmonic representation is explicitly given as a weighted linear combination of spherical harmonics. This provides a more natural statistical modeling framework. A validation study showing the improved performance of the weighted spherical harmonic representation over heat kernel smoothing was given in [76].

Comparing measurements defined at mesh vertices across different cortical surfaces is not a trivial task due to the fact no two cortical surfaces are identically shaped. In comparing measurements across different 3D whole brain images, 3D volume-based image registration is needed. However, 3D image registration techniques tend to misalign sulcal and gyral folding patterns of the cortex. Hence, 2D surface-based registration is needed in order to compare measurements across different cortical surfaces. Various surface registration methods have been proposed before [76, 363, 99, 248, 121]. These methods solve a complicated optimization problem of minimizing the measure of discrepancy between two surfaces. Unlike the previous computationally intensive methods, the weighted spherical harmonic representation provides a simple way of establishing surface correspondence between two surfaces in reducing the improper alignment of sulcal folding patterns without time consuming numerical optimization.

Once we establish surface correspondence between two surfaces, we also need to establish hemispheric correspondence within a subject for asymmetry analysis. However, it is not straightforward to establish a 2D surface-based hemispheric correspondence. Although there are many 3D volume-based brain hemisphere asymmetry analyses [25, 200], due to this simple reason, there is a lack of 2D surface-based asymmetry analyses. This will be the first unified mathematical framework on 2D cortical asymmetry. The inherent angular symmetry presented in the weighted spherical harmonic representation can be used to establish the inter-hemispheric correspondence. It turns out that the usual asymmetry index of (L-R)/(L+R) is expressed as the ratio between the sum of positive and negative order harmonics.

The novelty of our proposed method is that surface parameterization, surface-based smoothing, and within- and between- subject surface registration can be performed within a single unified mathematical framework that provides a more consistent modeling framework than previously available for cortical analysis.

11.2 Spherical Coordinates

Cortical thickness is measured at each vertex and used as a measure for characterizing cortical shape variation. There exists a bijective mapping between the cortical surface $\mathcal{M}$ and a unit sphere S^2 that is obtained via the deformable surface algorithm. Consider the parameterization of the unit sphere S^2 given by

$$(u_1, u_2, u_3) = (\sin\theta\cos\varphi, \sin\theta\sin\varphi, \cos\theta),$$

with $(\theta, \varphi) \in [0, \pi) \otimes [0, 2\pi)$. The polar angle θ is the angle from the north pole and the azimuthal angle φ is the angle along the horizontal cross-section. Then, using the bijective mapping, we can parameterize the Cartesian coordinates $v = (v_1, v_2, v_3)$ of each cortical mesh vertex in the cortical surface $\mathcal{M}$ with the spherical angles (θ, φ), i.e., $v = v(\theta, \varphi)$ (Figure 11.1). This enables us to represent cortical thickness measurements f with respect to the spherical coordinates, i.e., $f = f(\theta, \varphi)$. Each component of surface coordinates will be modeled independently as

$$v_i(\theta, \varphi) = h_i(\theta, \varphi) + \epsilon_i(\theta, \varphi), \tag{11.1}$$

where h_i is the unknown smooth coordinate function and ϵ_i is a zero mean random field, possibly Gaussian. We model cortical thickness f similarly as

$$f(\theta, \varphi) = g(\theta, \varphi) + e(\theta, \varphi),$$

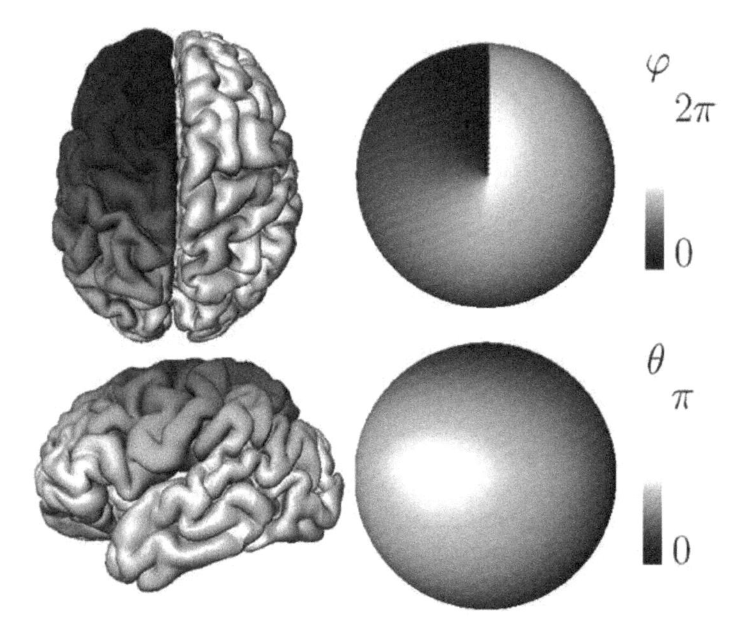

FIGURE 11.1
Parameterization of cortical surface using the spherical coordinates; the north and south poles are chosen in the plane, i.e. $u_2 = 0$, that separates the left and the right hemispheres.

where g is the unknown mean cortical thickness and e is a zero mean random field. We further assume $v_i, f \in \mathcal{L}^2(S^2)$, the space of square integrable functions on unit sphere S^2. The unknown signals h_i and g are then estimated in the finite subspace of $\mathcal{L}^2(S^2)$ spanned by harmonic basis functions in least squares fashion.

11.3 Spherical Harmonics

Definition 15 *The spherical harmonic Y_{lm} of degree l and order m is defined as*

$$Y_{lm} = \begin{cases} c_{lm} P_l^{|m|}(\cos\theta)\sin(|m|\varphi), & -l \le m \le -1, \\ \frac{c_{lm}}{\sqrt{2}} P_l^{|m|}(\cos\theta), & m = 0, \\ c_{lm} P_l^{|m|}(\cos\theta)\cos(|m|\varphi), & 1 \le m \le l, \end{cases}$$

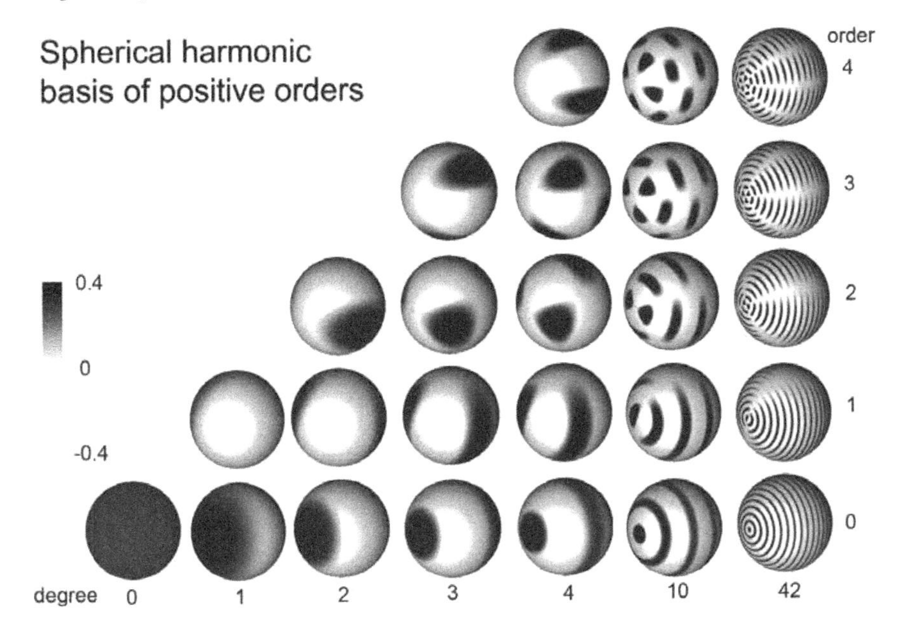

FIGURE 11.2
Spherical harmonics of positive orders. The negative orders are simply the rotation of the positive order harmonics.

where $c_{lm} = \sqrt{\frac{2l+1}{2\pi}\frac{(l-|m|)!}{(l+|m|)!}}$ *and* P_l^m *is the* associated Legendre polynomial *of order* m *[92, 372].*

The associated Legendre polynomial is given by

$$P_l^m(x) = \frac{(1-x^2)^{m/2}}{2^l l!}\frac{d^{l+m}}{dx^{l+m}}(x^2-1)^l, x \in [-1,1].$$

The first few terms of the spherical harmonics are

$$Y_{00} = \frac{1}{\sqrt{4\pi}}, Y_{1,-1} = \sqrt{\frac{3}{4\pi}}\sin\theta\sin\varphi,$$

$$Y_{1,0} = \sqrt{\frac{3}{4\pi}}\cos\theta, Y_{1,1} = \sqrt{\frac{3}{4\pi}}\sin\theta\cos\varphi.$$

Few representative spherical harmonics are shown in Figure 11.2. The spherical harmonics are orthonormal with respect to the inner product

$$\langle f_1, f_2\rangle = \int_{S^2} f_1(\Omega) f_2(\Omega)\, d\mu(\Omega),$$

where $\Omega = (\theta, \varphi)$ and the Lebesgue measure $d\mu(\Omega) = \sin\theta d\theta d\varphi$. The norm is then defined as

$$||f_1|| = \langle f_1, f_1\rangle^{1/2}. \tag{11.2}$$

Consider the subspace $\mathcal{I}_l$ spanned by the l-th degree spherical harmonics:

$$\mathcal{I}_l = \{ \sum_{m=-l}^{l} \beta_{lm} Y_{lm}(\Omega) : \beta_{lm} \in \mathbb{R} \}.$$

Then the subspace $\mathcal{H}_k$ spanned by up to k-th degree spherical harmonics is decomposed as the direct sum of $\mathcal{I}_0, \cdots, \mathcal{I}_k$:

$$\begin{aligned} \mathcal{H}_k &= \mathcal{I}_0 \oplus \mathcal{I}_1 \cdots \oplus \mathcal{I}_k. \\ &= \{ \sum_{l=0}^{k} \sum_{m=-l}^{l} \beta_{lm} Y_{lm}(\Omega) : \beta_{lm} \in \mathbb{R} \}. \end{aligned}$$

Traditionally, the coordinate functions h_i are estimated by minimizing the integral of the squared residual within $\mathcal{H}_k$:

$$\widehat{h}_i(\Omega) = \arg \min_{h \in \mathcal{H}_k} \int_{S^2} \Big| v_i(\Omega) - h(\Omega) \Big|^2 \, d\mu(\Omega). \tag{11.3}$$

It can be shown that the minimization is obtained when

$$\widehat{h}_i(\Omega) = \sum_{l=0}^{k} \sum_{m=-l}^{l} \langle v_i, Y_{lm} \rangle Y_{lm}(\Omega). \tag{11.4}$$

Representing an anatomical boundary via the Fourier series expansion (11.4) has been referred to as the *spherical harmonic representation* [135, 152, 323, 322]. This technique has been used in representing hippocampi [323], ventricles [135] and cortical surfaces [71, 152].

11.3.1 Weighted Spherical Harmonic Representation

The weakness of the traditional spherical harmonic representation is that it produces the Gibbs phenomenon (ringing artifacts) [71, 132] for discontinuous and rapidly changing continuous measurements. The Gibbs phenomenon can be effectively removed if the spherical harmonic representation converges fast enough as the degree goes to infinity. By weighting the spherical harmonic coefficients exponentially smaller, we can make the representation converge faster; this can be achieved by additionally weighting the squared residuals in equation (11.3) with the *heat kernel.*

Example 8 *Figure 11.3 demonstrates the severe Gibbs phenomenon in the traditional spherical harmonic representation (top row) on a hat-shaped 2D surface. The hat shaped step function is simulated as* $z = 1$ *for* $x^2 + y^2 < 1$ *and* $z = 0$ *for* $1 \leq x^2 + y^2 \leq 2$*. On the other hand the weighted spherical harmonic representation shows substantially reduced ringing artifacts. In both representations, we have used degree* $k = 42$*. For the weighted spherical harmonic representation, the bandwidth* $\sigma = 0.001$ *is used throughout the book.*

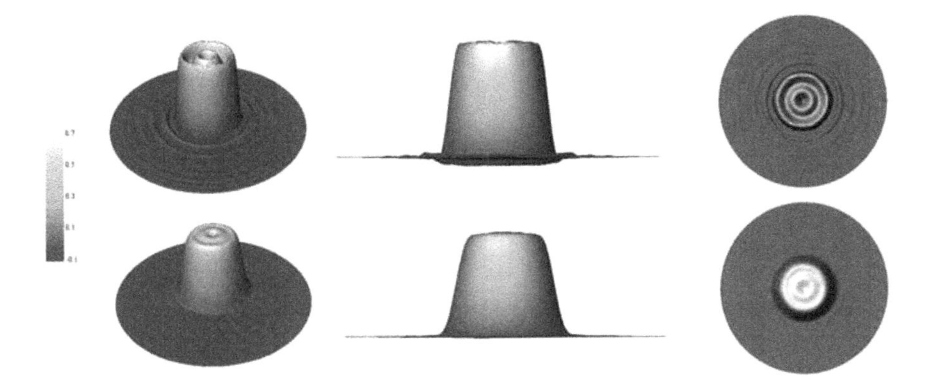

FIGURE 11.3
The Gibbs phenomenon on a hat shaped simulated surface showing the severe ringing effect on the traditional spherical harmonic representation (top) and reduced ringing effect on the weighted spherical harmonic representation (bottom). The degree $k = 42$ is used for the both cases and the bandwidth $\sigma = 0.001$ is used for the weighted spherical harmonic representation.

Due to very complex folding patterns, sulcal regions of the brain exhibit more abrupt directional change than the simulated hat surface(upward of 180 degree compared to 90 degree in the hat surface) so there is a need for reducing the Gibbs phenomenon in the traditional spherical harmonic representation.

The heat kernel is the generalization of the Gaussian kernel defined on Euclidean space to an arbitrary Riemannian manifold [76, 304]. On a unit sphere, the heat kernel is written as

$$K_\sigma(\Omega, \Omega') = \sum_{l=0}^{\infty} \sum_{m=-l}^{l} e^{-l(l+1)\sigma} Y_{lm}(\Omega) Y_{lm}(\Omega'), \tag{11.5}$$

where $\Omega = (\theta, \varphi)$ and $\Omega' = (\theta', \varphi')$. The heat kernel is symmetric and positive definite, and a probability distribution since

$$\int_{S^2} K_\sigma(\Omega, \Omega')\, d\mu(\Omega) = 1.$$

The parameter σ controls the dispersion of the kernel so we simply call it the *bandwidth*. The heat kernel satisfies

$$\lim_{\sigma \to \infty} K_\sigma(\Omega, \Omega') = \frac{1}{4\pi} \text{ and } \lim_{\sigma \to 0} K_\sigma(\Omega, \Omega') = \delta(\Omega - \Omega')$$

with δ as the Dirac-delta function. The heat kernel can be further simplified using the harmonic addition theorem [372] as

$$K_\sigma(\Omega, \Omega') = \sum_{l=0}^{\infty} \frac{2l+1}{4\pi} e^{-l(l+1)\sigma} P_l^0(\Omega \cdot \Omega'), \tag{11.6}$$

where $\cdot$ is the Cartesian inner product.

Let us define *heat kernel smoothing* [76] as

$$K_\sigma * f(\Omega) = \int_{S^2} K(\Omega, \Omega') f(\Omega')\, d\mu(\Omega'). \tag{11.7}$$

Then heat kernel smoothing has the following spectral representation, which can be easily seen by substituting (11.5) into equation (11.7) and rearranging the integral with the summation:

$$K_\sigma * f(\Omega) = \sum_{l=0}^{\infty} \sum_{m=-l}^{l} e^{-l(l+1)\sigma} \langle f, Y_{lm} \rangle Y_{lm}(\Omega), \tag{11.8}$$

The k-th degree finite series approximation of heat kernel smoothing is referred to as the *k-th degree weighted spherical harmonic representation.* The unknown mean coordinates h_i are estimated using the weighted spherical harmonic representation, which is the minimizer of the of the weighted squared distance between measurements v_i and a function h in $\mathcal{H}_k$ space. The unknown mean cortical thickness g is estimated similarly.

Theorem 7

$$\sum_{l=0}^{k} \sum_{m=-l}^{l} e^{-l(l+1)\sigma} \langle v_i, Y_{lm} \rangle Y_{lm}$$

$$= \arg \min_{h \in \mathcal{H}_k} \int_{S^2} \int_{S^2} K_\sigma(\Omega, \Omega') |v_i(\Omega') - h(\Omega)|^2 \, d\mu(\Omega')\, d\mu(\Omega)$$

Theorem 7 is proved as follows. Let $v_i = \sum_{l=0}^{k} \sum_{m=-l}^{l} \beta_{lm} Y_{lm}$. Let the inner integral be

$$I = \int_{\mathcal{M}} K_\sigma(\Omega, \Omega') \Big| v_i(\Omega') - \sum_{l=0}^{k} \sum_{m=-l}^{l} \beta_{lm} Y_{lm}(\Omega) \Big|^2 \, d\mu(\Omega').$$

Simplifying the expression, we obtain

$$I = \sum_{l=0}^{k} \sum_{m=-l}^{l} \sum_{l'=0}^{k} \sum_{m'=-l'}^{l'} Y_{lm}(\Omega) Y_{l'm'}(\Omega) \beta_{lm} \beta_{l'm'}$$

$$-2K_\sigma * v_i(\Omega) \sum_{l=0}^{k} \sum_{m=-l}^{l} Y_{lm}(\Omega) \beta_{lm} + K * v_i^2(\Omega).$$

Since I is an unconstrained positive semidefinite quadratic program (QP) in β_{lm}, there is no unique global minimizer of I without additional linear constraints. Integrating I further with respect to $\mu(\Omega)$, we collapses the QP to a positive definite QP, which yields a unique global minimizer as

$$\int_{S^2} I\, d\mu(\Omega) = \sum_{l=0}^{k} \sum_{m=-l}^{l} \beta_{lm}^2 - 2\sum_{i=0}^{k} e^{-l(l+1)\sigma} \langle v_i, Y_{lm} \rangle \beta_{lm} + \sum_{i=0}^{\infty} e^{-l(l+1)\sigma} \langle v_i^2, Y_{lm} \rangle.$$

The minimum of the above integral is obtained when all the partial derivatives with respect to β_j vanish.

$$\int_{S^2} \frac{\partial I}{\partial \beta_{lm}}\, d\mu(\Omega) = 2\beta_{lm} - 2e^{-l(l+1)\sigma} \langle v_i, Y_{lm} \rangle = 0.$$

Hence $\sum_{l=0}^{k} \sum_{m=-l}^{l} e^{-l(l+1)\sigma} \langle v_i, Y_{lm} \rangle Y_{lm}$ is the unique minimizer in $\mathcal{H}_k$.

We can also show that the weighted spherical harmonic representation is related to previously available surface-based isotropic diffusion smoothing [10, 53, 83, 76] via the following theorem.

Theorem 8

$$\sum_{l=0}^{k} \sum_{m=-l}^{l} e^{-l(l+1)\sigma} \langle v_i, Y_{lm} \rangle Y_{lm}(\Omega) = \arg \min_{h \in \mathcal{H}_k} \|h - h_0\|,$$

where h_0 satisfies isotropic heat diffusion

$$\frac{\partial h_0}{\partial \sigma} = \Delta h_0 = \frac{1}{\sin\theta} \frac{\partial}{\partial \theta} \Big(\sin\theta \frac{\partial h_0}{\partial \theta} \Big) + \frac{1}{\sin^2\theta} \frac{\partial^2 h_0}{\partial^2 \varphi}, \tag{11.9}$$

with the initial value condition $h_0(\Omega, \sigma = 0) = v_i(\Omega)$.

We first prove that heat kernel smoothing (11.7) and its spectral representation (11.8) are the solution of the heat equation (11.9). At each fixed σ, which serves as the physical time of the heat equation, the solution $h_0(\Omega, \sigma)$ belongs to $\mathcal{L}^2(S^2)$. Then the solution can be written as

$$h_0(\Omega, \sigma) = \sum_{l=0}^{\infty} \sum_{m=-l}^{l} c_{lm}(\sigma) Y_{lm}(\Omega). \tag{11.10}$$

Since the spherical harmonics are the eigenfunctions of the spherical Laplacian [372], we have

$$\Delta Y_{lm}(\Omega) = -l(l+1) Y_{lm}(\Omega). \tag{11.11}$$

Substituting (11.10) into (11.9) and using (11.11), we obtain

$$\frac{\partial c_{lm}(\sigma)}{\partial \sigma} = -l(l+1) c_{lm}(\sigma). \tag{11.12}$$

The solution of the ordinary differential equation (11.12) is given by $c_{lm}(\sigma) = b_{lm}e^{-l(l+1)\sigma}$ for some constant b_{lm}. Hence, we obtain the solution of the form

$$h_0(\Omega, \sigma) = \sum_{l=0}^{\infty} \sum_{m=-l}^{l} b_{lm} e^{-l(l+1)\sigma} Y_{lm}(\Omega).$$

When $\sigma = 0$, we have the initial condition

$$h_0(\Omega, 0) = \sum_{l=0}^{\infty} \sum_{m=-l}^{l} b_{lm} Y_{lm}(\Omega) = v_i(\Omega).$$

The coefficients b_{lm} must be the spherical harmonic coefficients, i.e. $b_{lm} = \langle v_i, Y_{lm} \rangle$. Then from the property of the generalized Fourier series [307], the finite expansion is the closest to the infinite series in $\mathcal{H}_k$:

$$\sum_{l=0}^{k} \sum_{m=-l}^{l} e^{-l(l+1)\sigma} \langle v_i, Y_{lm} \rangle Y_{lm}(\Omega) = \arg \min_{h \in \mathcal{H}_k} \Big|\Big| h - h_0(\Omega, \sigma) \Big|\Big|.$$

This proves the statement of the theorem.

11.3.2 Estimating Spherical Harmonic Coefficients

The spherical harmonic coefficients are estimated based on an iterative procedure that utilizes the orthonormality of spherical harmonics. We assume that coordinate functions are measured at n points $\Omega_1, \cdots, \Omega_n$. Then we have the normal equations

$$v_i(\Omega_j) = \sum_{l=0}^{k} \sum_{m=-l}^{l} e^{-l(l+1)\sigma} \langle v_i, Y_{lm} \rangle Y_{lm}(\Omega_j), j = 1, \cdots, n. \tag{11.13}$$

The normal equations (11.13) can be written in the matrix form as

$$\mathbf{V} = \underbrace{[\mathbf{Y}_0, e^{-1(1+1)\sigma}\mathbf{Y}_1, \cdots, e^{-k(k+1)\sigma}\mathbf{Y}_k]}_{\mathbf{Y}} \beta, \tag{11.14}$$

where the column vectors are

$$\begin{aligned} \mathbf{V} &= [v_i(\Omega_1), \cdots, v_i(\Omega_n)]' \\ \beta' &= (\beta_0', \beta_1', \cdots, \beta_k') \end{aligned}$$

with $\beta_l' = (\langle v_i, Y_{l,-l} \rangle, \cdots, \langle v_i, Y_{l,l} \rangle)$. The length of the vector β is

$$1 + (2 \cdot 1 + 1) + \cdots + (2 \cdot k + 1) = (k+1)^2.$$

Each submatrix $\mathbf{Y}_l$ is given by

$$\mathbf{Y}_l = \begin{bmatrix} Y_{l,-l}(\Omega_1), & \cdots & , Y_{l,l}(\Omega_1) \\ \vdots & \ddots & \vdots \\ Y_{l,-l}(\Omega_n), & \cdots & , Y_{l,l}(\Omega_n) \end{bmatrix}.$$

We may tempted to directly estimate β in least squares fashion as

$$\widehat{\beta} = (\mathbf{Y}'\mathbf{Y})^{-1}\mathbf{Y}'\mathbf{V}.$$

However, since the size of matrix $\mathbf{Y}'\mathbf{Y}$ becomes $(k+1)^2 \times (k+1)^2$, for large degree k, it may be difficult to directly invert the matrix. Instead of directly solving the normal equations, we project the normal equations into a smaller subspace $\mathcal{I}_l$ and estimate $2l+1$ coefficients in an iterative fashion.

At degree 0, we write $\mathbf{V} = \mathbf{Y}_0\beta_0 + \mathbf{r}_0$, where $\mathbf{r}_0$ is the residual vector of estimating $\mathbf{V}$ in subspace $\mathcal{I}_0$. Note that the residual vector $\mathbf{r}_0$ consists of residuals $r_0(\Omega_j)$ for all Ω_j. Then we estimate β_0 by minimizing the residual vector in least squares fashion:

$$\widehat{\beta}_0 = (\mathbf{Y}_0'\mathbf{Y}_0)^{-1}\mathbf{Y}_0'\mathbf{V} = \frac{\sum_{j=1}^n v_i(\Omega_j)Y_{00}(\Omega_j)}{\sum_{j=1}^n Y_{00}^2(\Omega_j)}.$$

At degree l, we have

$$\mathbf{r}_{l-1} = e^{-l(l+1)\sigma}\mathbf{Y}_l\beta_l + \mathbf{r}_l, \tag{11.15}$$

where the residual vector $\mathbf{r}_{l-1}$ is obtained from the previous estimation as

$$\mathbf{r}_{l-1} = \mathbf{V} - \mathbf{Y}_0\widehat{\beta}_0 \cdots - e^{-(l-1)l\sigma}\mathbf{Y}_{l-1}\widehat{\beta}_{l-1}.$$

The least squares minimization of $\mathbf{r}_l$ is then given by

$$\widehat{\beta}_l = e^{l(l+1)\sigma}(\mathbf{Y}_l'\mathbf{Y}_l)^{-1}\mathbf{Y}_l'\mathbf{r}_{l-1}.$$

The correctness of the algorithm can be easily seen from

$$\begin{aligned}&\sum_{m=-l}^{l} e^{-l(l+1)\sigma}\langle v_i, Y_{lm}\rangle Y_{lm}\\ = \;& \arg\min_{h\in\mathcal{I}_l}\int_{S^2} K_\sigma(\Omega,\Omega')\Big|r_{l-1}(\Omega') - h(\Omega)\Big|^2\, d\mu(\Omega'),\end{aligned}$$

where the residual is given by

$$r_l(\Omega') = v_i(\Omega') - \sum_{l'=0}^{l}\sum_{m=-l'}^{l'} e^{-l(l+1)\sigma}\langle v_i, Y_{lm}\rangle Y_{lm}(\Omega').$$

This iterative algorithm is referred to as the *iterative residual fitting (IRF)* algorithm [71] since we are iteratively fitting a linear equation to the residuals obtained from the previous iteration. The IRF algorithm is similar to the *matching pursuit method* [240] although the IRF was developed independently. The IRF algorithm was developed to avoid the computational burden of inverting a huge linear problem while the matching pursuit method was originally developed to compactly decompose a time frequency signal into a linear combination of a pre-selected pool of basis functions.

In the IRF algorithm, we minimize the residual component $\mathbf{r}_l$ in least squares fashion, i.e. minimizing the sum of squared residuals $\sum_{j=1}^{n} r_l^2(\Omega_j)$ over all mesh vertices. On the other hand, in the marching pursuit method, the norm $\|\mathbf{Y}_l\beta_l\|^2$ is maximized. Due to orthonormality, maximizing the norm is equivalent to minimizing the norm of the residual

$$\|\mathbf{r}_l\|^2 = \int_{S^2} r_l^2(\Omega)\, d\mu(\Omega).$$

So there is a slight difference in how the residual is minimized. Although there is no limitation estimating multiple coefficients simultaneously in the matching pursuit method, [240] only deals with the problem of estimating one coefficient at a time rather than multiple coefficients as in the IRF algorithm.

Although increasing the degree of the representation increases the goodness-of-fit, it also increases the number of estimated coefficients quadratically. So it is necessary to stop the iteration at the specific degree k, where the goodness-of-fit and the number of coefficients balance out. From (11.1), we can see that the k-th degree weighted spherical harmonic representation can be modeled as a linear model setting:

$$v_i(\Omega_j) = \sum_{l=0}^{k} \sum_{m=-l}^{l} e^{-l(l+1)\sigma} \beta_{lm}^i Y_{lm}(\Omega_j) + \epsilon_i(\Omega_j),$$

where the least squares estimation of β_{lm}^i is $\widehat{\beta_{lm}^i} = \langle v_i, Y_{lm} \rangle$. Then we stop the iteration at degree k by testing if the $2k+3$ coefficients at the next iteration vanish:

$$H_0 : \beta_{k+1,-(k+1)}^i = \beta_{k+1,-k}^i = \cdots = \beta_{k+1,k+1}^i = 0.$$

If we assume ϵ_i to be a Gaussian random field, the usual F test at the significant level $\alpha = 0.01$ can be used to determine the stopping degree. In our study, at bandwidth $\sigma = 0.001$, we stop the iteration at degree $k = 42$.

11.3.3 Validation Against Heat Kernel Smoothing

The weighted spherical harmonic representation is validated against heat kernel smoothing as formulated in [76]. Heat kernel smoothing was implemented as an iterated weighted averaging technique, where the weights are spatially adapted to follow the shape of heat kernel in discrete fashion along a surface mesh. The algorithm has been implemented in `MATLAB` and it is freely available at `www.stat.wisc.edu/~mchung/softwares/hk/hk.html`. Since its introduction in 2005, the method has been used in smoothing various cortical surface data: cortical curvatures [235, 130], cortical thickness [234, 38], hippocampus [324, 411], magnetoencephalography (MEG) [161] and functional-MRI [157, 186].

Definition 16 *The* n-th iterated heat kernel smoothing *of signal* $f \in L^2(S^2)$ *is*

$$K_\sigma^{(n)} * f(\Omega) = \underbrace{K_\sigma * \cdots * K_\sigma}_{n \ times} * f(\Omega).$$

Then we have the following theorem.

Theorem 9 3 $K_\sigma * f(\Omega) = K_{\sigma/n}^{(n)} * f(\Omega)$.

By letting $f = Y_{l'm'}$ in (11.8), and using the orthonormality of spherical harmonics, we obtain

$$K_\sigma * Y_{l'm'}(\Omega) = \int_{S^2} K_\sigma(\Omega, \Omega') Y_{l'm'}(\Omega')\, d\mu(\Omega') = e^{-(l'+1)l'\sigma} Y_{l'm'}(\Omega).$$

This is the restatement of the fact that $e^{-l(l+1)}\sigma$ and $Y_{l'm'}$ are eigenvalues and eigenfunctions of the above integral equation with heat kernel. By reapplying heat kernel smoothing to (11.8), we obtain

$$K_\sigma^{(2)} * f(\Omega) = \sum_{l=0}^{\infty} \sum_{m=-l}^{l} e^{-l(l+1)\sigma} \langle f, Y_{lm} \rangle K_\sigma * Y_{lm}(\Omega) \quad (11.16)$$

$$= \sum_{l=0}^{\infty} \sum_{m=-l}^{l} e^{-l(l+1)2\sigma} \langle f, Y_{lm} \rangle Y_{lm}(\Omega). \quad (11.17)$$

Then, arguing inductively, we obtain the spectral representation of the n-th iterated heat kernel smoothing as

$$K_\sigma^{(n)} * f(\Omega) = \sum_{l=0}^{\infty} \sum_{m=-l}^{l} e^{-l(l+1)n\sigma} \langle f, Y_{lm} \rangle Y_{lm}(\Omega).$$

The right side is the spectral representation of heat kernel smoothing with bandwidth $n\sigma$. This proves $K_\sigma^{(n)} * f(\Omega) = K_{n\sigma} * f(\Omega)$. Rescaling the bandwidth, we obtain the result.

Theorem 3 shows that heat kernel smoothing with large bandwidth σ can be decomposed into n repeated applications of heat kernel smoothing with smaller bandwidth σ/n. When the bandwidth is small, the heat kernel behaves like the Dirac-delta function and, using the *parametrix expansion* [304, 374], we can approximate it locally using the Gaussian kernel:

$$K_\sigma(\Omega, \Omega') = \frac{1}{\sqrt{4\pi\sigma}} \exp\big[-\frac{d^2(\Omega, \Omega')}{4\sigma}\big]\big[1 + O(\sigma^2)\big], \quad (11.18)$$

where $d(p, q)$ is the geodesic distance between p and q. For small bandwidth, all the kernel weights are concentrated near the center, so we need only to worry about the first neighbors of a given vertex in a surface mesh.

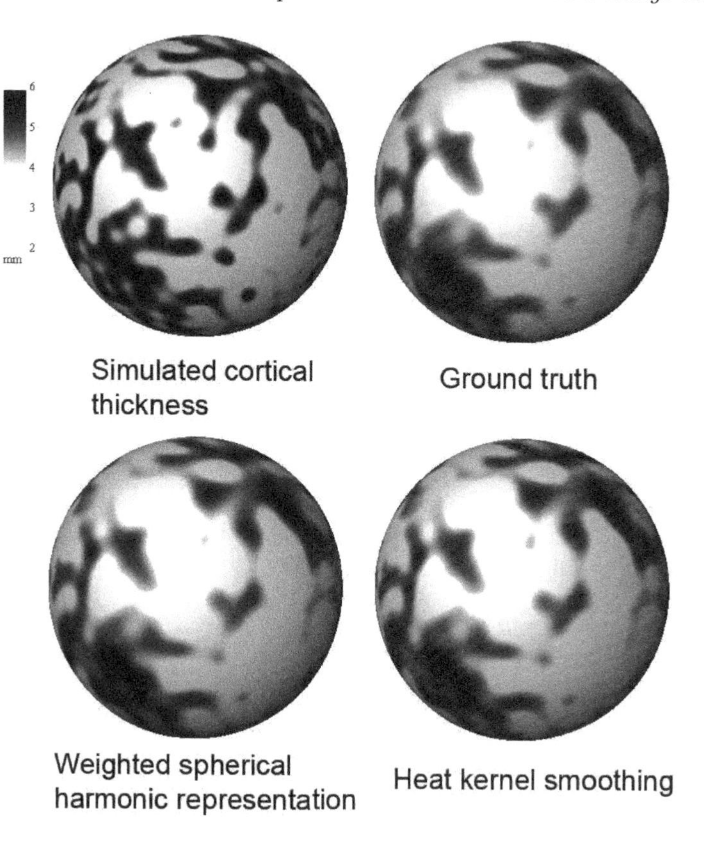

FIGURE 11.4
Cortical thickness is simulated from the sample cortical thickness. The ground truth is analytically constructed from the simulation. Then the weighted spherical harmonic representation and heat kernel smoothing of the simulated cortical thickness are compared against the ground truth for validation.

Let $\Omega_1, \cdots, \Omega_m$ be m neighboring vertices of vertex $\Omega = \Omega_0$ in the mesh. The geodesic distance between Ω and its adjacent vertex Ω_i is the length of edge between these two vertices in the mesh. Then the discretized and normalized heat kernel is given by

$$W_\sigma(\Omega, \Omega_i) = \frac{\exp\left(-\frac{d(\Omega,\Omega_i)^2}{4\sigma}\right)}{\sum_{j=0}^{m} \exp\left(-\frac{d(\Omega-\Omega_j)^2}{4\sigma}\right)}.$$

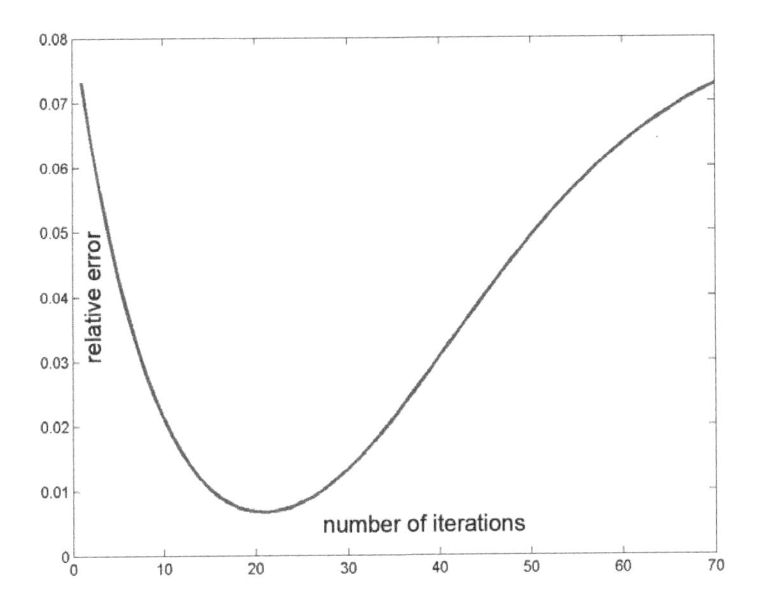

FIGURE 11.5
The plot is the relative error over the number of iterations for heat kernel smoothing against the ground truth.

Note that $\sum_{i=0}^{m} W_\sigma(\Omega, \Omega_i) = 1$. The discrete version of heat kernel smoothing on a triangle mesh is then defined as

$$W_\sigma * f(\Omega) = \sum_{i=0}^{m} W_\sigma(\Omega, \Omega_i) f(\Omega_i).$$

The discrete kernel smoothing should converge to heat kernel smoothing (11.7) as the mesh resolution increases. This is the form of the *Nadaraya-Watson estimator* [63] applied to surface data. Instead of performing a single kernel smoothing with large bandwidth $n\sigma$, we perform n iterated kernel smoothing with small bandwidth σ as follows $W_\sigma^{(n)} * f(\Omega)$.

For comparison between the weighted spherical harmonic representation and heat kernel smoothing, we used the sample cortical thickness data in constructing the analytical ground truth. Consider a surface measurement of the form

$$f(\Omega) = \sum_{l=0}^{k} \sum_{m=-l}^{l} \beta_{lm} Y_{lm}(\Omega) \tag{11.19}$$

for some given β_{lm}. Heat kernel smoothing of f is given as an exact analytic form, which serves as the ground truth for validation:

$$K_\sigma * f(\Omega) = \sum_{l=0}^{k} \sum_{m=-l}^{l} e^{-l(l+1)\sigma} \beta_{lm} Y_{lm}(\Omega). \tag{11.20}$$

Using the sample cortical thickness data, we simulated the measurement of the form (11.19) by estimating $\beta_{lm} = \langle f, Y_{lm} \rangle$ (Figure 11.4 top left). Then we compared the weighted spherical harmonic representation of f and the discrete version of heat kernel smoothing $W_{\sigma/n}^{(n)} * f$ against the the analytical ground truth (11.20) (Figure 11.4 top right) along the surface mesh.

For the weighted spherical harmonic representation, we used $\sigma = 0.001$ and the corresponding optimal degree $k = 42$ (Figure 11.4 bottom left). The relative error for the weighted spherical harmonic representation is up to 0.013 at a certain vertex and the mean relative error over all mesh vertices is 0.0012. For heat kernel smoothing, we used varying numbers of iterations, $1 \leq n \leq 70$, and the corresponding bandwidth $\sigma = 0.001/n$. The performance of heat kernel smoothing depended on the number of iterations, as shown in the plot of relative error over the number of iterations in Figure 11.5. The minimum relative error was obtained when 21 iterations were used. The relative error was up to 0.055 and the mean relative error was 0.0067. Our simulation result demonstrates that the weighted spherical harmonic representation performs better than heat kernel smoothing. The main problem with heat kernel smoothing is that the number of iterations needs to be predetermined, possibly using the proposed simulation technique. Even at the optimal iteration of 21, the weighted spherical harmonic representation provides a better performance.

11.4 Weighted-SPHARM Package

The cortical surface data we will use here as an example was first published in [76]. 16 high functioning autistic and 11 normal control subjects used in this study were screened to be right-handed males. There are 12 control subjects in [76]; however, one subject is removed due to image processing artifacts.

Age distributions for HFA and NC are 15.93±4.71 and 17.08±2.78 respectively. High resolution anatomical magnetic resonance images (MRI) were obtained using a 3-Tesla GE SIGNA (General Electric Medical Systems, Waukesha, WI) scanner with a quadrature head RF coil. A three-dimensional, spoiled gradient-echo (SPGR) pulse sequence was used to generate T_1-weighted images. The imaging parameters were TR/TE = 21/8 ms, flip angle = 30°, 240 mm field of view, 256x192 in-plane acquisition matrix (interpolated on the scanner to 256x256), and 128 axial slices (1.2 mm thick) covering the whole brain.

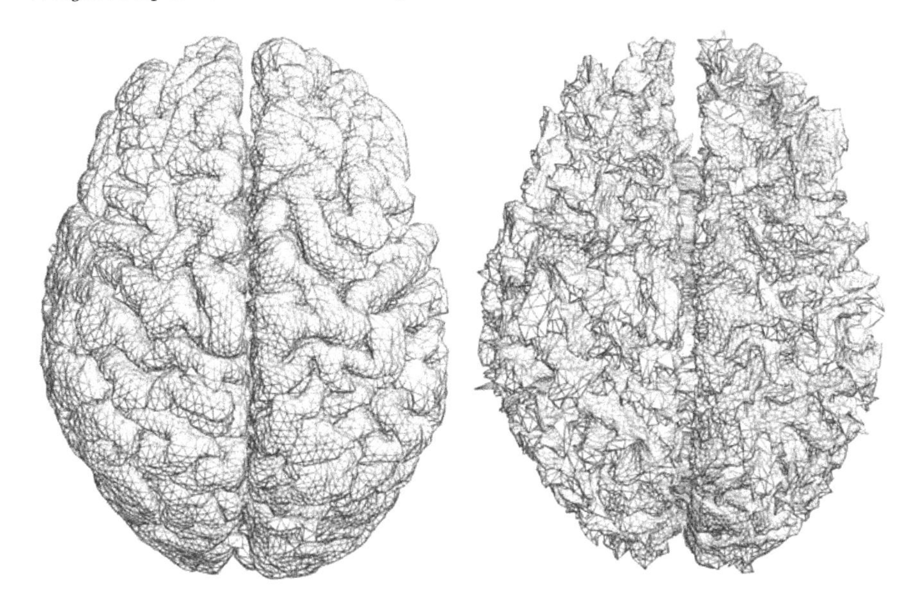

FIGURE 11.6
Outer and inner cortical surface meshes of a subject. Each surface consists of 40962 vertices and 81920 triangles. Vertices across subjects match anatomically so there is no need for the additional surface alignment.

Following image processing steps described in [76], both the outer and inner cortical surfaces were extracted for each subject via a deformable surface algorithm [237]. Surface normalization is performed by minimizing an objective function that measures the global fit of two surfaces while maximizing the smoothness of the deformation in such a way that the pattern of gyral ridges is matched smoothly [301, 76].

The surface data stored in `autism.surface.mat` consists of coordinates of the both inner and outer cortical surfaces of 27 subjects (16 autistic and 11 control) and ages in years. `autisminner` and `autismouter` are matrices of size $16 \times 3 \times 40962$ while `controlinner` and `controlouter` are matrices of size $11 \times 3 \times 40962$.

The `MATLAB` codes for performing the weighted-SPHARM representation are given in `http://brainimaging.waisman.wisc.edu/~chung/BIA`. The cortical surfaces of the first autistic subjects can be visualized using the commands (Figure 11.6)

```
load autism.surface.mat

surf.vertices=squeeze(autisminner(1,:,:))';
surf.faces=tri;
figure; figure_wire(surf, 'yellow', 'white');
```

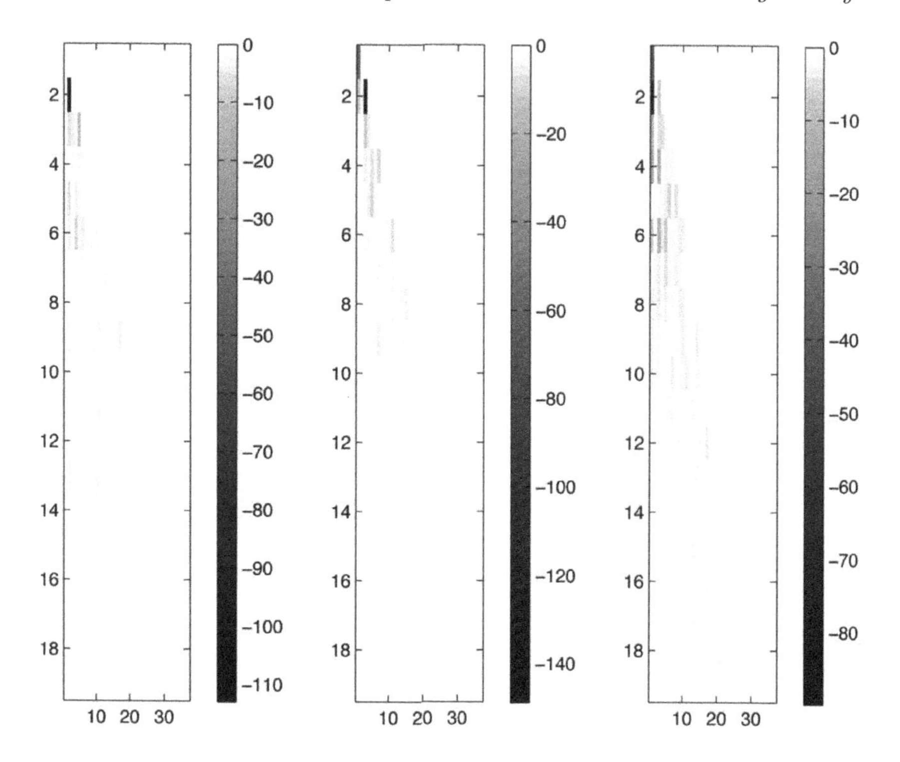

FIGURE 11.7
Spherical harmonic coefficients up to degree 18 are stored as a triangle matrix of size 19×37. The vertical axis is degree and the horizontal axis is order. The coefficient of degree l and order m is displayed in the position $(l, m + 19)$.

```
surf.vertices=squeeze(autismouter(1,:,:))';
surf.faces=tri;
figure; figure_wire(surf, 'yellow', 'white');
```

SPHARM representation requires a unit sphere mesh `sphere40962.mat` that corresponds to the cortical surfaces. Then we establish the spherical angles on the cortical surfaces using

```
load sphere40962.mat
[theta varphi]=SPHARMangles(sphere40962);
figure; figure_trimesh(surf,theta,'rwb');
figure; figure_trimesh(surf,varphi,'rwb');
```

The angles `theta` and `varphi` can be displayed either on the unit sphere or the cortical surface (Figure 11.6). Using the spherical angles, we construct the weighted-SPHARM.

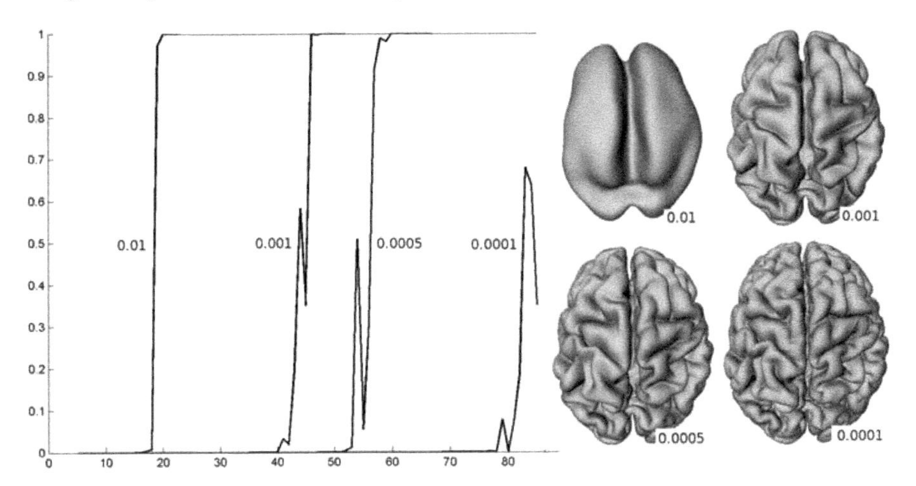

FIGURE 11.8
Cortical thickness projected onto the average outer cortex for various t and corresponding optimal degree: $k = 18(\sigma = 0.01), k = 42(\sigma = 0.001), k = 52(\sigma = 0.0005), k = 78(\sigma = 0.0001)$. The average cortex is constructed by averaging the coefficients of the weighted-SPHARM. The highly noisy first image shows thickness measurements obtained by computing the distance between two triangle meshes.

```
[surfsmooth, fourier]=SPHARMsmooth2(surf,sphere40962,18,0.01);
figure_wire(surfsmooth,'yellow','white');
```

This constructs the weighted-SPHARM representation of the outer cortical surface with $k = 18$ and $\sigma = 0.01$ (Figure 11.8). It requires exactly identical mesh topology between `surf` and `sphere4092`. `surfsmooth` is the smoothed surface of the form

```
surfsmooth =

    vertices: [40962x3 double]
       faces: [81920x3 double]
```

while `fourier` is a structured array of spherical harmonic coefficients for all x, y, and z-coordinates. The spherical harmonic coefficients can be extracted possibly for data reduction and classification purposes. For each representation, we have Fourier coefficients for x,y and z coordinates.

```
fourier =

    x: [19x37 double]
    y: [19x37 double]
```

```
    z: [19x37 double]

figure;
subplot(1,3,1)
imagesc(-abs(fourier.x)); colorbar
subplot(1,3,2)
imagesc(-abs(fourier.y)); colorbar;
subplot(1,3,3)
imagesc(-abs(fourier.z)); colorbar
colormap('gray'); figure_bg('white')
```

The first dimension in `fourier.x` is the degree between 0 and 18 and the second dimension is the order between -18 and 18. See Figure 11.7 for the display of the Fourier coefficients for subject 1. For the l-th degree, there are total $2l+1$ orders so not all elements in the matrix contain Fourier coefficients. Figure 11.7 shows the upper triangle elements are padded with zeros. Since it is difficult to handle the coefficients in a matrix form, we vectorize them using `SPHARMvectorize` command.

```
fourierv = SPHARMvectorize(fourier,18)

fourierv =

    x: [1x361 double]
    y: [1x361 double]
    z: [1x361 double]
```

There are total $1 + (2+1) + (2 \times 2 + 1) + ... + (2 \times 18 + 1) = (18+1)^2 = 361$ Fourier coefficients for each coordinate.

Using the estimated coefficients `fourier`, we can reconstruct smoothed cortical surfaces with the different amount of smoothing. Once the coefficients are estimated once, we can reuse them without reestimating them again. Using `SPHARMrepresent2.m`, 18 degree weighted-SPHARM representation with different bandwidth $\sigma = 0, 0.01, 0.1$ can be obtained (Figure 11.9).

```
surf=SPHARMrepresent2(sphere40962, fourier, 18,0);
figure; figure_wire(surf,'white','white');

surf=SPHARMrepresent2(sphere40962, fourier, 18,0.01);
figure; figure_wire(surf,'white','white');

surf=SPHARMrepresent2(sphere40962, fourier, 18,0.1);
figure; figure_wire(surf,'white','white');
```

The spherical harmonic representation can be used to establish surface correspondence between different subjects. The concept is introduced in [71].

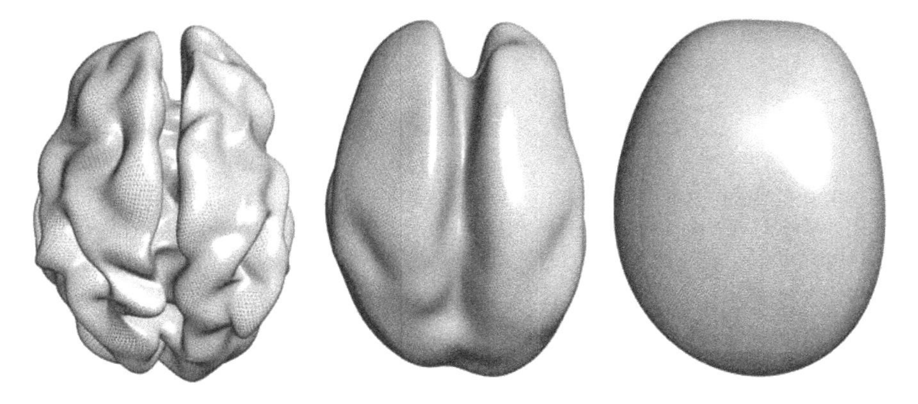

FIGURE 11.9
The weighted spherical harmonic representation with degree 18 and $\sigma = 0, 0.01, 0.1$. Even at the same degree, the changes in the bandwidth σ drastically change the shape of brain.

11.5 Surface Registration

Previously cortical surface normalization was performed by minimizing an objective function that measures the global fit of two surfaces while maximizing the smoothness of the deformation in such a way that the gyral patterns are matched smoothly [76, 301, 363]. In the spherical harmonic representation, the surface normalization is straightforward and does not require any sort of optimization explicitly but at least requires some initial alignment. A crude alignment can be done by coinciding the first order ellipsoid meridian and equator in the SPHARM-correspondence approach [135, 344]. For cortical meshes obtained using the anatomic segmentation using the proximities (ASP) algorithm [237], such alignments are not needed. An approximate surface alignment is done during the cortical surface extraction process. The algorithm generates 40,962 vertices and 81,920 triangles with the identical mesh topology for all subjects. The vertices indexed identically on two cortical meshes will have a very close anatomic homology and this defines the surface alignment. This provides the same spherical parameterization at identically indexed vertices across different cortical surfaces.

Consider a surface $h = (h_1, h_2, h_3)$ obtained from the coordinates v_i measured at point p:

$$h_i(p) = \sum_{l=0}^{k} \sum_{m=-l}^{l} \langle v_i, Y_{lm} \rangle (p).$$

Consider another surface j_i obtained from coordinate functions w_i:

$$j_i(p) = \sum_{l=0}^{k} \sum_{m=-l}^{l} \langle w_i, Y_{lm} \rangle (p).$$

Suppose the surface h_i is deformed to $h_i + d_i$ under the influence of the displacement vector field d_i. We wish to find d_i that minimizes the discrepancy between $h_i + d_i$ and j_i in the finite subspace $\mathcal{H}_k$. This can be easily done by noting that

$$\sum_{l=0}^{k} \sum_{m=-l}^{l} (w_{lm}^i - v_{lm}^i) Y_{lm}(p) = \arg \min_{d_i \in \mathcal{H}_k} \left| \left| \widehat{h_i} + d_i - \widehat{j_i} \right| \right|. \tag{11.21}$$

This implies that the optimal displacement in the least squares sense is obtained by simply taking the difference between two weighted spherical harmonic representations and matching coefficients of the same degree and order. Then a specific point $\widehat{h_i}(p_0)$ in one surface corresponds to $\widehat{j_i}(p_0)$ in the other surface. We refer to this point-to-point surface correspondence as the *spherical harmonic correspondence* [71]. The spherical harmonic correspondence shows that the optimal displacement in the least squares sense is obtained by simply taking the difference between two spherical harmonic representations. Unlike other surface registration methods used in warping surfaces between subjects [76, 301, 363], it is not necessary to consider an additional cost function that guarantees the smoothness of the displacement field since the displacement field $d = (d_1, d_2, d_3)$ is already a linear combination of smooth basis functions.

The previously available approaches for computing the cortical thickness in discrete triangle meshes produce noisy thickness measures [76, 120, 237]. So it is necessary to smooth the thickness measurements along the cortex via surface-based smoothing techniques [10, 54, 53, 83]. On the other hand, the weighted-SPHARM provides smooth functional representation of the outer and inner surfaces so that the distance measures between the surfaces should be already smooth. Hence, the weighted-SPHARM avoids the additional step of thickness smoothing done in most of thickness analysis literature [76, 83] while it is not necessary to perform data smoothing in the spherical harmonic formulation. The distance between the outer and inner cortical surfaces can be determined using the spherical harmonic correspondence. Given the outer surface h_i and the inner surface j_i, the cortical thickness is defined to be the Euclidean distance between the two representations:

$$\texttt{thick}(p) = \sqrt{\sum_{i=1}^{3} \Big[\sum_{l=0}^{k} \sum_{m=-l}^{l} \langle v_i - w_i, Y_{lm} \rangle \Big]^2}.$$

A similar approach has been proposed for measuring the closeness between two surfaces [135]. Figure 11.10 shows the comparison of cortical thickness computed from the traditional deformable surface algorithm [237] and the

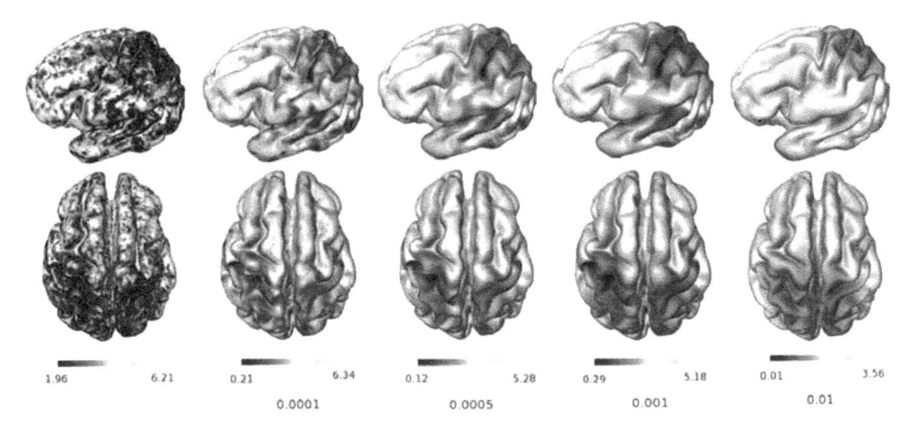

FIGURE 11.10
Cortical thickness of a subject projected onto a template. The cortical thickness is computed from the spherical harmonic correspondence with heat kernel weights. As the bandwidth increases from $\sigma = 0.0001$ to 0.01, the amount of smoothing also increases. The first image shows the cortical thickness obtained from the traditional deformable surface algorithm [237].

spherical harmonic correspondence. The cortical thickness obtained from the traditional approach introduces a lot of triangle mesh noise into its estimation while the spherical harmonic correspondence approach does not. The spatial smoothness of the thickness is explicitly incorporated via the bandwidth σ.

11.5.1 MATLAB Implementation

The SPHARM-correspondence will be explained using `autism.surface.mat`. We will first compute the cortical thickness of the autistic subjects 1 and 2.

```
load autism.surface.mat
surfout1.vertices=squeeze(autismouter(1,:,:))';
surfout1.faces=tri;
surfin1.vertices=squeeze(autisminner(1,:,:))';
surfin1.faces=tri;

surfout2.vertices=squeeze(autismouter(2,:,:))';
surfout2.faces=tri;
surfin2.vertices=squeeze(autisminner(2,:,:))';
surfin2.faces=tri;

thick1 = L2norm(surfout1.vertices-surfin1.vertices);
thick2 = L2norm(surfout2.vertices-surfin2.vertices);
```

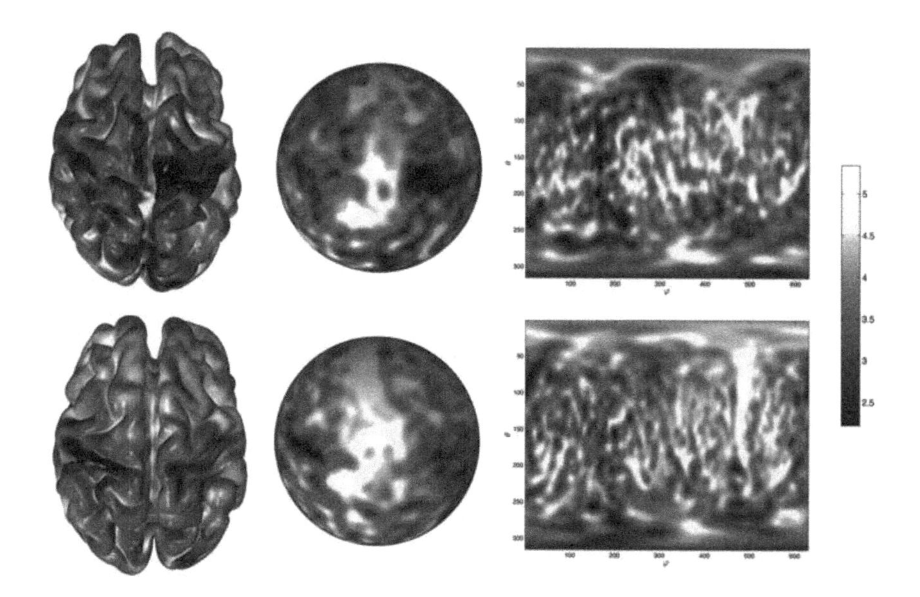

FIGURE 11.11
Cortical thickness of subjects 1 and 2. Cortical thickness has been smoothed using weighted-SPHARM with degree 42 and bandwidth $\sigma = 0.001$. The cortical surfaces are also smoothed using the same parameters. The cortical thickness maps are also projected onto a unit sphere for better visualization. SPHARM framework can be used to establish the point-wise correspondence between the surfaces.

`thick1` and `thick2` are the cortical thickness of subjects 1 and 2. This is simply computed as the Euclidean distance between the outer and inner cortical surfaces using `L2norm` function. Then we compute the weighted-SPHARM representation of the surfaces and their thickness with degree 42 and bandwidth $\sigma = 0.001$.

```
load sphere40962.mat
[surfout1,fourier1]=SPHARMsmooth2(surfout1,sphere40962,42,0.001);
[surfout2,fourier2]=SPHARMsmooth2(surfout2,sphere40962,42,0.001);

surfthick1.vertices=[thick1 thick1 thick1];
surfthick1.faces=tri;
[surfthick1,fourier]
 =SPHARMsmooth2(surfthick1,sphere40962,42,0.001);
thick1smooth = surfthick1.vertices(:,1);

surfthick2.vertices=[thick2 thick2 thick2];
```

```
surfthick2.faces=tri;
[surfthick2,fourier]
 =SPHARMsmooth2(surfthick2,sphere40962,42,0.001);
thick2smooth = surfthick2.vertices(:,1);
```

In order to smooth cortical thickness, `thick1` is treated as the coordinate of the mesh vertices in `surfthick1.vertices=[thick1 thick1 thick1]` and feeds into `SPHARMsmooth2` routine. The cortical thickness of subject 1 is visualized as Figure 11.11 using `figure_trimesh`.

```
figure;  figure_trimesh(surfout1,thick1smooth);
colormap('hot')
shading interp
view([0 90])
camlight headlight

figure;  figure_trimesh(sphere40962, thick1smooth);
colormap('hot')
shading interp
view([0 90])
camlight headlight
```

The subject 2 is displayed similarly. Cortical thickness is defined along a curved cortical surface. For flattening the thickness map onto a sphere, we can simply use the spherical mesh `sphere40962` that is topologically equivalent to the cortical mesh. However, if we want to flatten the spherical map further onto a rectangle, we need to interpolate the value of cortical thickness for each (θ, φ). However, the use of spherical harmonic expansion can avoid the interpolation problem. For this we need to discretize $[0, \pi] \otimes (0, 2\pi]$ into finite number of pixels.

```
theta_d=0:0.01:pi;
varphi_d=0:0.01:2*pi;
m=length(theta_d);
n=length(varphi_d);
```

There will be total `m` $\times$ `n` pixels in the flat map. Then for pixel value (θ, φ), we compute the basis $Y_{lm}(\theta, \varphi)$ and plot it. Here we show how to flatten the basis $Y_{3,1}$ and $Y_{10,2}$ (Figure 11.12).

```
theta=kron(ones(1,n),theta_d');
theta=reshape(theta,1,m*n);
varphi=kron(ones(m,1),varphi_d);
varphi=reshape(varphi,1,m*n);

Ylm= Y_lm(3,1,theta,varphi);
square=reshape(Ylm,m,n);
```

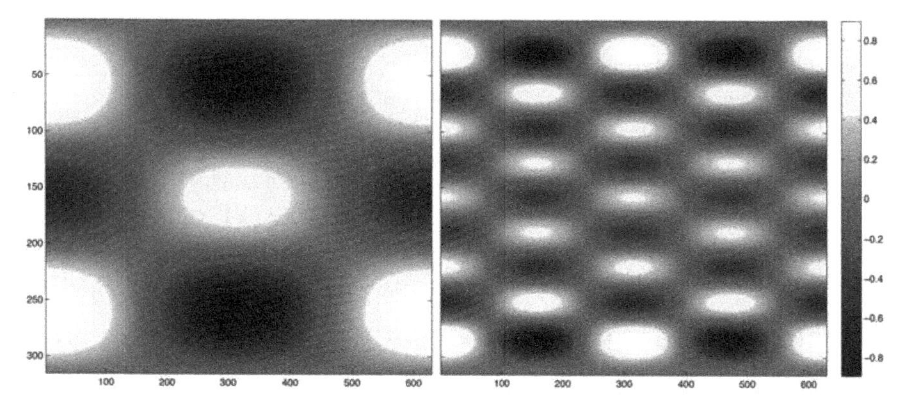

FIGURE 11.12
Spherical harmonic basis $Y_{3,1}$ and $Y_{10,2}$ are mapped onto a rectangle easily by discretizing the spherical angles (θ, φ). Any linear combination of spherical harmonics can then be easily mapped onto the rectangle.

```
figure;imagesc(square); colormap('hot'); colorbar
set(gcf,'Color','w')

Ylm= Y_lm(10,2,theta,varphi);
square=reshape(Ylm,m,n);
figure;imagesc(square); colormap('hot'); colorbar
set(gcf,'Color','w')
```

The same idea can be used to flatten any linear combination of spherical harmonics. Therefore, once we have weighted-SPHARM representation of any measurements on the cortical surface, it can be easily mapped onto a rectangle. We have written this as a function `SPHARM2square`. It requires the estimated spherical harmonic coefficients `fourier.x` of cortical thickness (Figure 11.11).

```
square=SPHARM2square(fourier.x,42, 0.001);
```

11.6 Encoding Surface Asymmetry

Given the weighted spherical harmonic representation, we need to establish surface correspondence between hemispheres and between subjects. This requires establishing anatomical correspondence using *surface registration*. The main motivation for the surface registration is to establish proper alignment for cortical thickness to be compared across subjects and between hemispheres. Previously, the cortical surface registration was performed by minimizing an objective function that measures the global fit of two surfaces while maximizing the smoothness of the deformation in such a way that the sulcal and gyral folding patterns are matched smoothly [363, 301, 76]. In the weighted spherical harmonic representation, surface registration is straightforward and does not require any sort of explicit time consuming optimization. Consider a surface $\widehat{h}_i$ obtained from coordinate functions v_i measured at points $\Omega_1, \cdots, \Omega_n$:

$$\widehat{h}_i(\Omega) = \sum_{l=0}^{k} \sum_{m=-l}^{l} e^{-l(l+1)\sigma} \langle v_i, Y_{lm} \rangle (\Omega).$$

Consider another surface $\widehat{j}_i$ obtained from coordinate functions w_i measured at points $\Omega'_1, \cdots, \Omega'_m$:

$$\widehat{j}_i(\Omega) = \sum_{l=0}^{k} \sum_{m=-l}^{l} e^{-l(l+1)\sigma} \langle w_i, Y_{lm} \rangle (\Omega).$$

Suppose the surface $\widehat{h}_i$ is deformed to $\widehat{h}_i + d_i$ under the influence of the displacement vector field d_i. We wish to find d_i that minimizes the discrepancy between $\widehat{h}_i + d_i$ and $\widehat{j}_i$ in the finite subspace $\mathcal{H}_k$. This can be easily done by noting that

$$\sum_{l=0}^{k} \sum_{m=-l}^{l} e^{-l(l+1)\sigma} (w^i_{lm} - v^i_{lm}) Y_{lm}(\Omega) = \arg \min_{d_i \in \mathcal{H}_k} \left|\left| \widehat{h}_i + d_i - \widehat{j}_i \right|\right|. \quad (11.22)$$

The proof of this statement is given in [71]. This implies that the optimal displacement in the least squares sense is obtained by simply taking the difference between two weighted spherical harmonic representations and matching coefficients of the same degree and order. Then a specific point $\widehat{h}_i(\Omega_0)$ in one surface corresponds to $\widehat{j}_i(\Omega_0)$ in the other surface. We refer to this point-to-point surface correspondence as the *spherical harmonic correspondence.*

The spherical harmonic correspondence can be further used to establish the inter-hemispheric correspondence by letting $\widehat{j}_i$ be the mirror reflection of $\widehat{h}_i$. The mirror reflection of $\widehat{h}_i$ with respect to the midsagittal cross section $u_2 = 0$ is simply given by

$$\widehat{j}_i(\theta, \varphi) = \widehat{h}_i^*(\theta, \varphi) = \widehat{h}_i(\theta, 2\pi - \varphi),$$

where * denotes the mirror reflection operation (Figure 11.13). The specific point $\widehat{h}_i(\theta_0, \varphi_0)$ in the left hemisphere will be mirror reflected to $\widehat{j}_i(\theta_0, 2\pi - \varphi_0)$ in the right hemisphere. The spherical harmonic correspondence of $\widehat{j}_i(\theta_0, 2\pi - \varphi_0)$ is $\widehat{h}_i(\theta_0, 2\pi - \varphi_0)$. Hence, the point $\widehat{h}_i(\theta_0, \varphi_0)$ in the left hemisphere corresponds to the point $\widehat{h}_i(\theta_0, 2\pi - \varphi_0)$ in the right hemisphere. This establishes the inter-hemispheric anatomical correspondence. The schematic

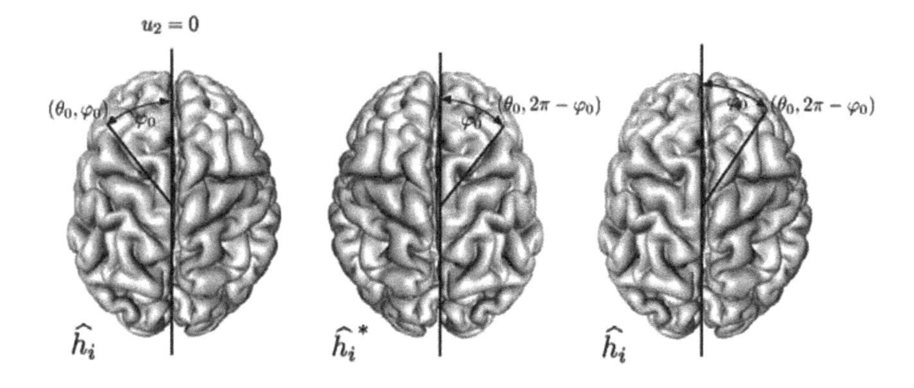

FIGURE 11.13
The point $\widehat{h}_i(\theta_0, \varphi_0)$ (left) corresponds to $\widehat{h}_i^*(\theta, 2\pi - \varphi_0)$ (middle) after mirror reflection with respect to the midsagittal cross section $u_2 = 0$. From the spherical harmonic correspondence, $\widehat{h}_i^*(\theta, 2\pi - \varphi_0)$ corresponds to $\widehat{h}_i(\theta, 2\pi - \varphi_0)$ (right). This establishes the mapping from the left hemisphere to the right hemisphere in least squares fashion.

of obtaining this inter-hemispheric correspondence is given in Figure 11.13. This inter-hemispheric correspondence is used to compare cortical thickness measurements f across the hemispheres. The weighted spherical harmonic representation of cortical thickness f is

$$\widehat{g}(\theta, \varphi) = \sum_{l=0}^{k} \sum_{m=-l}^{l} e^{-l(l+1)\sigma} \langle f, Y_{lm} \rangle Y_{lm}(\theta, \varphi).$$

At a given position $\widehat{h}_i(\theta_0, \varphi_0)$, the corresponding cortical thickness is $\widehat{g}(\theta_0, \varphi_0)$, which should be compared with the thickness $\widehat{g}(\theta_0, 2\pi - \varphi_0)$ at position $\widehat{h}_i(\theta_0, 2\pi - \varphi_0)$:

$$\widehat{g}(\theta_0, 2\pi - \varphi_0) = \sum_{l=0}^{k} \sum_{m=-l}^{l} e^{-l(l+1)\sigma} \langle f, Y_{lm} \rangle Y_{lm}(\theta, 2\pi - \varphi). \tag{11.23}$$

The equation (11.23) can be rewritten using the property of spherical harmonics:

$$Y_{lm}(\theta, 2\pi - \varphi) = \begin{cases} -Y_{lm}(\theta, \varphi), & -l \le m \le -1, \\ Y_{lm}(\theta, \varphi), & 0 \le m \le l, \end{cases}$$

$$\begin{aligned}\widehat{g}(\theta_0, 2\pi - \varphi_0) &= \sum_{l=0}^{k} \sum_{m=-l}^{-1} e^{-l(l+1)\sigma} \langle f, Y_{lm} \rangle Y_{lm}(\theta_0, \varphi_0) \\ &\quad - \sum_{l=0}^{k} \sum_{m=0}^{l} e^{-l(l+1)\sigma} \langle f, Y_{lm} \rangle Y_{lm}(\theta_0, \varphi_0).\end{aligned}$$

Comparing with the expansion for $\widehat{g}(\theta_0, \varphi_0)$, we see that the negative order terms are invariant while the positive order terms change sign. Hence we define the symmetry and asymmetry indices as follows.

Definition 17 *The symmetry index is defined as*

$$\begin{aligned}S(\theta, \varphi) &= \frac{1}{2}\Big[\widehat{g}(\theta, \varphi) + \widehat{g}(\theta, 2\pi - \varphi)\Big] \\ &= \sum_{l=0}^{k} \sum_{m=-l}^{-1} e^{-l(l+1)\sigma} \langle f, Y_{lm} \rangle Y_{lm}(\theta_0, \varphi_0),\end{aligned}$$

while the asymmetry index is defined as

$$\begin{aligned}A(\theta, \varphi) &= \frac{1}{2}\Big[\widehat{g}(\theta, \varphi) - \widehat{g}(\theta, 2\pi - \varphi)\Big] \\ &= \sum_{l=0}^{k} \sum_{m=0}^{l} e^{-l(l+1)\sigma} \langle f, Y_{lm} \rangle Y_{lm}(\theta_0, \varphi_0).\end{aligned}$$

We normalize the asymmetry index by dividing it by the symmetry index as

$$\begin{aligned}N(\theta, \varphi) &= \frac{\widehat{g}(\theta, \varphi) - \widehat{g}(\theta, 2\pi - \varphi)}{\widehat{g}(\theta, \varphi) + \widehat{g}(\theta, 2\pi - \varphi)} \\ &= \frac{\sum_{l=1}^{k} \sum_{m=-l}^{-1} e^{-1(l+1)\sigma} \langle f, Y_{lm} \rangle Y_{lm}(\theta, \varphi)}{\sum_{l=0}^{k} \sum_{m=0}^{l} e^{-l(l+1)\sigma} \langle f, Y_{lm} \rangle Y_{lm}(\theta, \varphi)}.\end{aligned}$$

We refer to this index as the *normalized asymmetry index*. The numerator is the sum of all negative orders while the denominator is the sum of all positive and the 0-th orders. Note that $N(\theta, 0) = N(\theta, \pi) = 0$. This index is intuitively interpreted as the normalized difference between cortical thickness in the left and the right hemispheres. Note that the larger the value of the index, the larger the amount of asymmetry. The index is invariant under the affine scaling of the human brain so it is not necessary to control for the global brain size difference in the later statistical analysis. Figure 11.14 shows the asymmetry index for three subjects.

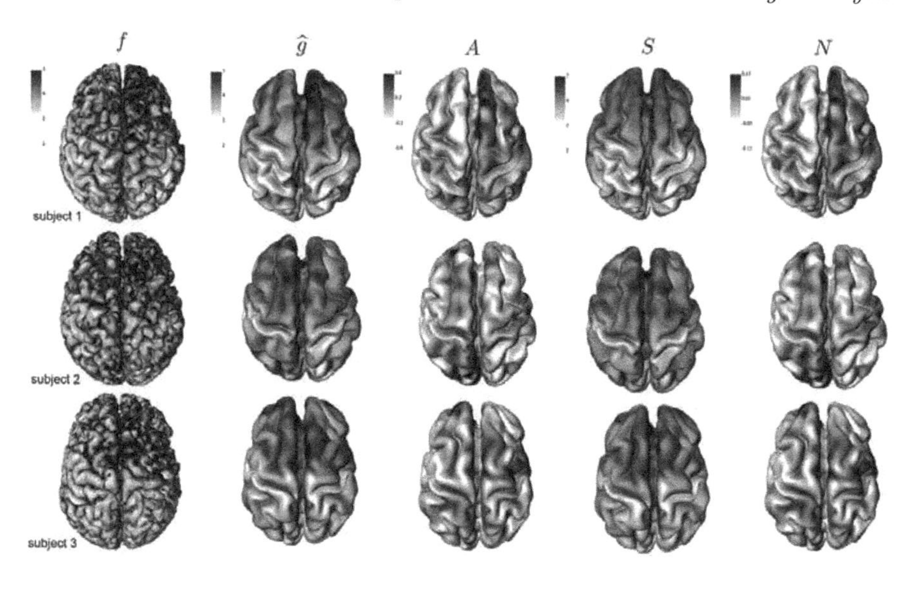

FIGURE 11.14
Three representative subjects showing cortical thickness (f), its weighted-SPHARM representation ($\widehat{g}$), asymmetry index (A), symmetry index (S) and normalized asymmetry index (N). The cortical thickness is projected onto the original brain surfaces while all other measurements are projected onto the 42-th degree weighed spherical harmonic representation.

11.7 Case Study: Cortical Asymmetry Analysis

As an application of the weighted spherical harmonic representation, we show how to perform cortical asymmetry analysis.

11.7.1 Descriptions of Data Set

Three Tesla T_1-weighted MR scans were acquired for 16 high functioning autistic and 12 control right handed males. The data set is first published in [76] with the detailed descriptions. The autistic subjects were diagnosed by a trained and certified psychologist at the Waisman Center at the University of Wisconsin-Madison [98]. The average ages were 17.1 ± 2.8 and 16.1 ± 4.5 for control and autistic groups respectively. Image intensity nonuniformity was corrected using a nonparametric nonuniform intensity normalization method and then the image was spatially normalized into the Montreal neurological institute stereotaxic space using a global affine transformation [85]. Afterwards,

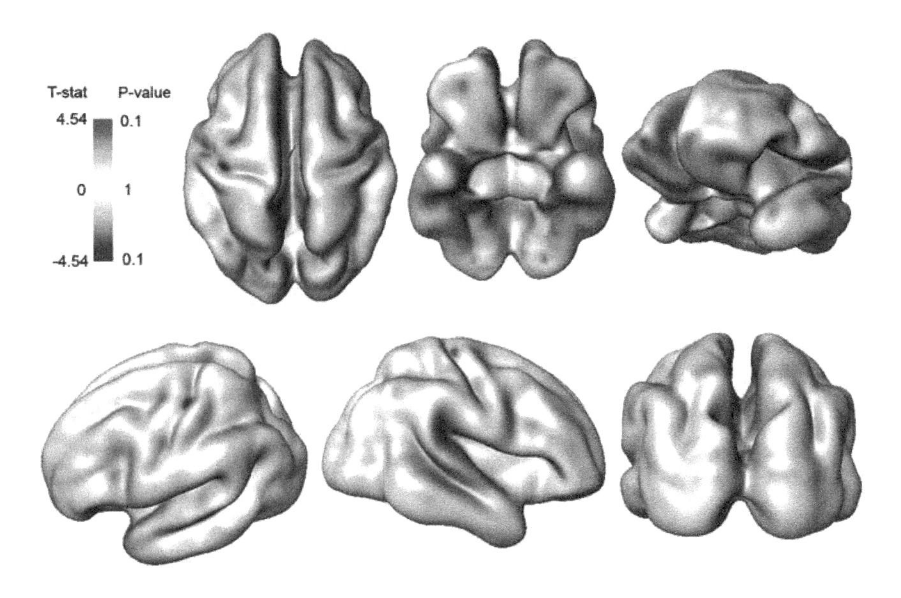

FIGURE 11.15
The statistically significant regions of cortical asymmetry thresholded at the corrected P-value of 0.1. The P-value has been corrected for multiple comparisons.

an automatic tissue-segmentation algorithm based on a supervised artificial neural network classifier was used to segment gray and white matters.

Triangle meshes for outer cortical surfaces were obtained by a deformable surface algorithm [237] and the mesh vertex coordinates v_i were obtained. At each vertex, cortical thickness f was also measured. Once we obtained the outer cortical surfaces of 28 subjects, the weighted spherical harmonic representations $\widehat{h}_i$ were constructed. We used bandwidth $\sigma = 0.001$ corresponding to $k = 42$ degrees. The weighted spherical harmonic representations for three representative subjects are given in Figure 11.14. The symmetry (S), asymmetry (A) and normalized asymmetry (N) indices are computed. The normalized asymmetry index is used in localizing the regions of cortical asymmetry difference between the two groups. These indices are projected on the average cortical surface (Figure 11.14). The average cortical surface is constructed by averaging the Fourier coefficients of all subjects within the same spherical harmonics basis following the spherical harmonic correspondence. The average surface serves as an anatomical landmark for displaying these indices as well as for projecting the final statistical analysis results in the next section.

11.7.2 Statistical Inference on Surface Asymmetry

For each subject, the normalized asymmetry index $A(\theta, \varphi)$ was computed and modeled as a Gaussian random field. The null hypothesis is that $A(\theta, \varphi)$ is identical in the both groups for all (θ, φ), while the alternate hypothesis is that there is a specific point (θ_0, φ_0) at which the normalized asymmetry index is different. The group difference on the normalized asymmetry index was tested using the T random field, denoted as $T(\theta, \varphi)$. Since we need to perform the test on every point on the cortical surface, it becomes a multiple comparison problem. We used the random field theory based t statistic thresholding to determine statistical significance [394]. The probability of obtaining false positives for the one-sided alternate hypothesis is given by

$$P\Big[\sup_{(\theta,\varphi)\in S^2} T(\theta, \varphi) > h\Big] \approx \sum_{d=0}^{2} R_d(S^2)\mu_d(h), \tag{11.24}$$

where R_d is the d-dimensional *Resel* of S^2, and ρ_d is the d-dimensional *Euler characteristic (EC) density* of the T-field [394, 396]. The Resels are

$$R_0(S^2) = 2, R_1(S^2) = 0, R_2(S^2) = \frac{4\pi}{\text{FWHM}^2},$$

where FWHM is the *full width at the half maximum* of the smoothing kernel. The FWHM of the heat kernel used in the weighted spherical harmonic representation is not given in a closed form, so it is computed numerically. From (11.6), the maximum of the heat kernel is obtained when $\Omega \cdot \Omega' = 1$. Then we numerically solve for $\Omega \cdot \Omega'$:

$$\frac{1}{2}\sum_{l=0}^{k} \frac{2l+1}{4\pi} e^{-l(l+1)\sigma} = \sum_{l=0}^{k} \frac{2l+1}{4\pi} e^{-l(l+1)\sigma} P_l^0(\Omega \cdot \Omega').$$

In previous surface data smoothing techniques [83, 76], a FWHM of between 20 to 30 mm was used for smoothing data directly along the brain surface. In our study, we used a substantially smaller FWHM since the analysis is performed on the unit sphere, which has a smaller surface area. The compatible Resels of the unit sphere can be obtained by using the bandwidth of $\sigma = 0.001$, which corresponds to a FWHM of 0.0968 mm. Then, based on the formula (11.24), we computed the multiple-comparison-corrected P-value and thresholded at $\alpha = 0.1$ (Figure 11.15). We found that the central sulci and the prefrontal cortex exhibit abnormal cortical asymmetry pattern in autistic subjects. The larger positive t statistic value indicates thicker cortical thickness with respect to the corresponding thickness at the opposite hemisphere.

11.8 Discussion

We have presented a novel cortical asymmetry technique called the weighted spherical harmonic representation that unifies surface representation, parameterization, smoothing, and registration in a unified mathematical framework. The weighed spherical representation is formulated as the least squares approximation to an isotropic heat diffusion on a unit sphere in such a way that the physical time of heat diffusion controls the amount of smoothing in the weighted spherical harmonic representation. The methodology is used in modeling cortical surface shape asymmetry. Within this framework the asymmetry index, that measures the amount of asymmetry presented in the cortical surface, was constructed as the ratio of the weighted spherical harmonic representation of negative and positive orders. The regions of the statistically different asymmetry index are localized using random field theory. As an illustration, the methodology was applied quantifying the abnormal cortical asymmetry pattern of autistic subjects. The weighted spherical harmonic representation is a very general surface shape representation so it can be used for any type of surface objects that are topologically equivalent to a unit sphere.

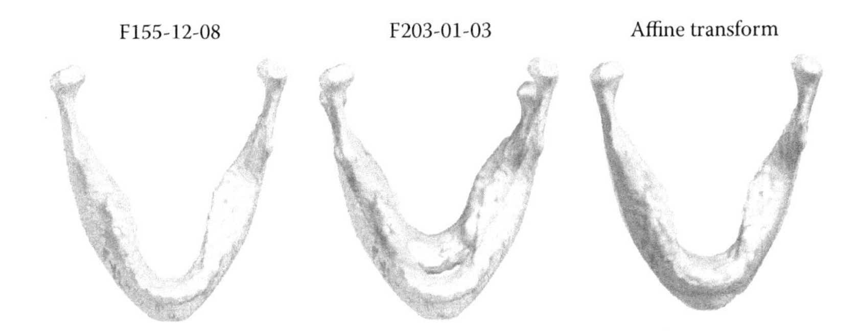

FIGURE 1.8
Mandible F155-12-08 (yellow) is used as a fixed template and other mandibles are affinely aligned to F155-12-08. For example, smaller F203-01-03 (gray) is aligned to larger F203-01-03 by affinely enlarging it.

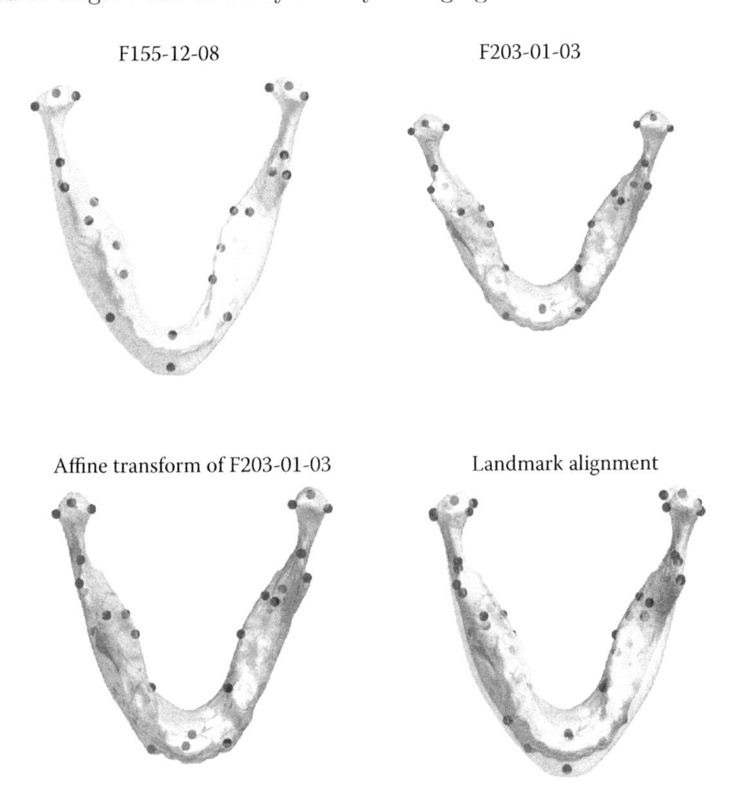

FIGURE 1.9
Mandible F155-12-08 (yellow) is used as a template and mandibles F203-01-03 is affinely aligned to F155-12-08 by matching 24 manually identified landmarks. The affine transform does not exactly match the landmarks perfectly but minimizes the distance between them in a least squares fashion.

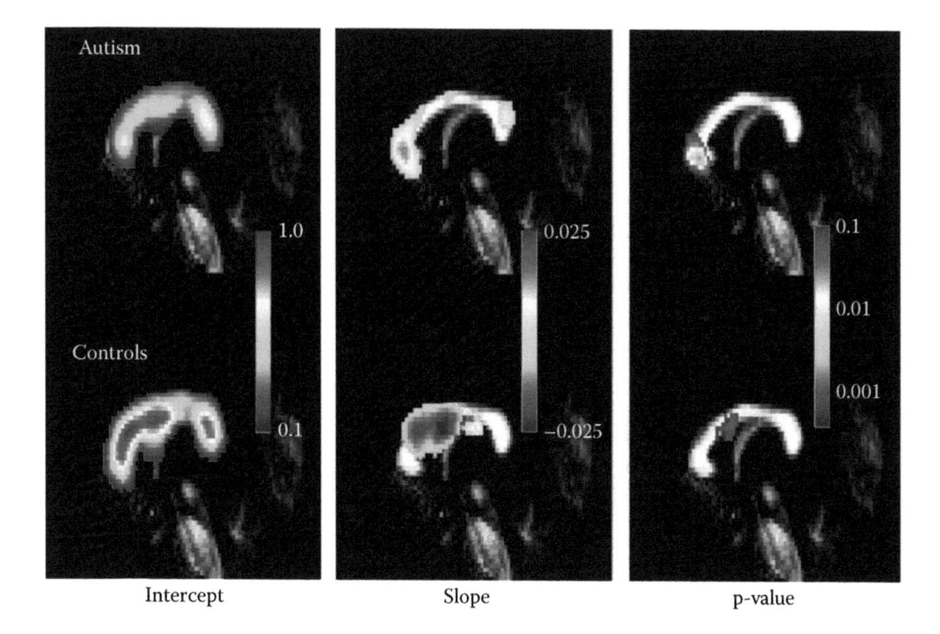

FIGURE 3.7
White matter density is linearly fitted over age and group at each voxel. The intercept and slopes of the linear regression is for each group. The autistic group shows lower white matter density compared to the control at lower age but gains white matter over time while the control group shows decreasing white matter density with age [69].

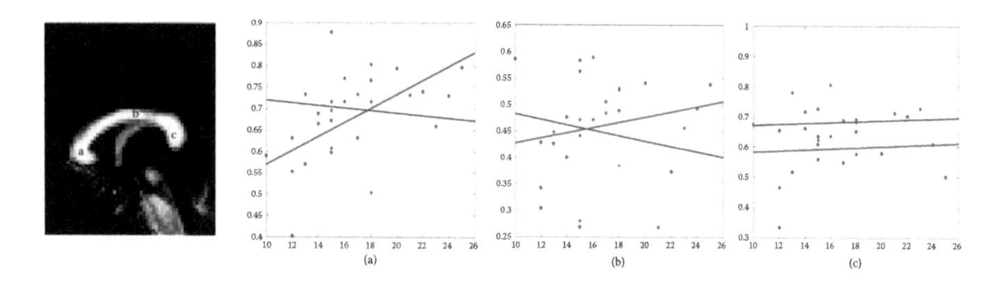

FIGURE 3.8
Linear regression fit for each group (red=autism, blue = control) (a) Genu (b) Midbody and (c) splenium of the corpus callosum.

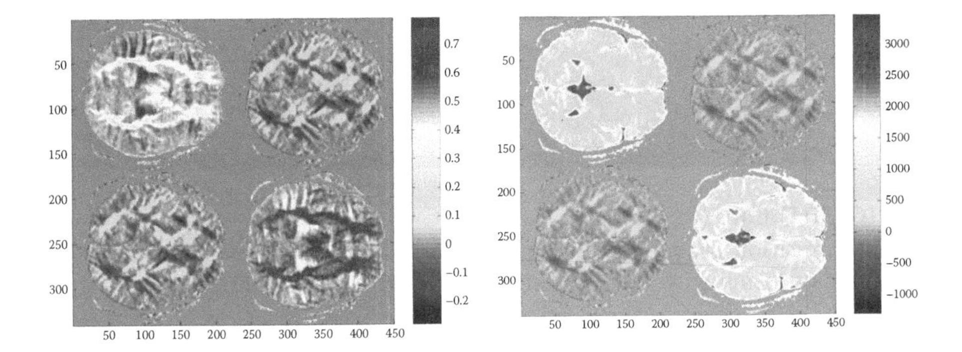

FIGURE 6.2
Left: the original diffusion tensor D_0 for 2D slice. Right: normalized diffusion tensor $D = D_0/\mathrm{tr}D_0$ for 2D. The diffusion tensor D gives a natural Riemannian metric tensor in DTI.

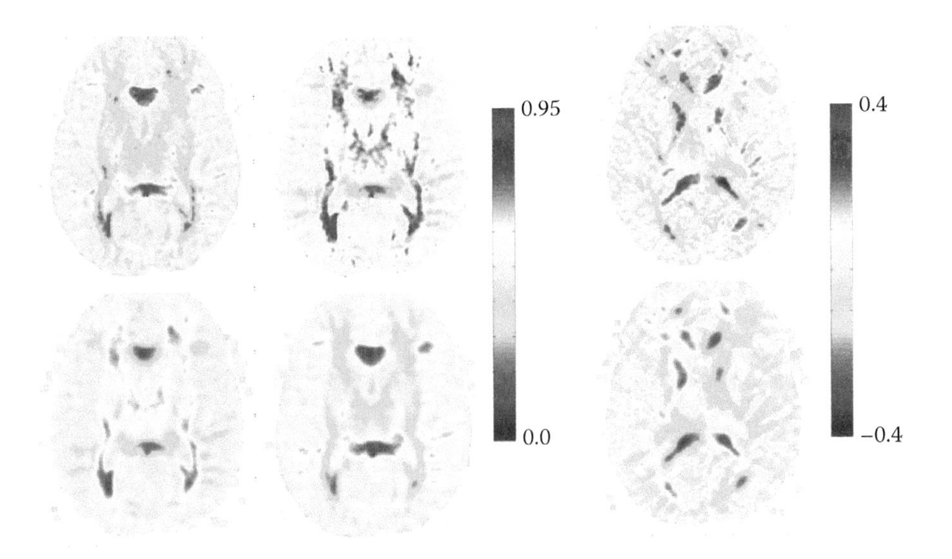

FIGURE 6.3
Top: the original diffusion tensors d_{11}, d_{22} and d_{12}. Bottom: the inverse of the diffusion tensor, d^{11}, d^{22} and d^{12}, which is needed for computing the anisotropic kernel. It has been smoothed and normalized by the trace of the inverse.

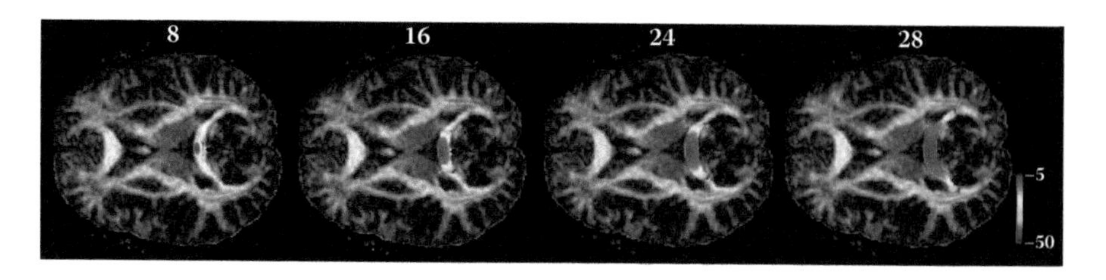

FIGURE 6.6
The transition probability in log scale from the seed at the splenium in the corpus callosum with $\Delta t = 0.1$ and $N = 8$, 16, 24, 28 iterations.

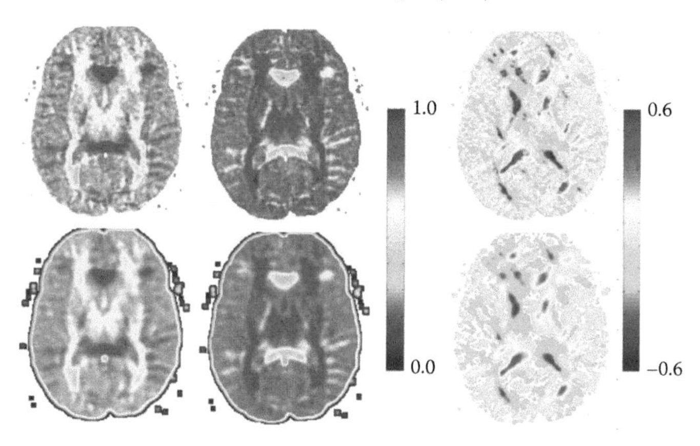

FIGURE 6.7
Top: Cholesky factors r_{11}, r_{22}, r_{12} of diffusion tensor D. Bottom: smoothed Cholesky factors with 8mm FWHM isotropic Gaussian kernel.

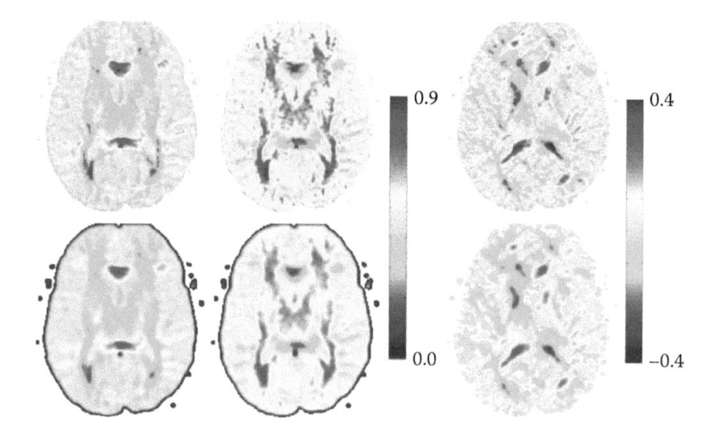

FIGURE 6.8
Top: normalized diffusion tensors d_{11}, d_{22}, d_{12} respectively. Bottom: smoothed diffusion tensors. We are smoothing the Cholesky factors r_{ij}.

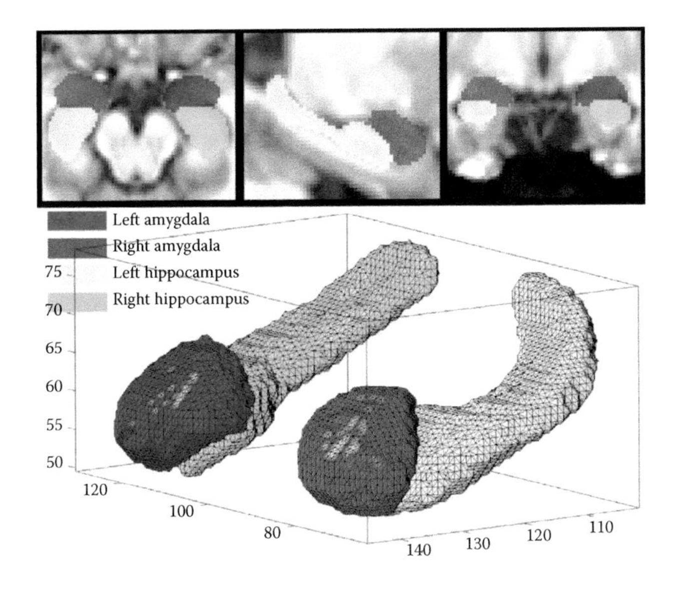

FIGURE 7.6
Subcortical structure template is superimposed on top of MRI [206, 205]. The figure is generated by Seung-Goo Kim of Seoul National University.

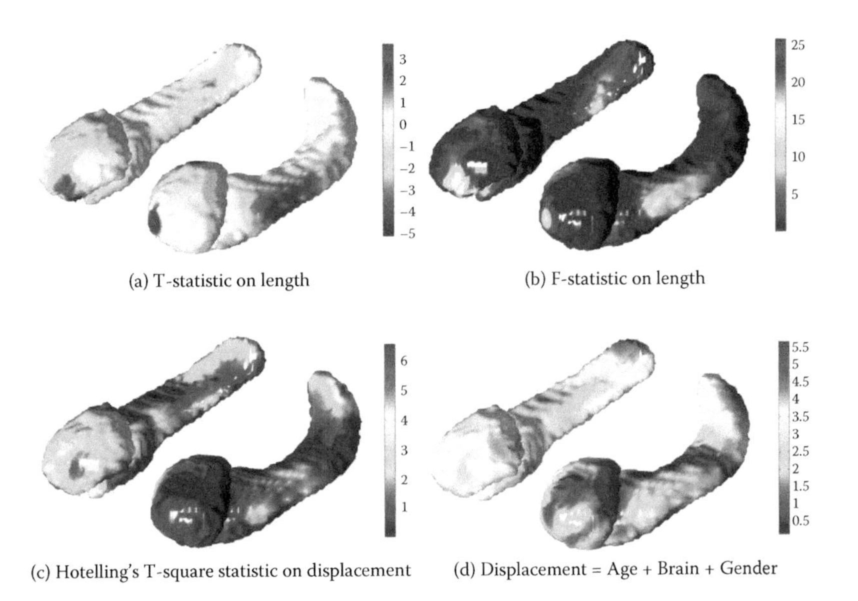

FIGURE 7.7
The gender effect on the length of displacement using (a) t-statistic and (b) F-statistic. The effect of gender on the displacement vector using (c) Hotelling's T^2 statistic and (d) Roy's maximum root.

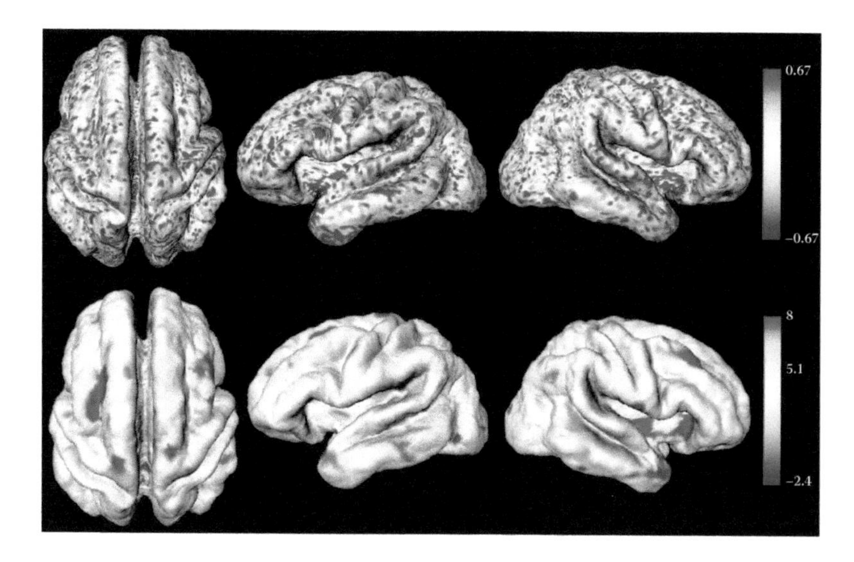

FIGURE 8.10
Top: Cortical thickness dilatation rate for a single subject. Due to such large measurement variations, surface-based smoothing is required. Bottom: T statistic map thresholded at the corrected p-value of 0.05.

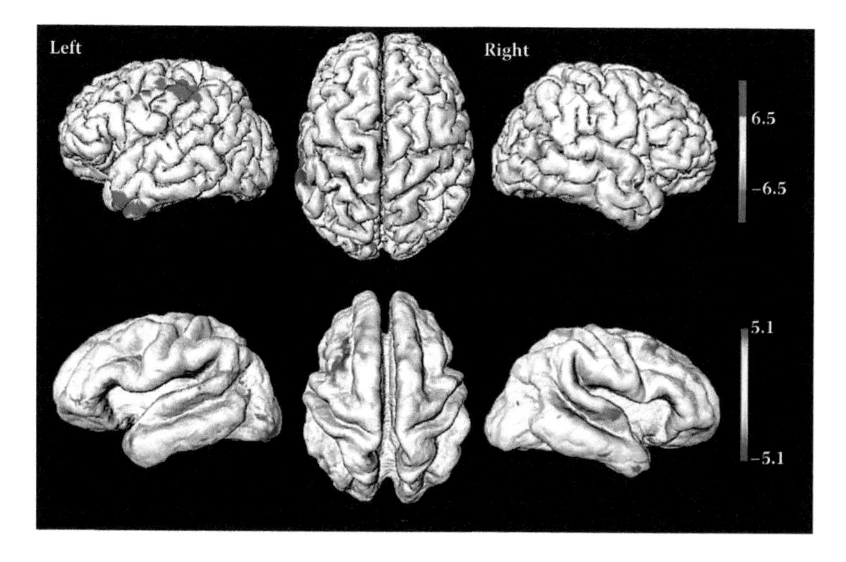

FIGURE 8.11
T-statistic map of local gray matter volume change (top) and cortical surface area dilatation rate (bottom) between ages 12 and 16

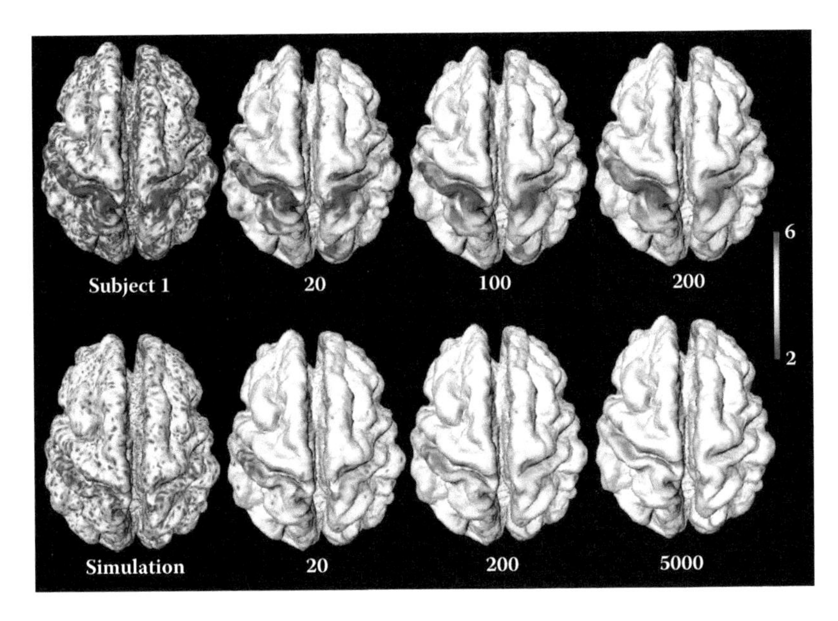

FIGURE 9.1
Heat kernel smoothing of cortical thickness (top) and simulated data (bottom) with $\sigma = 1$ and different number of iterations.

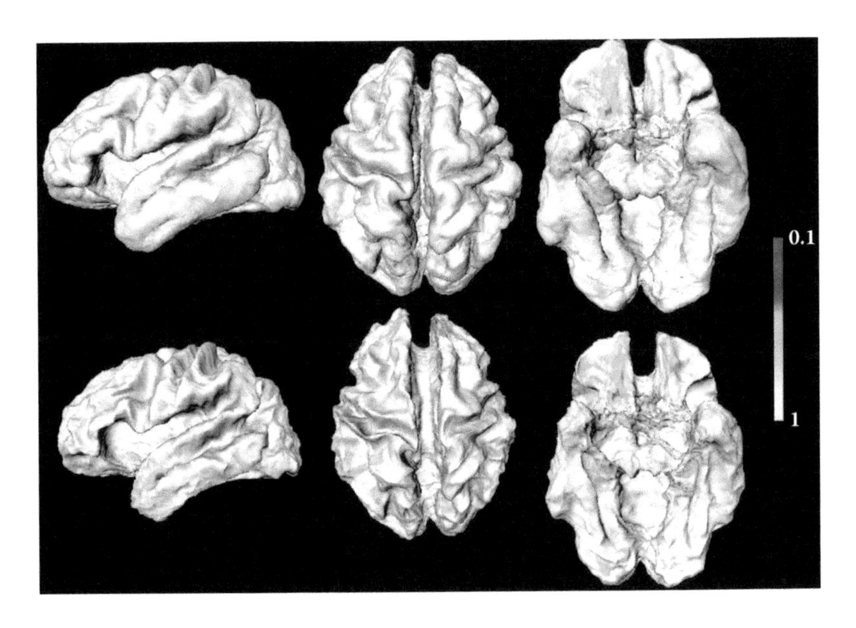

FIGURE 9.2
Corrected p-value maps of F-statistic removing the effect of age and relative gray matter volume difference projected onto the outer and inner surfaces.

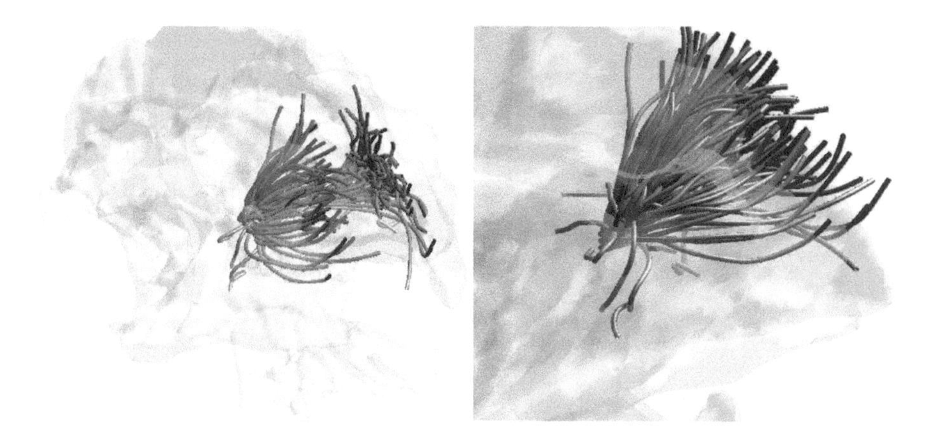

FIGURE 10.10
Each streamtube is the average tract in a subject. White matter fibers in controls (blue) are more clustered together with smaller spreading compared to autism (red).

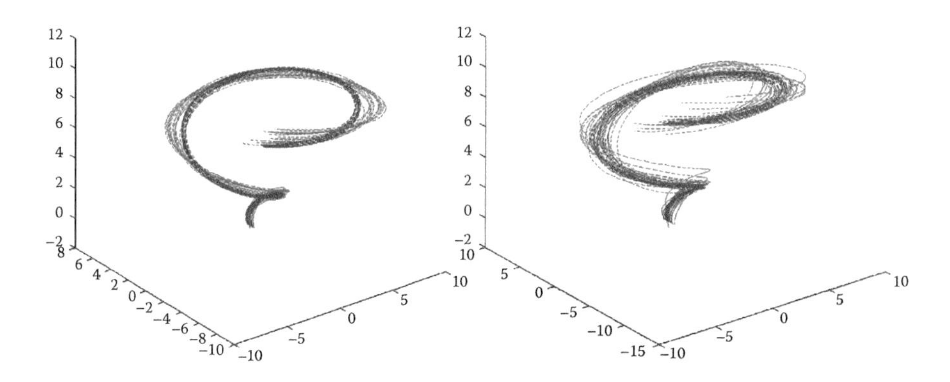

FIGURE 10.11
Simulated curves obtained from perturbing the basic curve shape $(x, y, z) = (s \sin s, s \cos s, s), s \in [0, 10]$. The first figure shows clear group separation while the second figure has too much overlap. We expect the cosine series representation to work extremely well for the first simulation while it may not work for the second simulation.

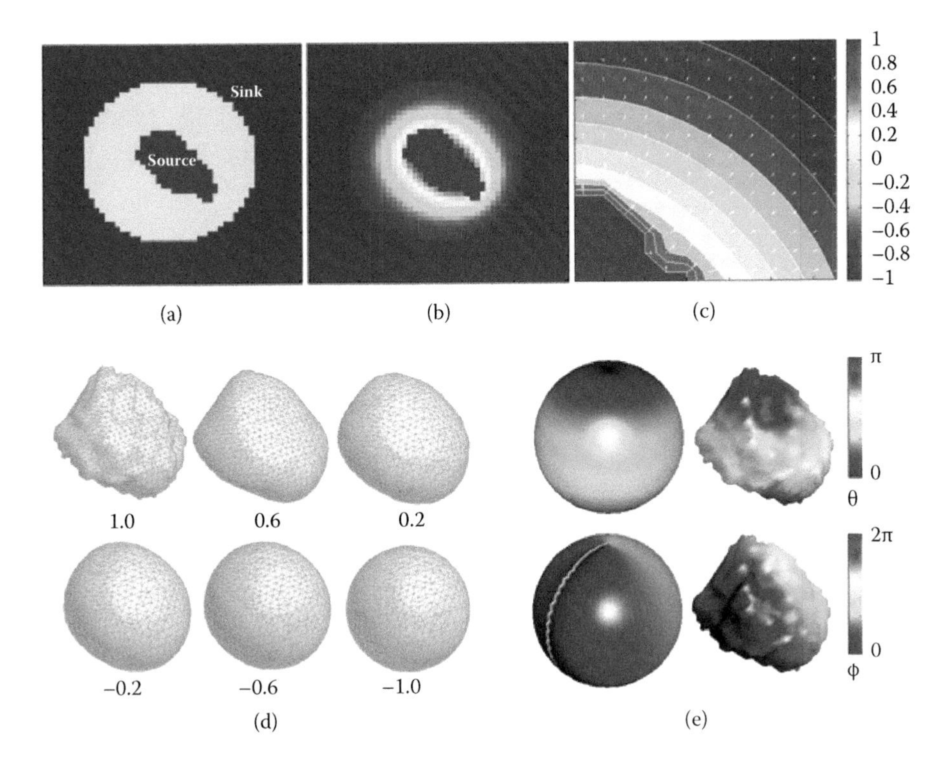

FIGURE 12.3
(a) The heat source (amygdala) and the heat sink are assigned values 1 and -1 respectively. The diffusion is solved with these boundary conditions. (b) After a sufficient number of iterations, the equilibrium state $f(x, \infty)$ is reached. (c) The gradient field shows the direction of heat propagation. The integral curve of the gradient field is computed by connecting one level set to the next level sets of $f(x, \infty)$. (d) Amygdala surface flattening is done by tracing the integral curve at each mesh vertex. The numbers $c = 1.0, 0.6, \cdots, -1.0$ correspond to the level sets $f(x, \infty) = c$. (e) Surface parameterization using the angles (θ, φ).

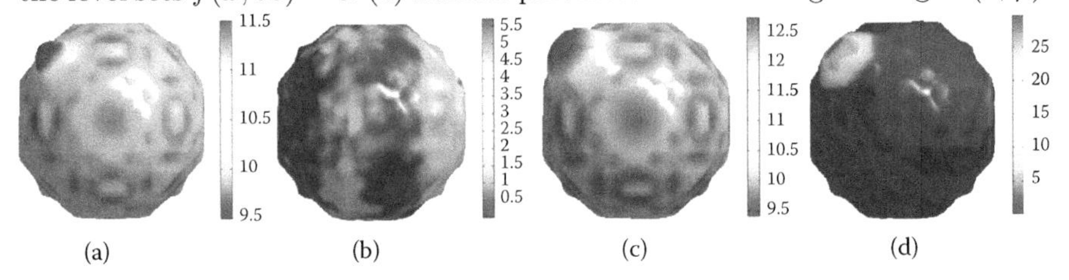

FIGURE 12.4
Simulation results. (a) small bump of height 1.5mm was added to a sphere of radius 10 mm. (b) T-statistic of comparing randomly simulated 20 spheres and 20 bumped spheres showing no group difference ($p = 0.35$). (c) small bump of height 3mm was added to a sphere of radius 10mm. (d) T-statistic of comparing randomly simulated 20 spheres and 20 bumped spheres showing significant group difference ($p < 0.0003$).

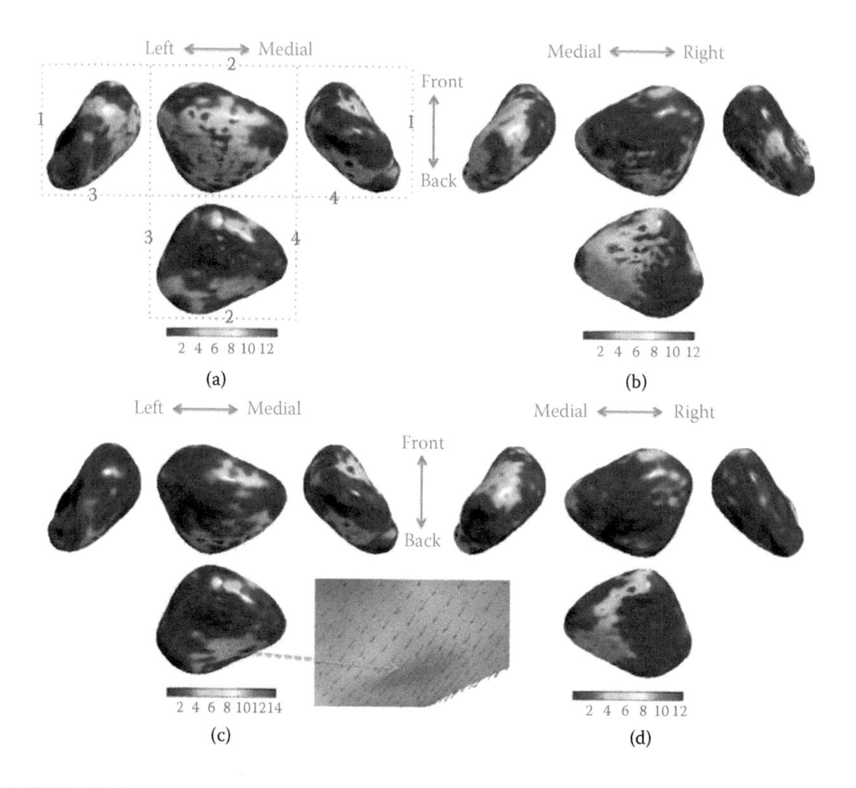

FIGURE 12.11
F statistic map of coordinate difference with and without accounting for age and the total brain volume for left (a,c) and right (b, d) amygdala.

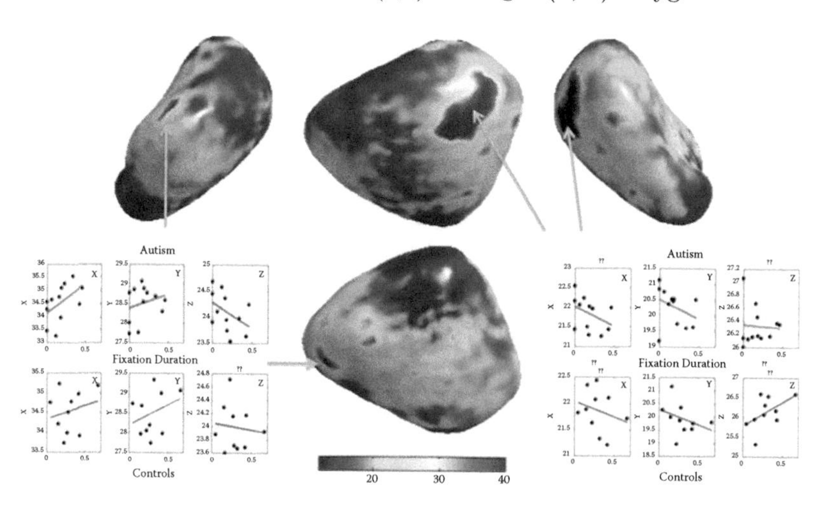

FIGURE 12.12
F-statistic of interaction between group and gaze fixation for right amygdala.

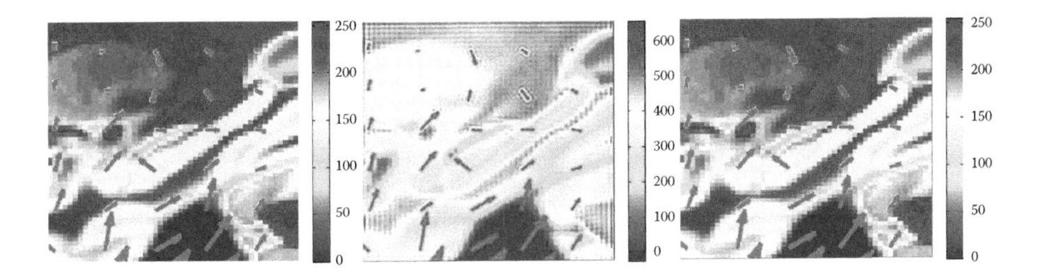

FIGURE 13.4
Left: original `deform.tif` image. Middle: 10000 basis expansion. Severe ringing artifacts are visible. Left: 90000 basis expansion.

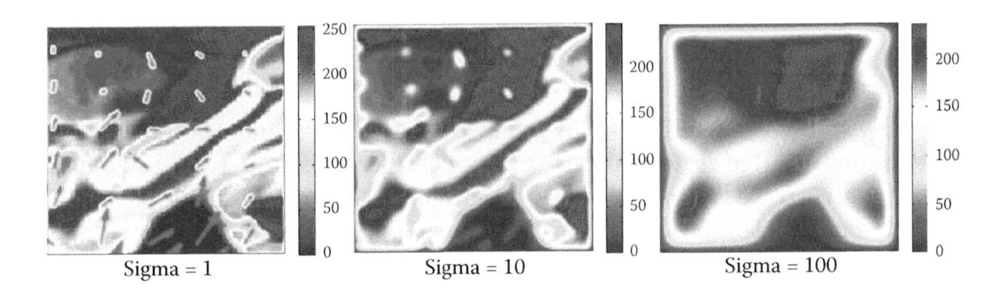

FIGURE 13.5
Heat kernel smoothing with 40000 basis and bandwidths 1, 10 and 100. Heat kernel smoothing does not exhibit ringing artifacts.

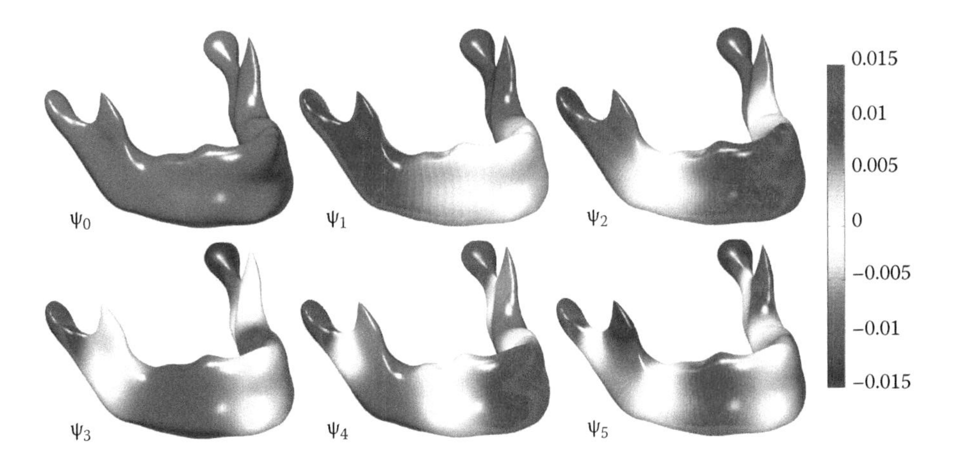

FIGURE 13.6
First six eigenfunctions on a mandible surface. The color scale is thresholded at ± 0.015 for better visualization. The figure was generated by Seongho Seo of Seoul National University.

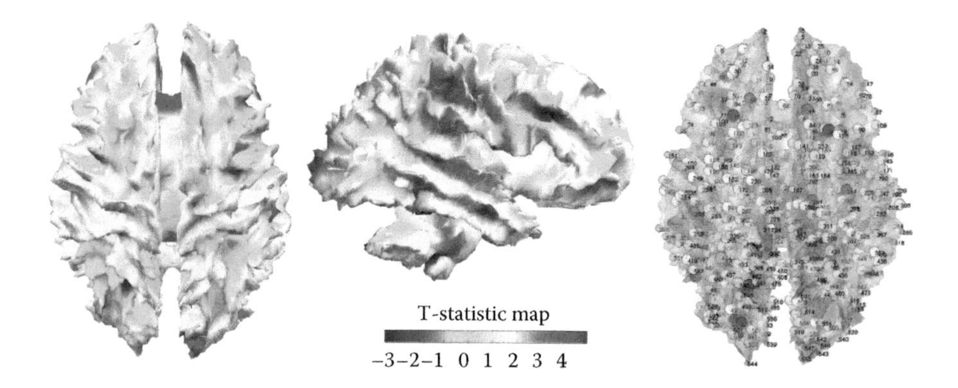

FIGURE 15.1
Left, Middle: T-statistic map of group differences (PI-controls) on Jacobian determinants. Red regions above 4.86 are considered as statistically significant at 0.05 (corrected) (see color insert). Right: 548 uniformly sampled nodes where multivariate-TBM will be performed. The nodes are sparsely sampled in the template to guarantee there is no spurious high correlation due to proximity between nodes.

FIGURE 15.2
548 uniformly sampled nodes along the white matter surface where multivariate-TBM will be performed. Exactly the same nodes used for Jacobian determinants are also selected.

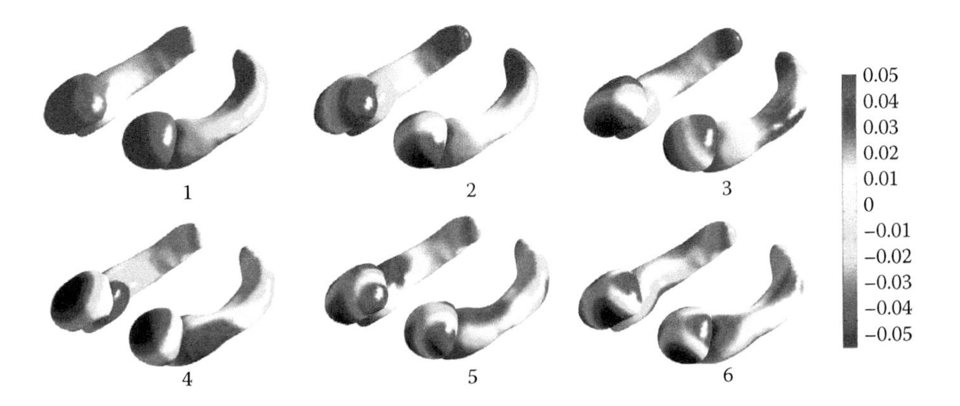

FIGURE 16.2
First six LB-eigenfunctions on amygdala and hippocampus surfaces.

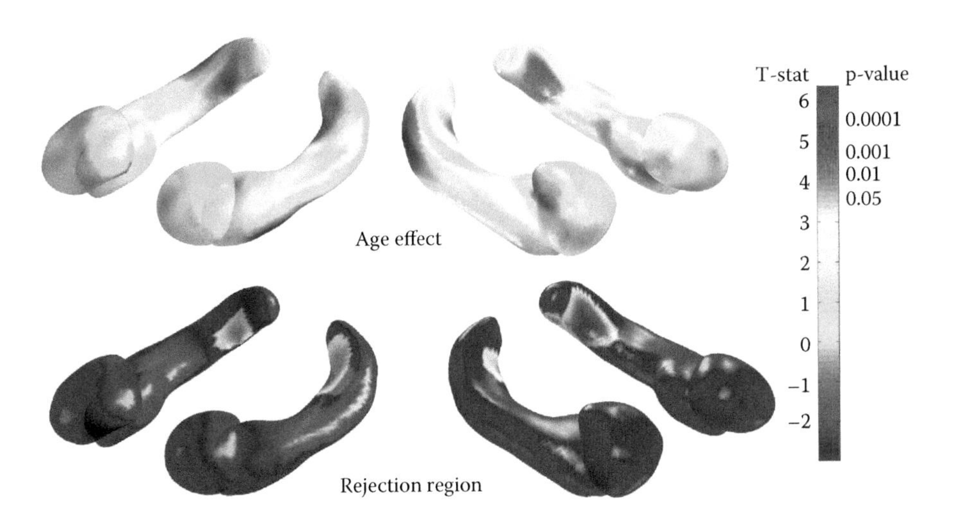

FIGURE 16.3
Age effect on hippocampus shape. The T-stat. and the corrected p-value are also shown. There is no age effect on amygdala. Rejection regions $\mathcal{M}_1$ corresponding to 0.05 level are also shown.

FIGURE 17.2
Stream tube representation of white matter fiber tracts.

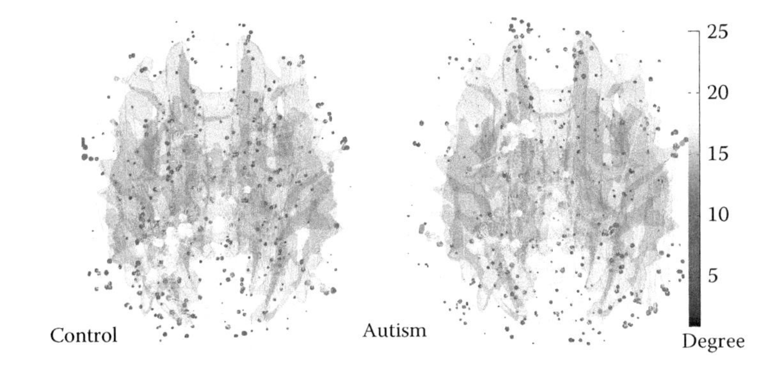

FIGURE 17.6
Node degrees are shown as colored spheres superimposed on top of the white matter surface. The size of a sphere at a node is proportional to the degree.

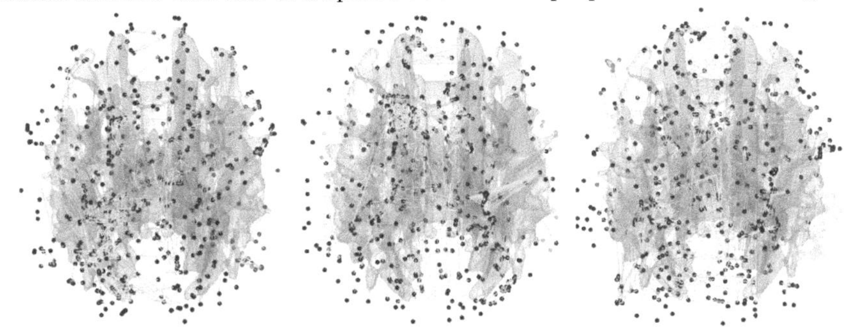

FIGURE 17.7
All nodes in the same connected component are colored identically. The brain network is characterized by a giant connected dominant component.

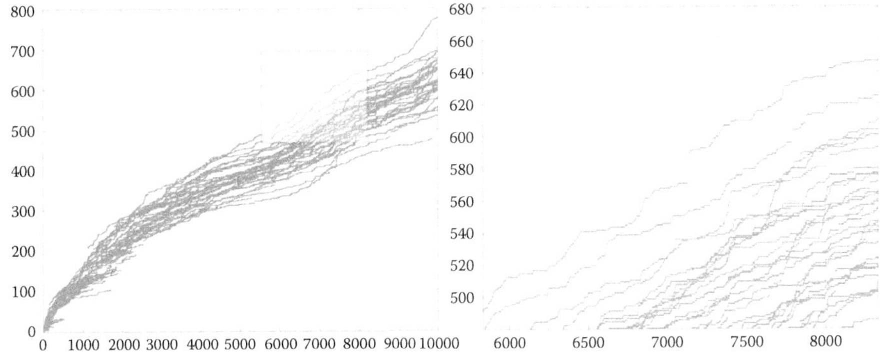

FIGURE 17.8
The size of the largest connected component (vertical) over the ϵ-filtration showing group difference (control = blue, autism= red).

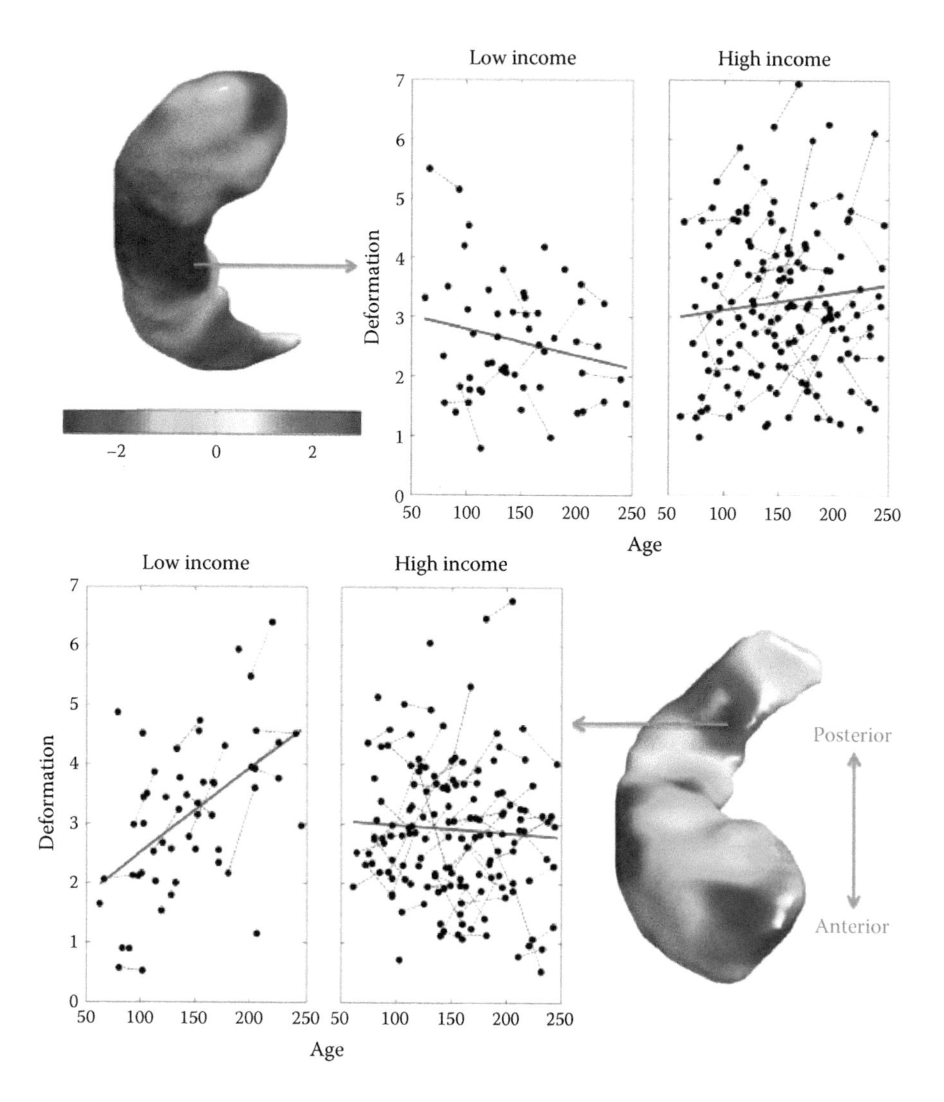

FIGURE 18.4
T-statistic map showing significant growth rate difference between children from low-income and high-income families. The significance of the growth rate difference is determined by the interaction term in a linear model. Highly focalized regions of group difference were detected in the right hippocampus (corrected p-value $=0.03$). The posterior region is enlarging while the midbody and the anterior parts are shrinking in children from low-income families. On the other hand the pattern is the opposite for children from high-income families.

12

Multivariate Surface Shape Analysis

Although there are many imaging studies on traditional ROI-based amygdala volumetry, there are very few studies on modeling amygdala shape variations. This chapter presents a unified computational and statistical framework for modeling amygdala shape variations in a clinical population.

Taking the anatomically corresponding surface coordinates across subjects as response variables, we set up a multivariate general linear model accounting for nuisance covariates such as age and brain size difference using `SurfStat` package that completely avoids the complexity of specifying design matrices. The methodology has been applied for quantifying abnormal local amygdala shape variations in 22 high functioning autistic subjects. This case study was first published in [80].

12.1 Introduction

Amygdala is an important brain substructure that has been implicated in abnormal functional impairment in autism [98, 257, 303]. Since structural abnormality might be the cause of the functional impairment, there have been many studies on amygdala volumetry. However, previous amygdala volumetry results have been inconsistent. [20] and [280] reported the amygdala volume was significantly smaller in the autistic subjects while [177] and [335] reported larger volume. [169] and [257] found no volume difference. [316] reported that age dependent amygdala volume difference in autistic children and indicated that the age dependency to be the cause of discrepancy. All these previous studies traced the amygdalae manually and by counting the number of voxels within the region of interest (ROI), the total volume of the amygdala was estimated. The limitation of the traditional ROI-based volumetry is that it can not determine if the volume difference is diffuse over the whole ROI or localized within specific regions of the ROI [81]. We present a novel computational and statistical framework that overcomes the limitation of the ROI-based volumetry.

Although there is extensive literature on local cortical shape analysis [76, 120, 195, 355, 363, 226, 235, 249], there is not much literature on amygdala shape analysis other than [61, 292, 201] mainly due to the difficulty of segmenting amgydala. On the other hand, there is extensive literature on shape modeling other subcortical structures using various techniques.

The medial representation [284] has been successfully applied to various subcortical structures including the cross sectional images of the corpus callosum [193] and hippocampus/amygdala complex [343], and ventricle and brain stem [284]. In the medial representation, the binary object is represented using the finite number of atoms and links that connect the atoms together to form a skeletal representation of the object. The medial representation is mainly used with the principal component analysis type of approaches for shape classification and group comparison.

Unlike the medial representation, which is in a discrete representation, there is a continuous parametric approach called the spherical harmonic representation [135, 152, 198, 323]. The spherical harmonic representation has been mainly used as a data reduction technique for compressing global shape features into a small number of coefficients. The main global geometric features are encoded in low degree coefficients while the noise will be in high degree spherical harmonics [152]. The method has been used to model various subcortical structures such as ventricles [135], hippocampi [323] and cortical surfaces [71]. The spherical harmonics have global support. So the spherical harmonic coefficients contain only the global shape features and it is not possible to directly obtain local shape information from the coefficients only. However, it is still possible to obtain local shape information by evaluating the representation at each fixed point, which gives the smoothed version of the coordinates of surfaces. In this fashion, the spherical harmonic representation can be viewed as mesh smoothing [71]. Instead of using the global basis of spherical harmonics, there have been attempts to use the local wavelet basis for parameterizing cortical surfaces [259, 405].

Other shape modeling approaches include distance transforms [228], deformation fields obtained by warping individual substructures to a template [248] and the particle-based method [61]. A distance transform is a function that for each point in the image is equal to the distance from that point to the boundary of the object [144]. The distance map approach has been applied in classifying a collection of hippocampi [144]. The deformation fields based approach has been somewhat popular and has been applied to modeling whole 3D brain volume [17, 81, 131], cortical surfaces [83, 361], hippocampus [195], and cingulate gyrus [96]. The particle-based method uses a nonparametric, dynamic particle system to simultaneously sample object surfaces and optimize correspondence point positions [61].

In this chapter, we use the *weighted spherical harmonic representation* for parameterization, surface smoothing and surface registration in a unified Hilbert space framework. [71] presented the underlying mathematical theory and a new iterative algorithm for estimating the coefficients of the representation for extremely large meshes such as cortical surfaces. Here we apply the method to real autism surface data in a truly multivariate fashion for the first time.

Our approach differs from the traditional spherical harmonic representation in many ways. Although the truncation of the series expansion in the spherical harmonic representation can be viewed as a form of smoothing, there is no direct equivalence to the *full width at half maximum* (FWHM) usually associated with kernel smoothing. So it is difficult to relate the unit of FWHM widely used in brain imaging to the degree of spherical harmonic representation. On the other hand, our new representation can easily relate to FWHM of smoothing kernel so we have a clear sense of how much smoothing we are performing beforehand.

The traditional representation suffers from the Gibbs phenomenon (ringing artifacts) [132] that usually happens in representing rapidly changing or discontinuous data with smooth periodic basis. Our new representation can substantially reduce the amount of Gibbs phenomenon by weighting the coefficients of the spherical harmonic expansion. The weighting has the effect of actually performing heat kernel smoothing, and thus reducing the ringing artifacts. We quantify the improved performance of our new representation in both the real and simulated data for the first time.

Since the proposed new representation requires a smooth map from amygdala surfaces to a sphere, we have developed a new and very fast surface flattening technique based on the propagation of heat diffusion. By tracing the integral curve of heat gradient from a heat source (amygdala) to a heat sink (sphere), we can obtain the flattening map. Since solving an isotropic heat equation in a 3D image volume is fairly straightforward, our proposed method offers a much simpler numerical implementation than available surface flattening techniques such as conformal mappings [12, 152, 180] quasi-isometric mappings [367] and area preserving mappings [49]. The established spherical mapping is used to parameterize an amygdala surface using two angles associated with the unit sphere. The angles serve as coordinates for representing amygdala surfaces using the weighted linear combination of spherical harmonics. The tools containing the weighted spherical harmonic representation and the surface flattening algorithm can be found in `www.stat.wisc.edu/~mchung/research/amygdala`. It should be pointed out that our representation and parameterization techniques are general enough to be applied to various brain structures such as hippocampus and caudate that are topologically equivalent to a sphere.

Based on the weighted spherical harmonic representation of amygdalae, various multivariate tests were performed to detect the group difference between autistic and control subjects. Most of the multivariate shape models on coordinates and deformation vector fields have mainly used the Hotelling's T-square as a test statistic [56, 81, 86, 131, 195, 359]. The Hotelling's T-square statistic tests for the equality of vector means without accounting for the additional covariates such as gender, brain size and age. Since the size of amygdala is dependent on brain size and possibly on age as well, there is a definite need for a model that is able to include these covariates explicitly.

The proposed multivariate linear model does exactly this by generalizing the Hotelling's T-square framework to incorporate additional covariates.

In order to simplify the computational burden of setting up the proposed multivariate linear models, we have developed the `SurfStat` package (`www.math.mcgill.ca/keith/surfstat`). that offers a unified statistical analysis platform for various 2D surface mesh and 3D image volume data. The novelty of `SurfStat` is that there is no need to specify design matrices that tend to baffle researchers not familiar with contrasts and design matrices. `SurfStat` supersedes fMRISTAT, and contains all the statistical and multiple comparison correction routines.

12.2 Surface Parameterization

Once the binary segmentation $\mathcal{M}_a$ of an object is obtained either manually or automatically, the marching cubes algorithm [232] was applied to obtain a triangle surface mesh $\partial\mathcal{M}_a$. The weighted spherical harmonic representation requires a smooth mapping from the surface mesh to a unit sphere S^2 to establish a coordinate system. We have developed a new surface flattening algorithm based on heat diffusion.

We start with putting a larger sphere $\mathcal{M}_s$ that encloses the binary object $\mathcal{M}_a$. Figure 12.3 shows an illustration with the binary segmentation of amygdala. The center of the sphere $\mathcal{M}_s$ is taken as the average of the mesh coordinates of $\partial\mathcal{M}_a$, which forms the surface mass center. The radius of the sphere $\mathcal{M}_s$ is taken in such a way that the shortest distance between the sphere to the binary object $\mathcal{M}_a$ is fixed(5mm for amygdala). The final flattening map is definitely affected by the perturbation of the position of the sphere but since we are fixing it to be the mass center of surface for all amygdala, we do not need to worry about the perturbation effect.

The binary object $\mathcal{M}_a$ is assigned the value 1 while the enclosing sphere is assigned the value -1, i.e.

$$f(\mathcal{M}_a, \sigma) = 1 \text{ and } f(\mathcal{M}_s, \sigma) = -1 \tag{12.1}$$

for all $\sigma \in [0, \infty)$. The parameter σ is the diffusion time. $\mathcal{M}_a$ and $\mathcal{M}_s$ serve as a heat source and a heat sink respectively. Then we solve isotropic diffusion

$$\frac{\partial f}{\partial \sigma} = \Delta f \tag{12.2}$$

with the given boundary condition (12.1). Δ is the 3D Laplacian. When $\sigma \to \infty$, the solution reaches the heat equilibrium state where the additional diffusion does not make any change in heat distribution. The heat equilibrium state is also obtained by letting $\frac{\partial f}{\partial \sigma} = 0$ and solving for the Laplace equation

$$\Delta f = 0 \tag{12.3}$$

with the same boundary condition. This will result in the equilibrium state denoted by $f(x, \sigma = \infty)$. Once we obtained the equilibrium state, we trace the path from the heat source to the heat sink for every mesh vertices on the isosurface of $\mathcal{M}_a$ using the gradient of the heat equilibrium $\nabla f(x, \infty)$. A similar formulation called the *Laplace equation method* has been used in estimating cortical thickness bounded by outer and inner cortical surfaces by establishing correspondence between two surfaces by tracing the gradient of the equilibrium state [401, 189, 226].

The heat gradients form vector fields originating at the heat source and ending at the heat sink (Figure 12.3). The integral curve of the gradient field at a mesh vertex $p \in \partial\mathcal{M}_a$ establishes a smooth mapping from the mesh vertex to the sphere. The integral curve τ is obtained by solving a system of differential equations

$$\frac{d\tau}{dt}(t) = \nabla f(\tau(t), \infty)$$

with $\tau(t = 0) = p$. The integral curve approach is a widely used formulation in tracking white matter fibers using diffusion tensors [30, 218]. These methods rely on discretizing the differential equations using the Runge-Kutta method, which is computation intensive. However, we avoided the Runge-Kutta method and solved using the idea of the propagation of level sets. Instead of directly computing the gradient field $\nabla f(x, \infty)$, we computed the level sets $f(x, \infty) = c$ of the equilibrium state corresponding to for varying c between -1 and 1. The integral curve is then obtained by finding the shortest path from one level set to the next level set and connecting them together in a piecewise fashion. This is done in an iterative fashion as shown in Figure 12.3, where five level sets corresponding to the values $c = 0.6, 0.2, -0.2, -0.6, -1.0$ are used to flatten the amygdala surface. Once we obtained the spherical mapping, we can then project the angles (θ, φ) onto $\partial\mathcal{M}_a$ and the two angles serve as the underlying parameterization for the weighted spherical harmonic representation.

For the proposed flattening method to work, the binary object has to be close to either star-shape or convex. For shapes with a more complex structure, the gradient lines that correspond to neighboring nodes on the surface will fall within one voxel in the volume, creating numerical singularities in mapping to the sphere. Other more complex mapping methods such as conformal mapping [12, 152, 180] can avoid this problem but they are numerically more demanding. On the other hand, our approach is simpler and more computationally efficient because it works for a limited class of shapes.

Since we are solving the steady state heat equation between an amygdala surface and a sphere, the level set of heat equilibrium near the sphere is expected to be close to be spherical. So even if we increase the radius of the sphere, the amygdala mesh vertices will be mapped to similar radial directions.

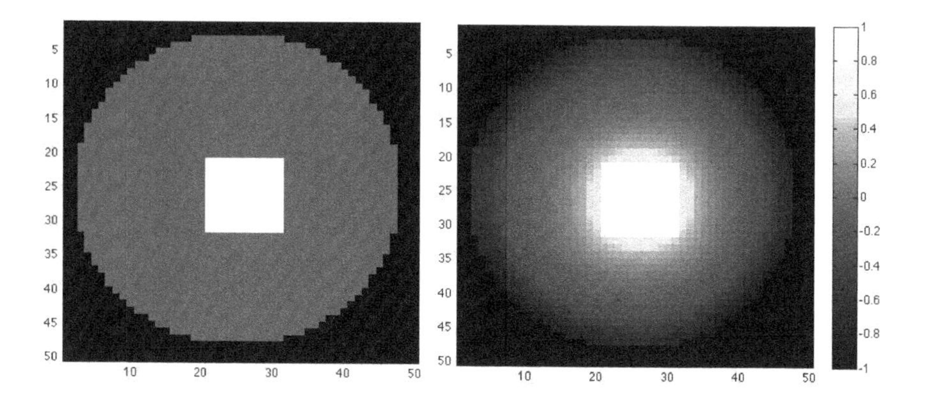

FIGURE 12.1
The process of flattening a simulated cube to a sphere by solving heat diffusion. The square is treated as a heat source assigned the value 1 while the sphere is treated as a heat sink and assigned the value -1. The system makes the heat to flow from the source to the sink introducing heat flow. By tracing the heat flow, we can flatten the cube to a sphere [80].

12.2.1 Flattening of Simulated Cube

We have simulated a cube of size $10 \times 10 \times 10$. Figures 12.1 and 12.2 show the simulated cube and how it is diffused to become a sphere.

```
vol=zeros(70,70,70);
vol(40:50,40:50,40:50)=1;

surf=isosurface(vol)
figure;figure_wire(surf,'yellow', 'white')
axis off
```

We enclose the cube with a larger sphere using `CREATEenclosedamyg.m` and assign the value 1 to the cube, which is considered as a heat source. The sphere is treated as a heat sink and assigned the value -1.

```
[amyg,sphere,amygsphere]=CREATEenclosedamyg(vol,surf);
figure;imagesc(squeeze(amygsphere(:,25,:)));
colorbar; colormap('hot')
```

The heat sink and source make the heat to flow from the source to the sink introducing heat flow. The heat equation is solved using `LAPLACE3Dsmooth.m` by performing 5 iterations.

```
n_steps=5;
```

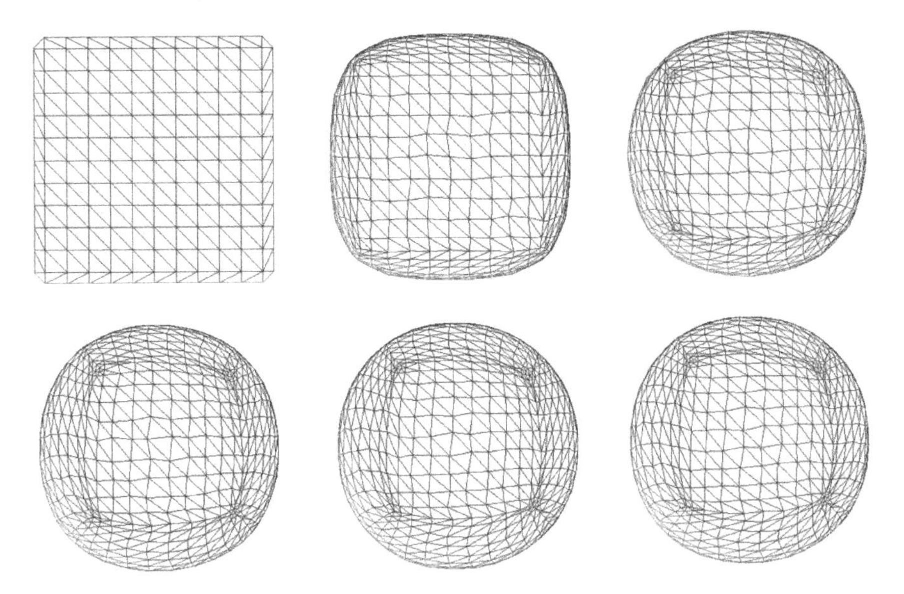

FIGURE 12.2
The process of flattening a simulated cube to a sphere by solving heat diffusion. The square is treated as a heat source assigned the value 1 while the sphere is treated as a heat sink and assigned the value -1. The system makes the heat to flow from the source to the sink introducing heat flow. By tracing the heat flow, we can flatten the cube to a sphere [80].

```
stream=LAPLACE3Dsmooth(amygsphere,amyg,-sphere, n_steps);
figure;imagesc(squeeze(stream(:,25,:)));
colorbar; colormap('hot')
```

The heat gradient is then obtained using the following commands. Again 5 iterations are used for flattening. The number of iterations is sufficient as shown in Figure 12.2 where the convergence to a sphere is visible only after three iterations.

```
sphere=isosurface(amyg);
for alpha=1:n_steps
    sphere=LAPLACEcontour(stream,sphere, 1 - 2*alpha/n_steps);
    sphere=REGULARIZEarea(sphere, 0.5);
    figure;figure_wire(sphere,'yellow', 'white'); axis off
end;
```

12.3 Weighted Spherical Harmonic Representation

The parameterized amygdala surfaces, in terms of spherical angles θ, φ, are further expressed using the weighted spherical harmonic representation [71], which expresses surface coordinate functions as a weighted linear combination of spherical harmonics. The automatic degree selection procedure was also introduced in the previous work but for the completeness of our paper, the method is briefly explained in section 12.3.1.

The mesh coordinates for the object surface $\partial \mathcal{M}_a$ are parameterized by the spherical angles $\Omega = (\theta, \varphi) \in [0, \pi] \otimes [0, 2\pi)$ as

$$p(\theta, \varphi) = (p_1(\theta, \varphi), p_2(\theta, \varphi), p_2(\theta, \varphi)).$$

The weighted spherical harmonic representation is given by

$$p(\theta, \varphi) = \sum_{l=0}^{k} \sum_{m=-l}^{l} e^{-l(l+1)\sigma} f_{lm} Y_{lm}(\theta, \varphi),$$

where

$$f_{lm} = \int_{\theta=0}^{\pi} \int_{\varphi=0}^{2\pi} p(\theta, \varphi) Y_{lm}(\theta, \varphi) \sin\theta d\theta d\varphi$$

are the spherical harmonic coefficient vectors and Y_{lm} are spherical harmonics of degree l and order m. The coefficients f_{lm} are estimated in a least squares fashion [71, 135, 323].

Many previous imaging and shape modeling literature have used the complex-valued spherical harmonics [52, 135, 152, 323], but we have only used real-valued spherical harmonics [92, 175] throughout the paper for the convenience in setting up a real-valued stochastic model. The relationship between the real- and complex-valued spherical harmonics is given in [42, 175]. The complex-valued spherical harmonics can be transformed into real-valued spherical harmonics using an unitary transform.

12.3.1 Optimal Degree Selection

Since it is impractical to sum the representation to infinity, we need a rule for truncating the series expansion. Given the bandwidth σ of heat kernel, we automatically determine if increasing degree k has any effect on the goodness of the fit of the representation. In all spherical harmonic literature [136, 135, 152, 322, 323], the truncation degree is simply selected based on a pre-specified error bound. On the other hand, our proposed statistical framework is based on a type-I error.

Although increasing the degree increases the goodness-of-fit of the representation, it also increases the number of coefficients to be estimated quadratically. It is necessary to find the optimal degree where the goodness-of-fit and

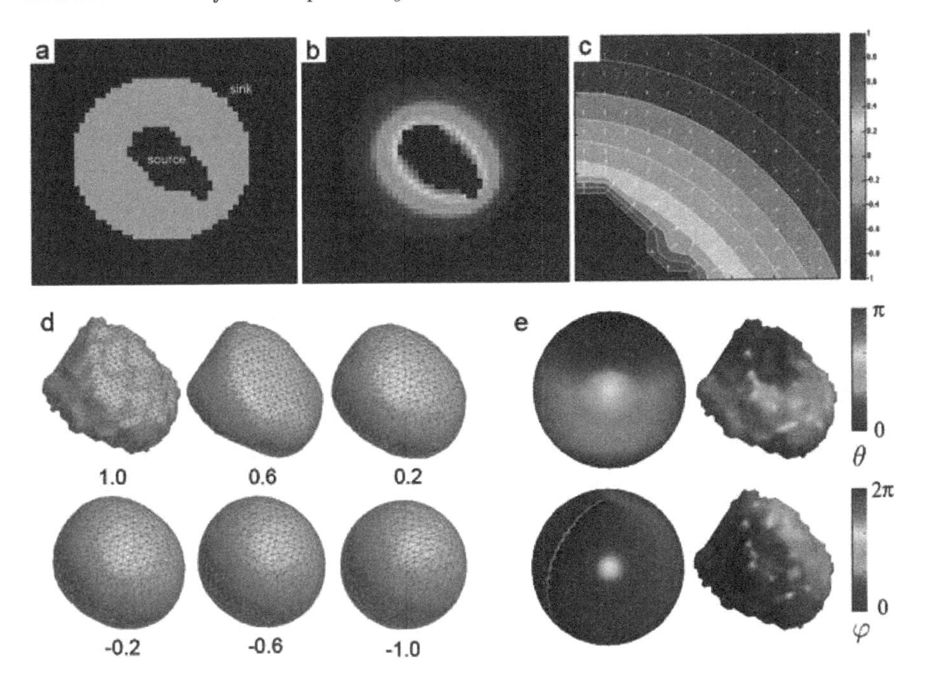

FIGURE 12.3
(a) The heat source (amygdala) and the heat sink are assigned values 1 and -1 respectively. The diffusion is solved with these boundary conditions. (b) After a sufficient number of iterations, the equilibrium state $f(x,\infty)$ is reached. (c) The gradient field shows the direction of heat propagation. The integral curve of the gradient field is computed by connecting one level set to the next level sets of $f(x,\infty)$. (d) Amygdala surface flattening is done by tracing the integral curve at each mesh vertex. The numbers $c = 1.0, 0.6, \cdots, -1.0$ correspond to the level sets $f(x,\infty) = c$. (e) Surface parameterization using the angles (θ, φ). (See color insert.)

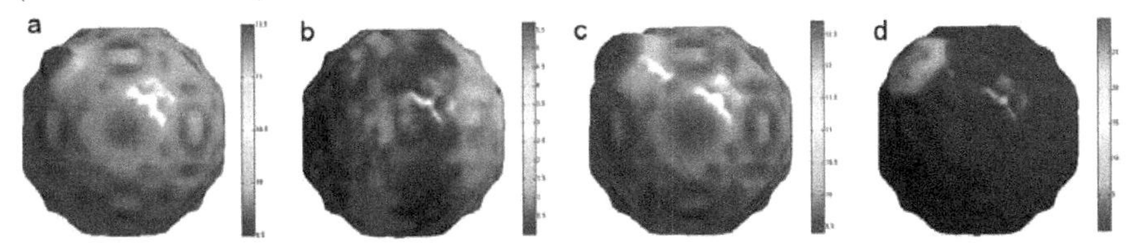

FIGURE 12.4
Simulation results. (a) small bump of height 1.5mm was added to a sphere of radius 10 mm. (b) T-statistic of comparing randomly simulated 20 spheres and 20 bumped spheres showing no group difference ($p = 0.35$). (c) small bump of height 3mm was added to a sphere of radius 10mm. (d) T-statistic of comparing randomly simulated 20 spheres and 20 bumped spheres showing significant group difference ($p < 0.0003$). (See color insert.)

the number of parameters balance out. Consider the k-th degree error model:

$$p(\Omega) = \sum_{l=0}^{k-1} \sum_{m=-l}^{l} e^{-l(l+1)\sigma} f_{lm} Y_{lm}(\Omega) + \sum_{m=-k}^{k} e^{-k(k+1)\sigma} f_{km} Y_{km}(\Omega) + \epsilon(\Omega),$$

where ϵ is a zero mean Gaussian random field. We test if adding the k-th degree terms to the $k-1$-th degree model is statistically significant by formally testing

$$H_0 : f_{k,-k} = f_{k,-k+1} = \cdots = f_{k,k-1} = f_{k,k} = 0.$$

This can be easily done using the F statistic $2k+1$ and $n-(k+1)^2$ degrees of freedom.

At each degree, we compute the corresponding p-value and stop increasing the degree if it is smaller than the pre-specified significance $\alpha = 0.01$. For bandwidths $\sigma = 0.01, 0.001$ and 0.0001, the approximate optimal degrees are 18, 42 and 78 respectively. In our study, we have used $k = 42$ degree representation corresponding to bandwidth $\sigma = 0.001$. The bandwidth 0.01 smoothes out too much local details while the bandwidth 0.0001 introduces too much voxel discretization error into the representation.

12.4 Gibbs Phenomenon in SPHARM

The weakness of the traditional spherical harmonic representation is that it produces the Gibbs phenomenon (ringing artifacts) for discontinuous and rapidly changing continuous measurements [132, 71].

Consider the finite Fourier series expansion of 1D piecewise smooth function $f \geq 0$ with discontinuity at c given by

$$S_k(u) = \sum_{j=0}^{k} f_j \psi_j(u),$$

where $f_j = \langle f, \psi_j \rangle$. The basis is the usual sin and cosine functions. Let

$$d = \lim_{u \to c^+} f(u) - \lim_{u \to c^-} f(u) > 0$$

be the size of jump. Let $u_\circ$ be the first local maximum. Then the amount of overshoot associated with the k-th series expansion is given by

$$S_k(u_\circ) - \lim_{u \to c^+} f(u).$$

Then we can show that the limit of the overshoot is

$$\lim_{k \to \infty} S_k(u_\circ) - \lim_{u \to c^+} f(u) = \frac{d}{2}(g-1),$$

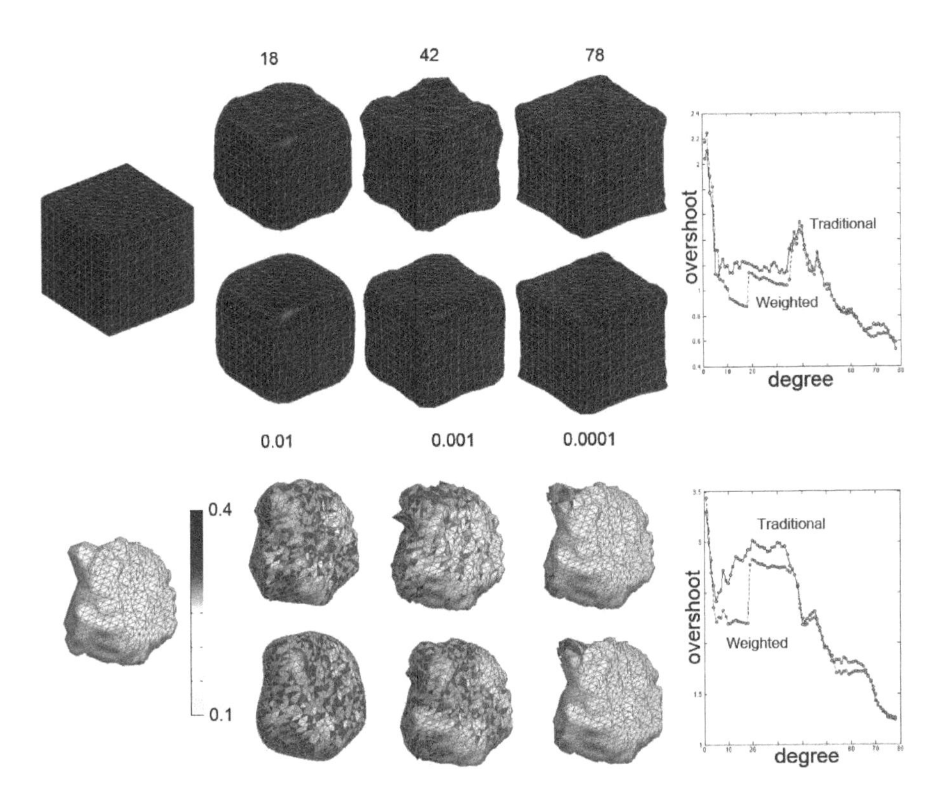

FIGURE 12.5
The first (third) row shows the significant Gibbs phenomenon in the spherical harmonic representation of a cube (left amygdala) for degrees $k = 18, 42, 78$. The second (fourth) row is the weighted spherical harmonic representation at the same degrees but with bandwidth $\sigma = 0.01, 0.001, 0.0001$ respectively. The color scale for amygdala is the absolute error between the original and reconstructed amygdala. In almost all degrees, the traditional spherical harmonic representation shows more prominent Gibbs phenomenon compared to the weighted version. The plots display the amount of overshoot for the traditional representation vs. the weighted version.

where the Gibbs constant g is given by

$$g = \frac{2}{\pi} \int_0^{\pi} \frac{\sin x}{x}\, dx = 1.17897974\cdots.$$

There are few available techniques for reducing Gibbs phenomenon [50, 148]. Most techniques are a variation on some sort of kernel methods. One of the standard methods is to use the Fejer kernel which is defined as

$$K_n(u) = \frac{1}{n} \sum_{j=0}^{n-1} D_j(u),$$

where D_j is the Dirichlet kernel

$$D_j = \sum_{k=-j}^{j} e^{iku}.$$

Then it can be shown that

$$K_n(u) = \frac{1}{n} \Big(\frac{\sin \frac{nu}{2}}{\sin \frac{u}{2}} \Big)^2.$$

The kernel is symmetric and positive. Then we have

$$K_n * f \to f$$

for any $f \in L^2([-\pi, \pi])$ as $n \to \infty$. Since the kernel is unimodal, it has the effect of smoothing the discontinuous signal f and in turn the convolution will not exhibit the ringing artifacts for sufficiently large n. Heat kernel smoothing and weighted Fourier representation behave similarly and can be used in reducing the Gibbs phenomenon.

The Gibbs phenomenon will likely arise in modeling arbitrary anatomical objects with possible sharp corners. The Gibbs phenomenon can be effectively removed if the spherical harmonic representation converges faster as the degree goes to infinity. By weighting the spherical harmonic coefficients exponentially smaller, we can make the representation converge faster. This can be achieved by additionally weighting the spherical harmonic coefficients with the heat kernel. Figure 12.6 demonstrates the severe Gibbs phenomenon in the traditional spherical harmonic representation (top) on a hat shaped 2D surface. The hat shaped surface is simulated as $z = 1$ for $x^2 + y^2 < 1$ and $z = 0$ for $1 \leq x^2 + y^2 \leq 2$. On the other hand the weighted spherical harmonic representation (bottom) shows substantially reduced ringing artifacts. Due to very complex folding patterns, sulcal regions of the brain exhibit more abrupt directional change than the simulated hat surface (upward of 180 degrees compared to 90 degrees in the hat surface) so there is a need for reducing the Gibbs phenomenon in the traditional spherical harmonic representation.

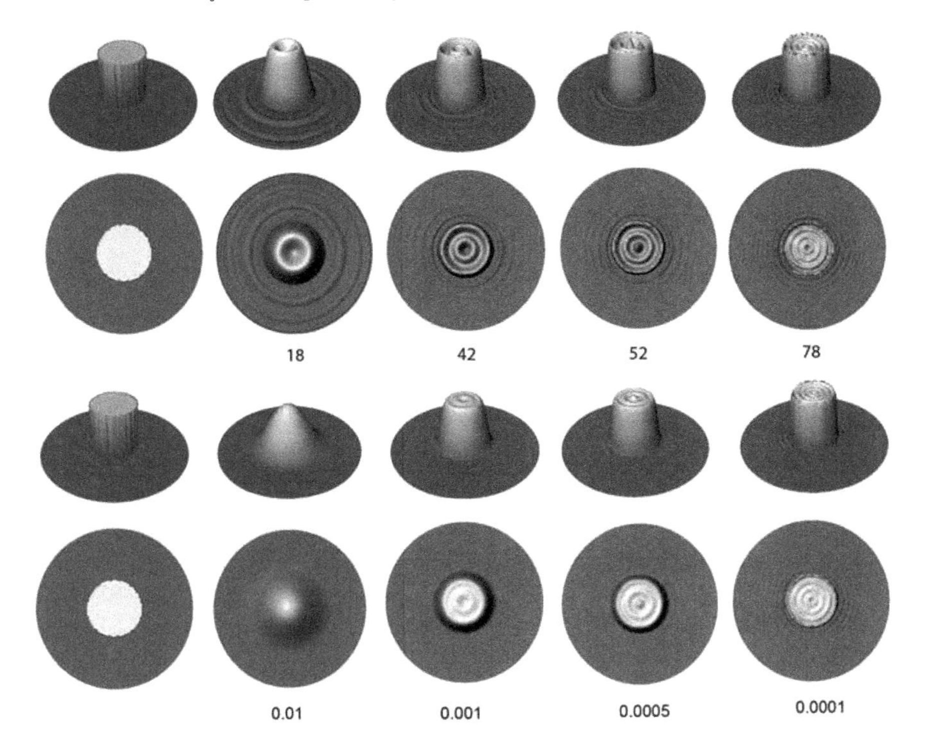

FIGURE 12.6
Gibbs phenomenon on a hat shaped simulated surface. The SPHARM representation (top) of degrees 18, 42, 52 and 78 show severe ringing artifacts. One the other hand, the weighted-SPHARM representation (bottom) with bandwidths 0.01, 0.001, 0.0005, 0.0001 shows less ringing artifacts. The optimal degrees for the weighed representation are determined by the model selection procedure and found to be 18, 42, 52 and 78 respectively.

In Figure 12.7, a different example is given for Gibbs phenomenon. Discontinuous measurements are constructed as a step function of value 1 in the circular band $\frac{1}{8} < \theta < \frac{1}{4}$ and 0 outside of the band on a unit sphere. The SPAHRM representation of the step function resulted in significant ringing artifacts even for fairly high degrees up to $k = 78$. In comparison, the weighted-SPHARM representation does not exhibit any serious ringing artifacts. The superior performance of the weighted-SPHARM can be easily explained in terms of convergence. The weighted-SPHARM representation additionally weights Fourier coefficients with exponentially decaying weights, which contributes to more rapid convergence even for discontinuous measurements. This robustness of weighted-SPHARM is also related to the fact that

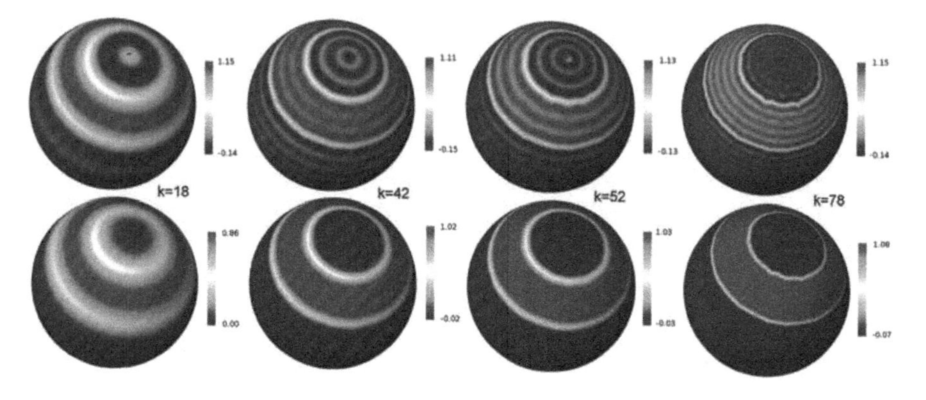

FIGURE 12.7
Gibbs phenomenon in the SPHARM representation for degrees 18, 42, 52 and 78. The traditional SPHARM (top) and the weighted representations (bottom) are performed on the discontinuous measurements on a unit sphere, which are defined as 1 in region $\frac{1}{8} < \theta < \frac{1}{4}$ and 0 in other regions. The SPHARM representation shows severe ringing artifacts while the weighted-SPHARM shows the negligible ringing effect.

it is a PDE-based data smoothing technique while the traditional SPHARM is more of an interpolation or reconstruction technique.

12.4.1 Overshoot in Gibbs Phenomenon

To numerically quantify the amount of overshoot, we define the *overshoot* as the maximum of L_2 norm of the residual difference between the original and the reconstructed surface as

$$\sup_{(\theta,\varphi)\in S^2} \left|\left| p(\theta,\varphi) - \sum_{l=0}^{k} \sum_{m=-l}^{l} e^{-l(l+1)\sigma} f_{lm} Y_{lm}(\theta,\varphi) \right|\right|.$$

If surface coordinates are abruptly changing or their derivatives are discontinuous, the Gibbs phenomenon will severely distort the surface shape and the overshoot will never converge to zero. We have reconstructed a cube with various degree presentations and the bandwidth showing more ringing artifacts and overshoot in the traditional representation compared to the proposed weighted version (Figure 12.5). The exponentially decaying weights make the representations converge faster and reduce the Gibbs phenomenon significantly. The plots in Figure 12.5 display the amount of overshoot for the traditional representation (black) and the weighted version (red). The weighted spherical harmonic representation shows smaller overshoot compared to the traditional representation.

12.4.2 Simulation Study

We have performed two simulation studies to determine if the proposed pipeline can detect a small artificial bump. A similar bump test was done in [405] for testing the effectiveness of a spherical wavelet representation. In the first simulation, we have generated the binary mask of a sphere with radius 10mm. Then we obtained the weighted spherical harmonic representation (11.8) of the sphere with $\sigma = 0.001$ and degree $k = 42$. Taking the estimated coefficients f_{lm} as the ground truth, we simulated 20 spheres (group A) by putting noise $N(f_{lm}, (f_{lm}/20)^2)$ in the spherical harmonic coefficients. The standard deviation is taken as the 20th of the estimated coefficient. We have also given a bump of height 1.5mm to the sphere and simulated 20 bumped sphere (Figure 12.4 -(a)). Two groups of surfaces are fed into the multivariate linear model testing for the group effect. The T-statistic map is projected on the average of 40 simulated surfaces (Figure 12.4-(b)). Since the bump is so small with respect to the noise level, we did not detect any the bump ($p = 0.35$).

In the second simulation, we increased the height of the bump to 3mm (Figure 12.4-(c)) and repeated the first simulation. The resulting T-statistic map is projected on the average of 40 simulated surfaces (Figure 12.4-(d)). Unlike the first simulation study, we have detected the bump in yellow and red regions ($p < 0.0003$). These experiments demonstrate that the proposed framework works for detecting sufficiently large shape differences, and further demonstrate that what we detected in the real data is of sufficiently large shape difference. Otherwise, we simply wouldn't detect the signal in the first place.

12.5 Surface Normalization

MRIs were first reoriented manually to the pathological plane for the manual binary segmentation of amygdala [89]. The images then further underwent a 6-parameter rigid-body alignment with manual landmarking for the subsequent anterior commissure (AC) and posterior commissure (PC) alignment [257]. The aligned left amygdala are displayed in Figure 12.8 showing an approximate initial alignment. The proposed weighted spherical harmonic representations were then obtained. The additional alignment beyond the rigid-body alignment was done by matching the weighted spherical harmonic representations. Note we are not trying to match the original noisy surfaces but rather their smooth analytic representations. Once object surfaces are represented with weighted spherical harmonics, we need to establish surface correspondence across different surfaces for the subsequent statistical analysis. The correspondence is established by matching the coefficient of spherical harmonics at the same degree and order. This guarantees the sum of squares errors to be minimum in the following sense. Consider two surface coordinates p and q given by the representations

$$p(\Omega) = \sum_{l=0}^{k} \sum_{m=-l}^{l} e^{-l(l+1)\sigma} f_{lm} Y_{lm}(\Omega)$$

and

$$q(\Omega) = \sum_{l=0}^{k} \sum_{m=-l}^{l} e^{-l(l+1)\sigma} g_{lm} Y_{lm}(\Omega),$$

where f_{lm} and g_{lm} are Fourier vectors. Suppose the surface p is deformed to $p + d$ under the influence of the displacement vector field d. We wish to find $d = (d_1, d_2, d_3)$ that minimizes the discrepancy between $p+d$ and q in the finite subspace $\mathcal{H}_k$, which is spanned by up to degree k spherical harmonics. The restriction of the search space to the finite subspace simplifies the computation as follows:

$$\sum_{l=0}^{k} \sum_{m=-l}^{l} e^{-l(l+1)\sigma} (g_{lm} - f_{lm}) Y_{lm}(\Omega) = \arg \min_{d_1, d_2, d_3 \in \mathcal{H}_k} \|p + d - q\|^2. \quad (12.4)$$

The proof is given in [71]. The optimal displacement in the least squares sense is obtained by simply taking the difference between two weighted spherical harmonic representations and matching coefficients of the same degree and order. (12.4) can be used to establish the correspondence across different meshes with different mesh topology, i.e. mesh connectivity. For instance, the first surface in Figure 12.8-(a) has 1270 vertices and 2536 faces while the second surface has 1302 vertices and 2600 faces. We establish correspondence between topologically different meshes by matching a specific point $p(\Omega_0)$ in one surface to $q(\Omega_0)$ in the other surface and it is optimal in the least squares fashion. Since the representation is continuously defined in any $\Omega \in [0, \pi] \otimes [0, 2\pi)$, it is possible to resample surface meshes using a topologically different spherical mesh. We have uniformly sampled the unit sphere and constructed a spherical mesh with 2563 vertices and 5120 faces. This spherical mesh serves as a common mesh topology for all surfaces. After the resampling, all surfaces will have the identical mesh topology as the spherical mesh, and the identical vertex indices will correspond across different surfaces (Figure 12.8-(c)). This is also illustrated in Figure 12.8-(d), where the pattern of basis Y_{22} corresponds across different amygdala. A similar idea of uniform mesh topology has been previously used for establishing MNI cortical correspondence [83, 76, 237, 226, 355, 396].

Denote the surface coordinates corresponding to the i-th surface as p^i. Then we have the representation

$$p^i(\Omega) = \sum_{l=0}^{k} \sum_{m=-l}^{l} e^{-l(l+1)\sigma} f_{lm}^i Y_{lm}(\Omega). \quad (12.5)$$

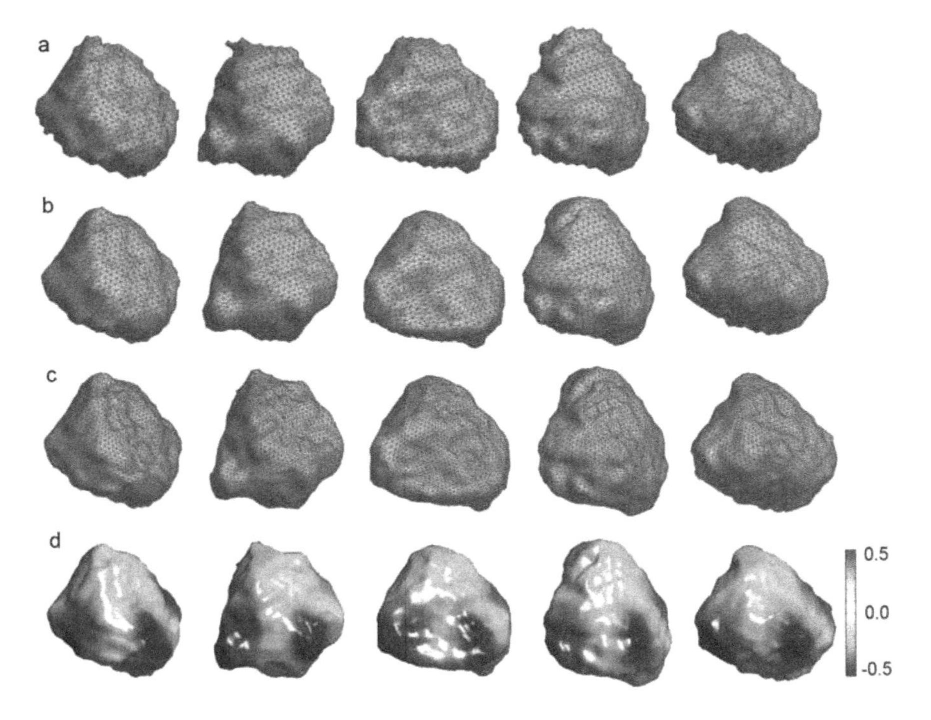

FIGURE 12.8
(a) Five representative left amygdala surfaces. (b) 42 degree weighted spherical harmonic representation. Surfaces have different mesh topology. (c) However, meshes can be resampled in such a way that all meshes have identical topology with exactly 2562 vertices and 5120 faces. Identically indexed mesh vertices correspond across different surfaces in the least squares fashion. (d) Spherical harmonic basis Y_{22} is projected on each amygdala to show surface correspondence.

There are total $(k+1)^2 \times 3$ coefficients to be estimated. Assume there are total n surfaces, the average surface $\overline{p}$ is given as

$$\overline{p} = \frac{1}{n} \sum_{i=1}^{n} \sum_{l=0}^{k} \sum_{m=-l}^{l} e^{-l(l+1)\sigma} f_{lm}^{i} Y_{lm}. \tag{12.6}$$

In our study, the average left and right amygdala templates are constructed by averaging the spherical harmonic coefficients of all 24 control subjects. The template surfaces serve as the reference coordinates for projecting the subsequent statistical parametric maps (Figure 12.11 and 12.12).

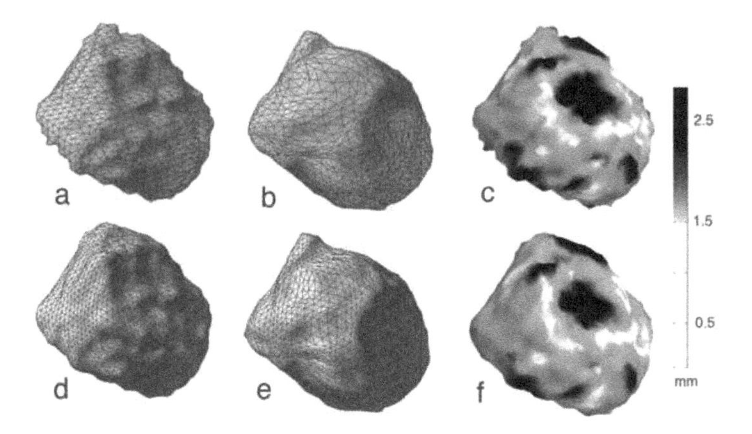

FIGURE 12.9
(a) (b) Simulated surfaces with the known displacement field between them. (c) The displacement in mm. (d) (e) Corresponding weighted spherical harmonic representation (f) The estimated displacement from the weighted spherical harmonic representations.

12.5.1 Validation

The methodology is validated in simulated surfaces where the ground truth is exactly known. In order not to bias the result, we have used an intrinsic geometric method using the Laplace-Beltrami eigenfunctions as a way to simulate surfaces with the known ground truth [229]. For the surface coordinates p, we have the Laplace-Beltrami operator Δ and its eigenfunctions ψ_j satisfying

$$\psi_j = \lambda_j \Delta \psi_j$$

where

$$0 = \lambda_0 < \lambda_1 \leq \lambda_2 \leq \cdots .$$

Then each surface can be represented as a linear combination of the Laplace-Beltrami eigenfunctions:

$$p = \sum_{j=0}^{\infty} f_j \psi_j,$$

where $f_j = \langle p, \psi_j \rangle$. Note that low degree coefficients represent global shape features and high degree coefficients represent high frequency local shape features. So by changing the high degree coefficients a bit, we can simulate new surfaces with similar global features but with the exact surface correspondence.

For the first simulated surface, we simply used the left amygdala surface of a randomly selected subject with 1000 basis ψ_j (Figure 12.9-(a)). Now if we reuse the first five coefficients f_j while changing the remaining coefficients to g_j, we can obtain the second simulated surface given by

$$q = \sum_{j=0}^{4} f_j \psi_j + \sum_{j=5}^{999} g_j \psi_j.$$

This is shown in Figure 12.9-(b) where the global shape is similar to (a) but local shape features differ substantially. The high degree coefficients g_j were obtained from the remaining 45 amygdala surfaces to generate 45 simulated surfaces. This process generates one fixed surface which serves as a template and 45 matched surfaces with the known displacement fields. The simulated surface went through the proposed processing pipeline and the weighted spherical harmonic representations were computed. The displacement between the representations is given by the minimum distance (12.4). Figure 12.9-(f) shows the estimated displacement which exhibits a smoother pattern than the ground truth. This is expected since the ground truth is the distance between noisy surfaces while the estimated displacement is the distance between smooth functional representations. However, the pattern of estimation does follow the pattern of the ground truth sufficiently well. In fact the mean relative error over each surface is 0.116 ± 0.011.

12.6 Image and Data Acquisition

High resolution T1-weighted magnetic resonance images (MRI) were acquired with a GE SIGNA 3-Tesla scanner with a quadrature head coil with 240 $\times$ 240 mm field of view and 124 axial sections. Details on image acquisition parameters are given in [98] and [257]. T2-weighted images were used to smooth out inhomogeneities in the inversion recovery-prepared images using FSL (`www.fmrib.ox.ac.uk/fsl`). Total 22 high functioning autistic and 24 normal control MRI were acquired. Subjects were all males aged between 8 and 25 years. The Autism Diagnostic Interview-Revised [231] was used for diagnoses by trained researchers K.M. Dalton and B.M. Nacewicz [98].

MRIs were first reoriented to the pathological plane for optimal comparison with anatomical atlases [89]. Image contrast was matched by alignment of white and gray matter peaks on intensity histograms. Manual segmentation was done by a trained expert B.M. Nacewicz who has been blind to the diagnoses [257]. The manual segmentation also involves refinement through plane-by-plane comparison with ex vivo atlas sections [238]. The reliability of the manual segmentation protocol was validated by two raters on 10 amygdala volumes resulting in interclass correlation of 0.95 and the spatial reliability (intersection over union) average of 0.84. Figure 12.10 shows the manual segmentation of an amygdala in three different cross sections. The amygdala

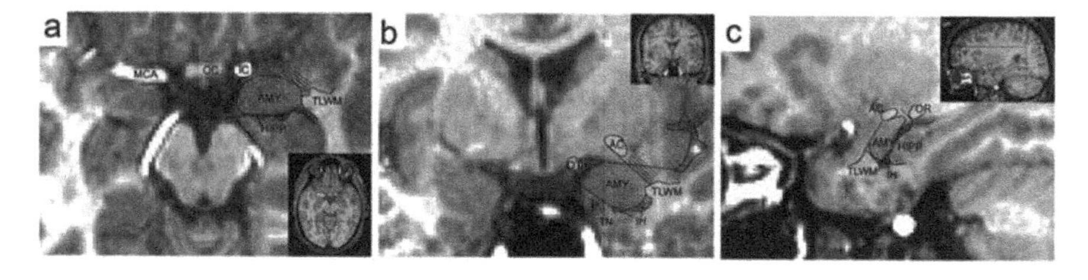

FIGURE 12.10
Amygdala manual segmentation at (a) axial (b) coronal and (c) midsagittal sections. The amygdala (AMY) was segmented using adjacent structures such as anterior commissure (AC), hippocampus (HIPP), inferior horn of lateral ventricle (IH), optic radiations (OR), optic tract (OT), temporal lobe white matter (TLWM) and tentorial notch (TN). The figure was generated by Brendon Nacewicz of University of Wisconsin-Madison.

(AMY) was traced in detail using various adjacent structures such as anterior commissure (AC), hippocampus (HIPP), inferior horn of lateral ventricle (IH), optic radiations (OR), optic tract (OT), temporal lobe white matter (TLWM) and tentorial notch (TN).

The total brain volume was also computed using an automated threshold-based connected voxel search method, and manually edited afterwards to ensure proper removal of CSF, skull, eye regions, brainstem and cerebellum using in-house software Spamalize [263, 308, 257]. The brain volumes are 1224 ± 128 and 1230 ± 161 cm^3 for autistic and control subjects. The volume difference is not significant ($p = 0.89$).

A subset of subjects (10 controls and 12 autistic) went through a face emotion recognition task consisting of showing 40 standardized pictures of posed facial expressions (8 each of happy, angry and sad, and 16 neutral) [98]. Subjects were required to press a button distinguishing neutral from emotional faces. The faces were black and white pictures taken from the Karolinska Directed Emotional Faces set [236]. The faces were presented using E-Prime software (`www.pstnet.com`) allowing for the measurement of response time for each trial. iView system with a remote eye-tracking device (SensoMotoric Instruments, `www.smivision.com`) was used at the same time to measure gaze fixation duration on eyes and faces during the task. The system records eye movements as the gaze position of the pupil over a certain length of time along with the amount of time spent on any given fixation point. It has been hypothesized that subjects with autism should exhibit diminished eye fixation duration relative to face fixation duration. If there is no confusion, we will simply refer to *gaze fixation* as the ratio of durations fixed on eyes over faces. Note that this is a unitless measure. Our study enables us to show that abnormal gaze fixation duration is correlated with amygdala shape in spatially localized regions.

12.7 Results

12.7.1 Amygdala Volumetry

We have counted the number of voxels in amygdala segmentation and computed the left and right amygdala volumes. The volumes for control subjects ($n = 22$) are left $1892 \pm 173\text{mm}^3$, right $1883 \pm 171\text{mm}^3$. The volumes for autistic subjects ($n = 24$) are left $1858 \pm 182\text{mm}^3$, right $1862 \pm 181\text{mm}^3$. The volume difference between the groups is not statistically significant based on the two-sample t-test ($p = 0.52$ for left and 0.69 for right). Previous amygdala volumetry studies in autism have been inconsistent [20, 169, 257, 280, 316, 335]. [20] and [280] reported significantly smaller amygdala volume in the autistic subjects while [177] and [335] reported larger volume. [169] and [257] found no volume difference. These inconsistency might be due to the lack of control for brain size and age in statistical analyses [316].

12.7.2 Local Shape Difference

From the amygdala volumetry result, it is still not clear if shape difference might be still present within amygdala. It is possible to have no volume difference while having significant shape difference. So we have performed multivariate linear modeling on the weighted spherical harmonic representation. We have tested the effect of `group` variable in the model

$$\texttt{P} = \texttt{1} + \texttt{Group},$$

which resulted in the threshold of 26.99 at $\alpha = 0.1$. On the other hand the maximum F statistic value is 13.55 (Figure 12.11 (a)). So we could not detect any shape difference in the left amygdala. For the right amygdala, the threshold is 26.64 which is far larger than the maximum F statistic value of 12.11. So again there is no statistically significant shape difference in the right amygdala.

We have also tested the effect of `Group` variable while accounting for age and the total brain volume in the `SurfStat` model form

$$\texttt{P} = \texttt{Age} + \texttt{Brain} + \texttt{Group}. \tag{12.7}$$

The maximum F statistics are 14.77 (left) and 12.91 (right) while the threshold corresponding to the $\alpha = 0.1$ is 14.58 (left) and 14.61 (right). Hence, we still did not detect group difference in the right amygdala (Figure 12.11-(d)) while there seems to be a bit weak group difference in the left amygdala (Figure 12.11-(c)). However, they did not pass the $\alpha = 0.01$ test so our result is inconclusive. The enlarged area in Figure 12.11 shows the average surface coordinate difference (autism - control) in the region of the maximum F value.

Head circumference and brain enlargement are linked to autism [104, 349]

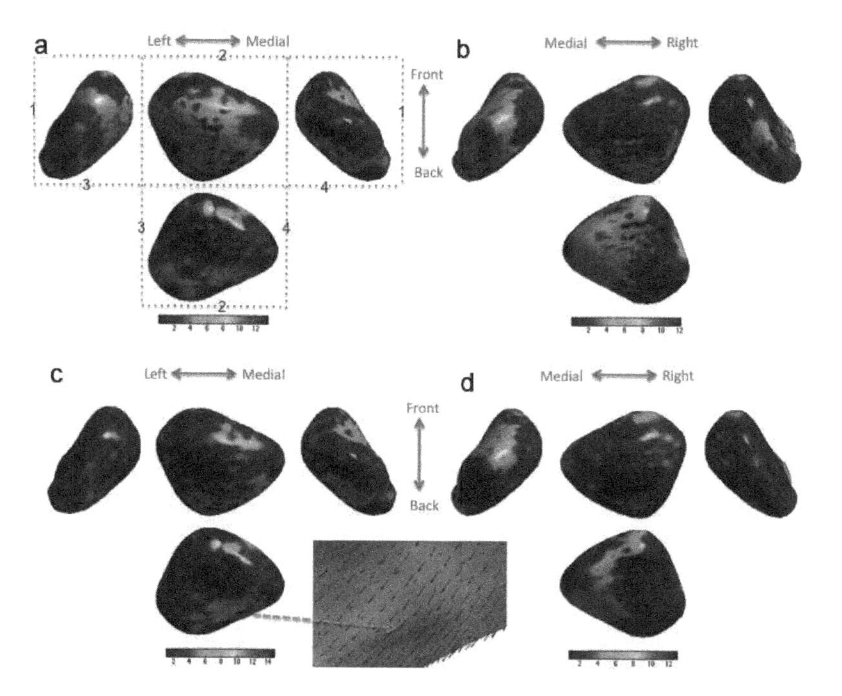

FIGURE 12.11
F statistic map of coordinate difference with and without accounting for age and the total brain volume for left (a,c) and right (b, d) amygdala. (See color insert.)

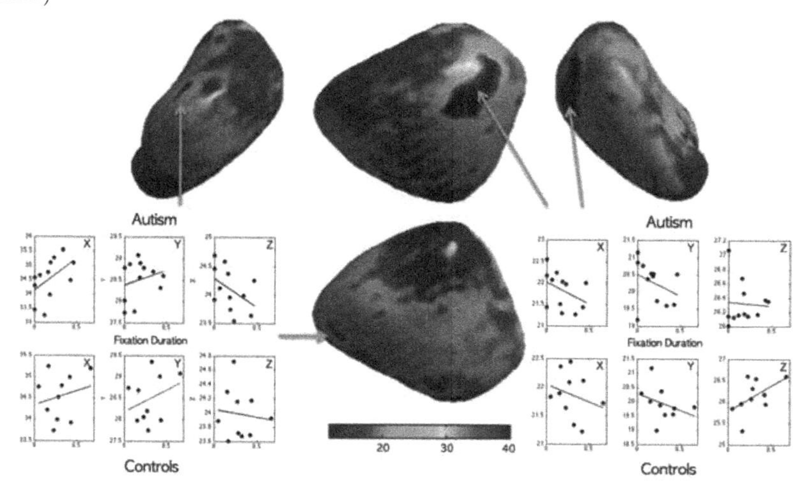

FIGURE 12.12
F-statistic of interaction between group and gaze fixation for right amygdala. (See color insert.)

and thus the covariate `Brain` in the model (12.7) may introduce a scaling related effect that was originally not present in the data. However, we did not find significant brain volume difference between the groups ($p = 0.89$). The brain size difference does not significantly compound our result. From Figure 12.12, we can see that the results between with and without covariating `Brain` are not much different (they are all statistically insignificant). Therefore, `Brain` in the model mostly accounts for subject-specific brain size difference rather than the group-specific brain size difference.

12.7.3 Brain and Behavior Association

Among total 46 subjects, 10 control and 12 autistic subjects went through the face emotion recognition task and gaze fixation (`Fixation`) was observed. The gaze fixations are 0.30 ± 0.17 (control) and 0.18 ± 0.16 (autism). Note that these are unitless measures. [257] showed the gaze fixation duration correlates differently with amygdala volume between the two groups; however, it was not clear if the association difference is local or diffuse over all regions of the amygdala. So we have tested the significance of the interaction between `Group` and `Fixation` using multivariate linear models. The reduced model is

$$\texttt{P} = \texttt{Age} + \texttt{Brain} + \texttt{Group} + \texttt{Fixation}$$

while the full model is

$$\texttt{P} = \texttt{Age} + \texttt{Brain} + \texttt{Group} + \texttt{Fixation} + \texttt{Group} * \texttt{Fixation} \tag{12.8}$$

and we tested for the significance of the interaction `Group*Fixation`.

We have obtained regions of significant interaction in the both left ($p < 0.05$) and right ($p < 0.02$) lateral nuclei in amygdala (Figure 12.12). The largest cluster in the right amygdala shows highly significant interaction ($\max F = 65.68, p = 0.003$). The color bar in Figure 12.12-(b) has been thresholded at 40 for better visualization. The scatter plots of the z-coordinate of the displacement vector field vs. `Fixation` are shown at the two most significant clusters in each amygdala. The red lines are linear regression lines. The significance of interaction implies difference in regression slopes between groups in a multivariate fashion. Note that there are three different slopes corresponding to x, y and z coordinates but due to the space limitation, we did not show other coordinates.

The total number of unknown parameters in our most complicated model (12.8) is $6 \times 3 = 18$ including the constant terms. This is a large number of parameters to estimate if (12.8) was a univariate linear model. However, in our multivariate setting, it is a reasonable number of parameters since we are also tripling the number of measurements as well. Note that Roy's maximum root statistic is based on maximizing an F-statistic with 1 and $n - 1 - 5$ degrees of freedom. Since the number of subjects is $n = 22 + 24$, we have sufficient degrees of freedom not to worry about the over-fitting problem. Unfortunately, practical power approximation for Roy's maximum root statistic does not exist although that of Lawley-Hotelling trace is available [28, 264] so the discussion of the parameter over-fitting is still an open statistical problem.

12.8 Discussion

This section discussed a unified multivariate linear modeling approach for a collection of binary neuroanatomical objects. The unified framework is applied to amygdala shape analysis in autism. The surfaces of the binary objects are flattened using a new technique based on heat diffusion. The coordinates of amygdala surfaces are smoothed and normalized using the weighted spherical harmonic representation. The multivariate linear models accounting for nuisance covariates are used using a newly developed `SurfStat` package.

Since surface data is inherently multivariate, traditionally Hotelling's T-square approach has been used on surface coordinates in a group comparison that can not account for nuisance covariates. On the other hand, the proposed multivariate linear model generalizes the Hotelling's T-square approach so that we can construct more complicated statistical models while accounting for additional covariates. The model formula based multivariate linear modeling tool `SurfStat` has been developed for this purpose and publicly available. We have applied the proposed methods to 22 autistic subjects to test if there is localized shape difference within an amygdala. We were able to localize regions, mainly in the right amygdala, that show differential association of gaze fixation with anatomy between the groups.

12.8.1 Anatomical Findings

Many MRI-based volumetric studies have shown inconsistent results in determining if there are any abnormal amygdala volume differences [20, 177, 169, 280, 316, 335, 257]. These studies focus on the total volume difference of amygdala obtained from MRI and was unable to determine if the volume difference is locally focused within the subregions of amygdala or diffuse over all regions.

Although we did not detect statistically significant shape difference within amygdala at 0.01 level, we detected significant group difference of shape in relation to the gaze fixation duration mostly in both the lateral nuclei (largest clusters in Figure 12.12). The lateral nucleus receives information from the thalamus and cortex, and relays it to other subregions within the amygdala. Our finding is consistent with literature that reports that autistic subjects fail to activate the amygdala normally when processing emotional facial and eye expressions [24, 95, 23]. There are two anatomical studies that additionally support our findings. A post-mortem study shows there are increased neuron-packing density of the medial, cortical and central nuclei, and medial and basal lateral nuclei of the amygdala in five autopsy cases [93]. Further, reduced fractional anisotropy is found in the temporal lobes approaching the amygdala bilaterally in a diffusion tensor imaging study [23].

The inconsistent amygdala volumetry results seem to be caused by the local volume and shape difference of the lateral nuclei that may or may not contribute to the total volume of amygdala. Further diffusion tensor imaging studies on the white matter fiber tracts connecting the lateral nuclei would shed a light on the abnormal nature of lateral nucleus of the amygdala and its structural connection to other parts of the brain.

12.9 Numerical Implementation

We illustrate `SurfStat` package by showing the step-by-step command lines for multivariate linear models used in the study. The detailed description of the `SurfStat` package can be found in `www.math.mcgill.ca/keith/surfstat`. The SurfStat is a general purpose surface analysis package and it requires additional codes for amygdala specific analysis. The extension to amygdala shape modeling can be found in `www.stat.wisc.edu/~mchung/research/amygdala`.

Given an amygdala mesh `surf`, which is, for instance, given as a structured array of the form

```
surf =
    vertices: [1270x3 double]
       faces: [2536x3 double]
```

the amygdala flattening algorithm will generate the corresponding unit sphere mesh `sphere` that has identical topology as `surf`. The weighted spherical harmonic representation `P` with degree $k = 42$ and the bandwidth $\sigma = 0.001$ is computed from

```
>[P,coeff]=SPHARMsmooth(surf,sphere,42,0.001);
```

The coordinates of the weighted spherical harmonic representation have been read into an array of size 46 (subjects) $\times$ 2562 (vertices) $\times$ 3 (coordinates) `P`. Brain size (`brain`), age (`age`), group variable (`group`) are read into 46 (subjects) $\times$ 1 vectors. The group categorical variable consists of strings `'control'` and `'autism'`. We now convert these to terms that can be combined into a multivariate linear model as follows:

```
>Brain = term( brain );
>Age = term( age );
>Group = term ( group );
```

```
>Group
autism  control
---------------
   0       1
   0       1
   1       0
   1       0
   .        .
   .        .
   .        .
```

To test the effect of group, the linear model of the from `P = 1 + Group` is fitted by

```
>E = SurfStatLinMod( P,1 + Group, Avg );
```

where `Avg` is the average surface obtained from the weighted spherical harmonic representation.

We specify a group contrast and calculate the T-statistic:

```
>contrast = Group.autism - Group.control
contrast =
    -1
    -1
     1
     1
     .
     .
     .
LM = SurfStatT( E, contrast );
```

`LM.t` gives the vector of 2562 T-statistic values for all mesh vertices. Instead of using the contrast and T-statistic, we can test the effect of group variable using the F-statistic as well:

```
>E0 = SurfStatLinMod( P,1 );
>LM = SurfStatF( E,E0 );
```

`E0` contains the information about the sum of squared residual of the reduced model `P = 1` in `E0.SSE` while `E` contains that of the full model `P = 1 + Group`. Based on the ratio of the sum of squared residuals, `SurfStatF` computes the F-statistics. To display the F-statistic value on top of the average surface, we use `FigureOrigami( Avg, LM.t )` which produces Figure 12.11.

We can determine the random field based thresholding corresponding to $\alpha = 0.01$ level:

```
>resels = SurfStatResels(LM);
>stat_threshold( resels, length(LM.t),1,LM.df,0.01,[],[],[],LM.k)
```

```
peak_threshold =
   26.9918
```

`resels` computes the resels of the random field and `peak_threshold` is the threshold corresponding to 0.01 level.

We can construct a more complicated model that includes the brain size and age as covariates:

```
>E0 = SurfStatLinMod(P,Age+Brain);
>E = SurfStatLinMod(P,Age+Brain+Group,Avg);
>LM = SurfStatF(E,E0);
```

`LM.t` contains the F-statistic of the significance of group variable while accounting for age and brain size. We can also test for interaction between gaze fixation `Fixation` and group variable:

```
>E0=SurfStatLinMod(P,Age+Brai +Group+Fixation);
>E=SurfStatLinMod(P,Age+Brain+Group+Fixation+Group*Fixation,Avg);
>LM=SurfStatF(E,E0);
```

13

Laplace-Beltrami Eigenfunctions for Surface Data

We present a novel surface data smoothing framework using the Laplace-Beltrami eigenfunctions. The Green's function of an isotropic diffusion equation on a manifold is analytically represented using the eigenfunctions of the Laplace-Beltrami operator. The Green's function is then used in explicitly constructing heat kernel smoothing as a series expansion of the eigenfunctions. Unlike many previous approaches involving surface diffusion, our approach represents the solution of diffusion analytically reducing numerical inaccuracy. The numerical implementation is validated against the spherical harmonic representation of heat kernel smoothing on a unit sphere, and compared against widely used iterative kernel smoothing. As an illustration, we have applied the method in localizing the regions of mandible growth between age 0 and 20.

13.1 Introduction

In medical image analysis, anatomical surfaces obtained from MRI and CT are often represented as triangular meshes. Image segmentation and surface extraction process themselves are likely to introduce noise to the mesh coordinates. It is imperative to reduce the mesh noise while preserving the geometric details of the anatomical structures for various applications.

Diffusion equations have been widely used in image processing as a form of noise reduction starting with Perona and Malik in 1990 [275]. Although numerous techniques have been developed for surface fairing and mesh regularization [333, 239, 351, 353]; only a few have tried to smooth out surface measurements for the purpose of statistical analysis [10, 79, 53, 54, 77, 191]. Particularly in brain imaging, isotropic heat diffusion on surfaces has been introduced for subsequent statistical analysis involving the random field theory that assumes an isotropic covariance function as a noise model [10, 79, 53, 54]. Since then, isotropic diffusion has been mainly used as a standard smoothing technique. Such diffusion approaches mainly use finite element or finite difference schemes which are known to suffer numerical instability if the forward Euler scheme is used.

Iterated kernel smoothing is also a widely used method in approximately solving diffusion equations on surfaces [77, 77, 162]. Iterated kernel smoothing is often used in smoothing various cortical surface data: cortical curvatures [235, 130], cortical thickness [234, 38], hippocampus [324, 411], magnetoencephalography (MEG) [161] and functional-MRI [157, 186]. In iterated kernel smoothing, kernel weights are spatially adapted to follow the shape of the heat kernel in a discrete fashion along a manifold. In the tangent space of the manifold, the heat kernel can be approximated linearly using the Gaussian kernel for small bandwidth. A kernel with large bandwidth is then constructed iteratively applying the kernel with small bandwidth. However, this process compounds the linearization error at each iteration as we demonstrate in this book.

In this paper, we propose a new smoothing framework that constructs the heat kernel analytically using the eigenfunctions of the Laplace-Beltrami operator, avoiding the need for the linear approximation [77, 77, 162]. Although solving for the eigenfunctions of the Laplace-Beltrami operator requires the finite element method, the proposed method is analytic in a sense that heat kernel smoothing is formulated as a series expansion explicitly. The proposed method represents isotropic heat diffusion analytically as a series expansion so it avoids the numerical instability associated with solving the diffusion equations numerically [10, 79, 191]. Our framework is radically different in that it bypasses the various numerical problems that are associated with previous approaches including numerical instability, slow convergence, and accumulated linearization error.

Although there are many studies on solving diffusion equations on arbitrary triangular meshes [10, 79, 191, 352], there is no study that solves and validates heat diffusion by explicitly constructing the heat kernel analytically as the eigenfunctions of the Laplace-Beltrami operator, and then use it as a form of data smoothing on surfaces. Although there recently have been a few studies that introduce heat kernel in computer vision, they mainly use heat kernel to compute shape descriptors [347, 51]; or to define a multi-scale metric [101]. In other words, these studies did not use heat kernel to smooth out data to increase signal-to-noise ratio.

13.2 Heat Kernel Smoothing

Consider a functional measurement Y observed on a closed compact manifold $\mathcal{M} \subset \mathbb{R}^3$. We assume the following additive model on Y:

$$Y(p) = \theta(p) + \epsilon(p), \tag{13.1}$$

where $\theta(p)$ is the unknown mean signal to be estimated and $\epsilon(p)$ is a zero-mean Gaussian random field. We may assume further $Y \in L^2(\mathcal{M})$, the space of square integrable functions on $\mathcal{M}$ with the inner product

$$\langle f, g \rangle = \int_{\mathcal{M}} f(p)g(p)\, d\mu(p), \tag{13.2}$$

where μ is the Lebesgue measure such that $\mu(\mathcal{M})$ is the total area of $\mathcal{M}$.

Various functional data such as electroencephalography (EEG), magnetoencephalography (MEG) [161] and functional-MRI [157, 186], and anatomical data such as cortical curvatures [235, 130], cortical thickness [234, 38] and surface coordinates [77] can be considered as possible functional measurements.

Let Δ be the Laplace-Beltrami operator Δ on $\mathcal{M}$. Solving the eigenvalue equation

$$\Delta\psi_j = -\lambda\psi_j, \tag{13.3}$$

we order eigenvalues

$$0 = \lambda_0 < \lambda_1 \leq \lambda_2 \leq \cdots,$$

and corresponding eigenfunctions $\psi_0, \psi_1, \psi_2, \cdots$ [77, 229, 304, 325]. Then, the eigenfunctions ψ_j form an orthonormal basis in $L^2(\mathcal{M})$. There is extensive literature on the use of eigenvalues and eigenfunctions of the Laplace-Beltrami operator in medical imaging and computer vision [229, 290, 298, 296, 408, 409]. The eigenvalues have been used in caudate shape discriminators [262]. Qiu et al. used eigenfunctions in constructing splines on cortical surfaces [290]. Reuter used the topological features of eigenfunctions [296]. Shi et al. used the Reeb graph of the second eigenfunction in shape characterization and landmark detection in cortical and subcortical structures [326, 325]. Lai et al. used the critical points of the second eigenfunction as anatomical landmarks for colon surfaces [215]. Since the direct application of eigenvalues and eigenfunctions as features of interest is the beyond the scope of this book, we will not pursue the issue in detail here.

Using the eigenfunctions, *heat kernel* $K_\sigma(p, q)$ is defined as

$$K_\sigma(p, q) = \sum_{j=0}^{\infty} e^{-\lambda_j \sigma}\psi_j(p)\psi_j(q), \tag{13.4}$$

where σ is the bandwidth of the kernel. Figure 13.1 shows examples of a heat kernel with different bandwidths. Then *heat kernel smoothing* of functional measurement Y is defined as

$$K_\sigma * Y(p) = \sum_{j=0}^{\infty} e^{-\lambda_j \sigma}\beta_j\psi_j(p), \tag{13.5}$$

where $\beta_j = \langle Y, \psi_j \rangle$ are Fourier coefficients [77]. Kernel smoothing $K_\sigma * Y$ is taken as the estimate for the unknown mean signal θ.

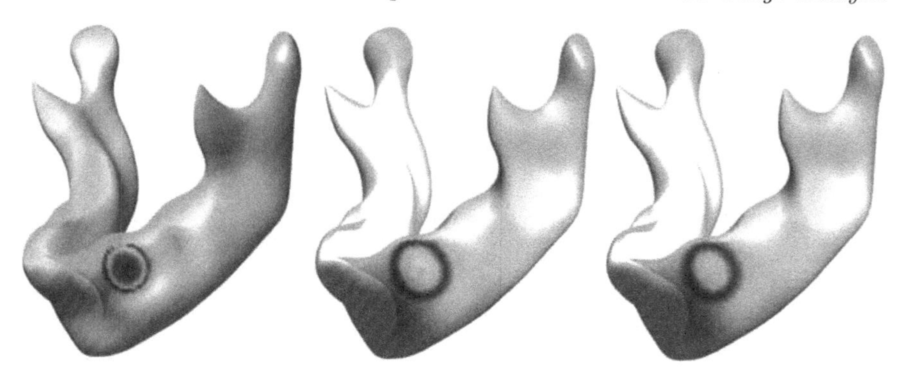

FIGURE 13.1
Heat kernel shape with bandwidths 0.025, 1.25 and 5 on a mandible surface. The level sets of the heat kernel form geodesic circles. The kernel is constructed by Seongho Seo of Seoul National University. All the heat kernels are constructed on the surface smoothed by the proposed heat kernel smoothing with $\sigma = 0.025$ and $k = 132$ for better visualization.

The heat kernel smoothing $K_\sigma * Y$ is the unique solution of the isotropic diffusion equation [77, 304]:

$$\frac{\partial f}{\partial \sigma} = \Delta f, \ f(p, \sigma = 0) = Y(p), \tag{13.6}$$

where the bandwidth σ is interpreted as diffusion time. This can be easily seen as follows. A Green's function or a fundamental solution of the Cauchy problem (13.6) is given by the solution of the following equation

$$\frac{\partial f}{\partial \sigma} = \Delta f, \ f(p, \sigma = 0) = \delta(p), \tag{13.7}$$

where δ is the Dirac delta function. It can be shown that the heat kernel K_σ is a Green's function of (13.7) [118]. Since the operators are linear in (13.7), we can further convolve the terms with the initial data Y so that we have

$$\frac{\partial}{\partial \sigma}(K_\sigma * Y) = \Delta(K_\sigma * Y), \ K_\sigma * Y(p, \sigma = 0) = Y(p).$$

Hence $K_\sigma * Y$ is a solution of (13.6).

Unlike previous approaches to heat diffusion [10, 79, 191, 352], our proposed method avoids the direct numerical discretization of the diffusion equation. Instead we discretize the basis functions of the given manifold $\mathcal{M}$ by solving for the eigensystem (13.3) and obtain λ_j and ψ_j.

13.2.1 Heat Kernel Smoothing in 2D Images

As an illustration, we will show how to construct the scale-space representation of 2D images analytically via heat kernel smoothing. The Laplace-Beltrami

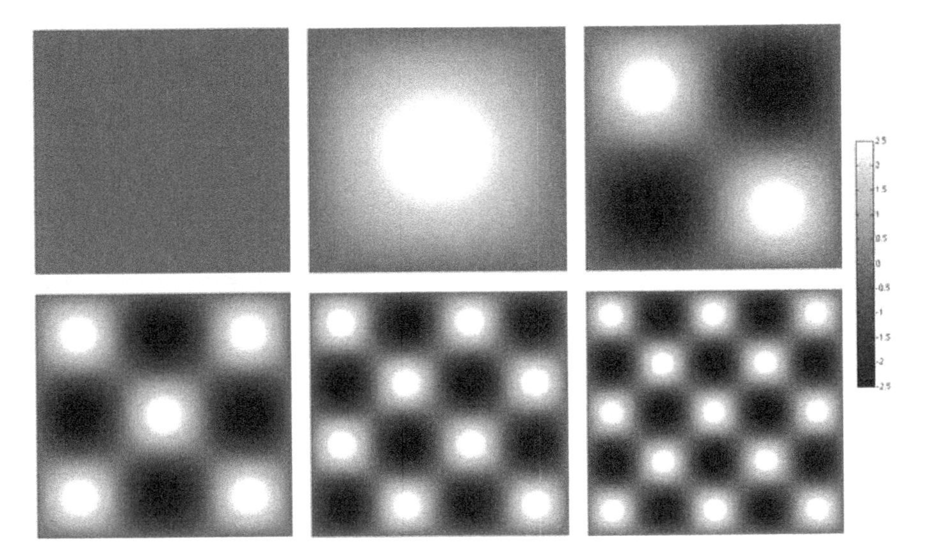

FIGURE 13.2
The Laplace-Beltrami eigenfunctions in 300×300 2D image domain.

eigenfunctions in the Euclidean space are simply given in terms of the product of sine and cosine functions. We first estimate the Fourier coefficients corresponding to the basis functions. Then we need to change the bandwidth of the heat kernel in the computation to have another scale-space. This is an extremely powerful framework.

In 2D rectangle of size 300×300, we first construct the eigenfunctions of degree l. The first six eigenfunctions in a rectangular domain are given in Figure 13.2. The first eigenfunction is a constant.

```
L=[300 300];
for j=0:5
    l=[j j];
    psi=eigenfunction_box(l,L);
    figure; imagesc(psi); colorbar;
end;
```

Let us perform the eigenfunction expansion of an image. The `deform.tif` image shows the part of T1 weighted MRI showing the deformation field (blue arrows) around the hippocampus (elongated greenish yellow structure) (Figure 13.4).

```
A = imread('deform.tif');
A=double(A);
imagesc(A);colorbar
```

If we perform the Fourier series expansion using the 10000 eigenfunctions, we obtain Figure 13.4. With 90000 basis expansion, the Fourier series expansion is almost like the original image.

```
expansion=zeros(300,300);
fourier=zeros(100,100);
L=[300 300]
degree=100;

for j=0:degree-1
    for k=0:degree-1
        l=[j k];
        psi=eigenfunction_box(l,L);
        coeff=sum(sum(psi.*A))/300^2;
        fourier(j+1,k+1) = coeff;
        expansion=expansion + coeff*psi;
    end;
end;
figure; imagesc(expansion); colorbar
```

Once we obtained the Fourier coefficient, heat kernel smoothing is done by simply changing the bandwidth (Figure 13.5) as follows.

```
sigma=10;
degree=100;
expansion=zeros(300,300);

for j=0:degree-1
    for k=0:degree-1
        l=[j k];
        psi=eigenfunction_box(l,L);
        lambda= (j*pi/L(1))^2 + (k*pi/L(2))^2;
        coeff = fourier(j+1,k+1);
        expansion=expansion + exp(-sigma*lambda)*coeff*psi;
    end;
end;
figure; imagesc(expansion); colorbar
```

13.3 Generalized Eigenvalue Problem

13.3.1 Finite Element Method

Since the closed form expression for the eigenfunctions of the Laplace-Beltrami operator on an arbitrary surface is unknown, the eigenfunctions are numer-

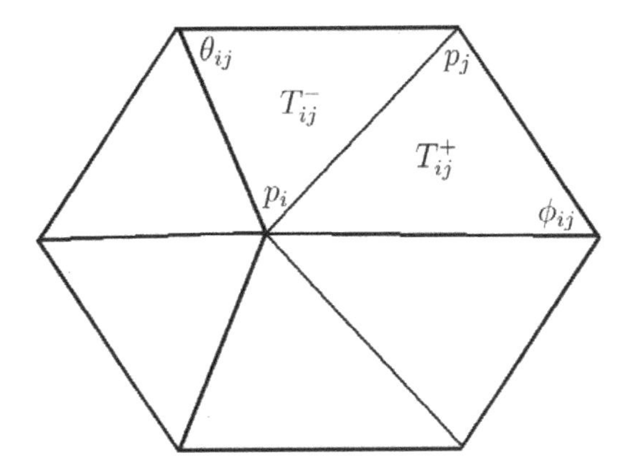

FIGURE 13.3
A typical 1-ring neighbor of a mesh vertex p_i. θ_{ij} and ϕ_{ij} are the angles opposite to the edge p_ip_j. T_{ij}^- and T_{ij}^+ are triangles sharing the edge p_ip_j.

ically computed by discretizing the Laplace-Beltrami operator. To solve the eigensystem (13.3), we need to discretize it on mandible triangular meshes using the cotan formulation [79, 325, 290, 309, 378, 370].

The cotan formulation has been often used in geometry processing in computer vision. Since the discretization of the Laplace-Beltrami operator in (13.3) results in the eigenvalue problem involving a nonsymmetric matrix, the eigenfunctions may not be real and orthonormal [229, 370, 408]. However, the finite element method (FEM) discretization avoids this problem by solving the generalized eigenvalue problem that is numerically more stable [79, 229, 290, 370, 297, 298].

Let N_T be the number of triangles in the mesh that approximate the underlying manifold $\mathcal{M}$. We seek a piecewise differentiable solution f_i in the i-th triangle T_i such that the solution $f_i(x)$ is continuous across neighboring triangles. The solution f for the whole mesh is then

$$f(x) = \sum_{i=1}^{N_T} f_i(x).$$

Let $p_{i_1}, p_{i_2}, p_{i_3}$ be the vertices of element T_i. In T_i, we estimate f_i linearly as

$$f_i(x) = \sum_{k=1}^{3} \xi_{i_k}(x) f(p_{i_k}),$$

where nonnegative ξ_{i_k} are given by the *barycentric coordinates* [79, 351]. Any point $x \in T_i$ is uniquely determined by two conditions:

$$x = \sum_{k=1}^{3} \xi_{i_k}(x) p_{i_k}, \quad \sum_{k=1}^{3} \xi_{i_k}(x) = 1.$$

Let g be an arbitrary piecewise linear function given by

$$g(x) = \sum_{i=1}^{N_T} \sum_{k=1}^{3} \xi_{i_k}(x) g_{i_k}, \tag{13.8}$$

where $g_{i_k} = g(p_{i_k})$ are the values of function g evaluated at vertices p_{i_k} of T_i. For the function f, we can represent similarly as $f_{i_k} = f(p_{i_k})$. Since the Laplace-Beltrami operator is *self-adjoint*, we have

$$\int g \Delta f \, d\mu = -\int \langle \nabla f, \nabla g \rangle \, d\mu = \int f \Delta g \, d\mu. \tag{13.9}$$

Then the integral version of the eigensystem $\Delta f = -\lambda f$ in the triangle T_i can be written as

$$\int_{T_i} g \lambda f \, d\mu = \int_{T_i} \langle \nabla f, \nabla g \rangle \, d\mu. \tag{13.10}$$

The left-hand term in (13.10) can be written further as

$$\int_{T_i} g \lambda f \, d\mu = \sum_{k,l=1}^{3} g_{i_k} \lambda f_{i_l} \int_{T_i} \xi_{i_k} \xi_{i_l} \, d\mu \tag{13.11}$$

$$= \lambda \mathbf{G}_i' \mathbf{A}^i \mathbf{F}_i, \tag{13.12}$$

where $\mathbf{G}_i = (g_{i_1}, g_{i_2}, g_{i_3})', \mathbf{F}_i = (f_{i_1}, f_{i_2}, f_{i_3})'$ and 3×3 mass matrix

$$\mathbf{A}^i = (A^i_{kl}), \; A^i_{kl} = \int_{T_i} \xi_{i_k} \xi_{i_l} \, d\mu.$$

It can be shown that

$$\mathbf{A}^i = \frac{|T_i|}{12} \begin{pmatrix} 2 & 1 & 1 \\ 1 & 2 & 1 \\ 1 & 1 & 2 \end{pmatrix},$$

where $|T_i|$ is the area of the triangle T_i [312, 313]. Similarly, the right-hand term in (13.10) is

$$\int_{T_i} \langle \nabla f, \nabla g \rangle \, d\mu = \sum_{k,l=1}^{3} g_{i_k} f_{i_l} \int_{T_i} \langle \nabla \xi_{i_k}, \nabla \xi_{i_l} \rangle \, d\mu \tag{13.13}$$

$$= \mathbf{G}_i' \mathbf{C}^i \mathbf{F}_i, \tag{13.14}$$

where 3×3 matrix $\mathbf{C}^i$ is given by

$$\mathbf{C}^i = (C^i_{kl}), \; C^i_{kl} = \int_{T_i} \langle \nabla \xi_{i_k}, \nabla \xi_{i_l} \rangle \, d\mu.$$

Since T_i is planar, the gradient $\nabla \xi_{i_k}$ is the standard planar gradient. The matrix $\mathbf{C}^i$ can be further written as [312, 313, 328]

$$\frac{1}{2}\begin{pmatrix} \cot\theta_{i_2} + \cot\theta_{i_3} & -\cot\theta_{i_3} & -\cot\theta_{i_2} \\ -\cot\theta_{i_3} & \cot\theta_{i_1} + \cot\theta_{i_3} & -\cot\theta_{i_1} \\ -\cot\theta_{i_2} & -\cot\theta_{i_1} & \cot\theta_{i_1} + \cot\theta_{i_2} \end{pmatrix},$$

where θ_{i_k} is the incident angle of vertex p_{i_k} in triangle T_i. By equating (13.12) and (13.14), we obtain

$$\mathbf{A}^i \lambda \mathbf{F}_i = \mathbf{C}^i \mathbf{F}_i. \tag{13.15}$$

We solve (13.15) by assembling all triangles. To simplify the indexing, we will use slightly different notations from now on. Let $N(p_i)$ be the set of neighboring vertices around p_i, and let T_{ij}^- and T_{ij}^+ denote two triangles sharing the vertex p_i and its neighboring vertex $p_j \in N(p_i)$. Then, both T_{ij}^- and T_{ij}^+ constitute incident triangles around p. Let two angles opposite to the edge containing p_i and p_j be ϕ_{ij} and θ_{ij} respectively for T_{ij}^+ and T_{ij}^- (Figure 13.3). Then, the assembled sparse matrices $\mathbf{A} = (A_{ij})$ are computed as follows. The diagonal entries are

$$A_{ii} = \frac{1}{12} \sum_{p_j \in N(p_i)} (T_{ij}^+ + T_{ij}^-),$$

and the off-diagonal entries are

$$A_{ij} = \frac{1}{12}(T_{ij}^+ + T_{ij}^-),$$

if p_i and p_j are adjacent, and $A_{ij} = 0$ otherwise. The global coefficient matrix $\mathbf{C} = (C_{ij})$, which is the assemblage of individual element coefficients is given similarly using the cotan formulation. The diagonal entries are

$$C_{ii} = \frac{1}{2} \sum_{p_j \in N(p_i)} (\cot\theta_{ij} + \cot\phi_{ij}),$$

and the off diagonal entries are

$$C_{ij} = -\frac{1}{2}(\cot\theta_{ij} + \cot\phi_{ij}),$$

if p_i and p_j are adjacent, and $C_{ij} = 0$ otherwise. When we construct $\mathbf{A}$ and $\mathbf{C}$ matrices, we compute the off-diagonal elements first and the diagonal elements next by summing the off-diagonal terms in the first ring neighbors. Finally, we can obtain the following generalized eigenvalue problem:

$$\mathbf{C}\psi = \lambda \mathbf{A}\psi. \tag{13.16}$$

Since $\mathbf{C}$ and $\mathbf{A}$ are large sparse matrices, we have solved (13.16) using the *Implicitly Restarted Arnoldi Method* [172, 224] without consuming a large amount of memory and time for sparse entries. The MATLAB code for computing $\mathbf{C}$ and $\mathbf{A}$ is given in brainimaging.waisman.wisc.edu/~chung/lb.

Figure 13.6 shows the first few eigenfunctions for a sample mandible surface. The first eigenfunction is trivially given as $\psi_0 = 1/\sqrt{\mu(\mathcal{M})}$ and $\lambda_0 = 0$ for a closed compact surface. It is possible to have multiple eigenfunctions corresponding to a single eigenvalue. The multiplicity of the eigenvalues of the Laplace-Beltrami operator is known although the exact number of multiplicity is unknown for arbitrary manifolds [174]. For smooth genus zero surfaces, the multiplicity m is bounded by

$$m(\lambda_k) \leq 2k - 3 \text{ for } k \geq 2.$$

Suppose $\psi_{k1}, \cdots \psi_{kk_m}$ are k_m eigenfunctions corresponding to eigenvalue λ_k. Then any linear combination of ψ_{kj} is also an eigenfunction. Hence, within the same degree, the space of eigenfunctions form a vector space. The eigenfunctions form a complete orthonormal basis in the space of square integrable functions, $L^2(\mathcal{M})$, so all other possible orthonormal bases are a linear combination of eigenfunctions.

13.3.2 Fourier Coefficients Estimation

Once we obtain the eigenfunctions numerically, we construct the subspace $\mathcal{H}_k$, which is spanned by up to k-th degree basis. Then we approximate the functional data Y in $\mathcal{H}_k$ by minimizing the sum of squared residual:

$$\arg\min_{f \in \mathcal{H}_k} \|f - K_\sigma * Y\|^2 = \sum_{j=0}^{k} e^{-\lambda_j \sigma} \beta_j \psi_j(p), \tag{13.17}$$

where

$$\beta_j = \langle Y, \psi_j \rangle, \tag{13.18}$$

are Fourier coefficients to be estimated.

The least squares method is used in estimating the Fourier coefficients [323, 344, 73]. By letting $\sigma = 0$ in (13.17), we can write (13.17) as the normal equation

$$\mathbf{Y} = \mathbf{\Psi} \boldsymbol{\beta}, \tag{13.19}$$

where $\mathbf{\Psi} = (\mathbf{\Psi}_0, \cdots, \mathbf{\Psi}_k)$ and $\boldsymbol{\beta} = (\beta_0, \cdots, \beta_k)'$. The coefficients $\boldsymbol{\beta}$ are then estimated least squarely as

$$\widehat{\boldsymbol{\beta}} = (\mathbf{\Psi}'\mathbf{\Psi})^{-1}\mathbf{\Psi}'\mathbf{Y}. \tag{13.20}$$

The advantage of the least squares method is that it does not require knowing the mass matrix $\mathbf{A}$. So the least squares method has been more often used in the case where the basis functions are already known. For instance, on a unit sphere, the eigenfunctions of the Laplace-Beltrami operator are explicitly given as spherical harmonics so there is no need to compute the mass matrix **A**. Thus, for the spherical harmonic representation, the least squares method has been the most widely used [323, 344, 73].

13.4 Numerical Implementation

The concept of heat kernel smoothing along an arbitrary manifold has been first introduced in [76, 77]. The Gaussian kernel weights observations according to their Euclidean distance. When the observations lie on a convoluted brain surface and arbitrary manifolds, it is more natural to assign the weight based on the geodesic distance along the manifolds. On the curved manifold, a straight line between two points is not the shortest distance so one may incorrectly assign less weights to closer observations. Therefore, smoothing data residing on manifolds requires constructing a kernel that is isotropic along the geodesic curves.

The original implementation given in [76, 77] was formulated as iterated kernel convolutions that approximate Gaussian kernel smoothing in the tangent space locally. The `MATLAB` code is given in `www.stat.wisc.edu/~mchung/softwares/hk/hk.html`. To remedy the confounding numerical error over each iteration, we introduced a new smoothing framework that uses the eigenfunctions of the Laplace Beltrami operator in this chapter. This implementation of heat kernel smoothing probably solves an isotropic heat diffusion on the manifolds accurately without the problem of divergence. Our new method performs surface data smoothing by constructing the series expansion of the eigenfunctions of the Laplace-Beltrami operator [319]. This new analytic framework improves upon the previous iterated kernel smoothing formulation with improved numerical accuracy and stability.

For demonstration, we will use the hippocampus surface data. The left and right hippocampus surfaces are saved as triangular mesh formats and displayed in Figure 13.7.

```
load hippocampus.mat;
```

It contains the left and right hippocampi of the format:

```
hippoleft =

    vertices: [2338x3 double]
       faces: [4672x3 double]


hipporight =
```

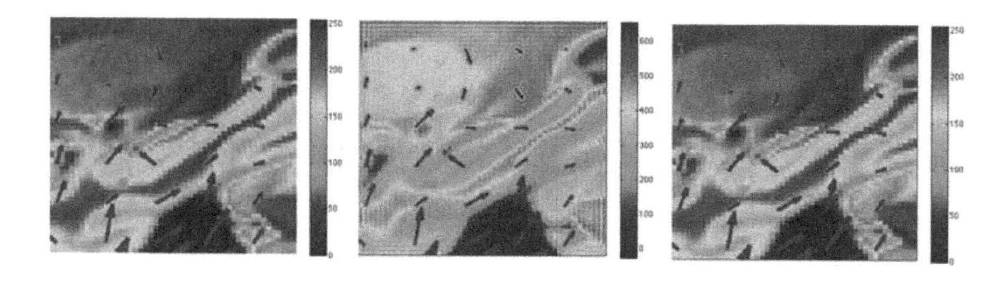

FIGURE 13.4
Left: original `deform.tif` image. Middle: 10000 basis expansion. Severe ringing artifacts are visible. Left: 90000 basis expansion. (See color insert.)

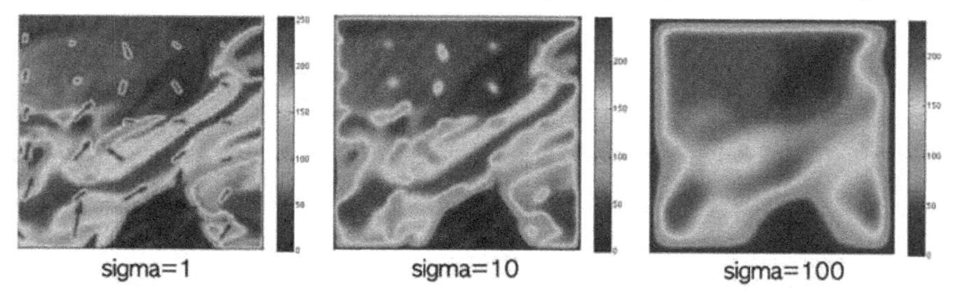

FIGURE 13.5
Heat kernel smoothing with 40000 basis and bandwidths 1, 10 and 100. Heat kernel smoothing does not exhibit ringing artifacts. (See color insert.)

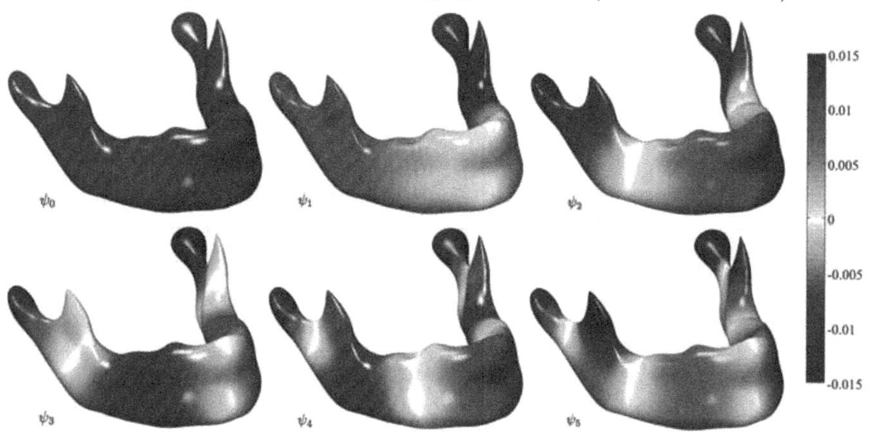

FIGURE 13.6
First six eigenfunctions on a mandible surface. The color scale is thresholded at ± 0.015 for better visualization. The figure was generated by Seongho Seo of Seoul National University. (See color insert.)

```
    vertices: [2312x3 double]
       faces: [4620x3 double]
```

```
figure;figure_patch(hippoleft,[0.7 0.7 0.6],0.5)
```

To compute the eigenfunctions of the Laplace-Beltrami operator on the hippocampus, we need to discretize the operator using the finite element method (FEM). The FEM discretization of the Laplace-Beltrami operator was originally given in [64]. Our cotan implementation follows notation and implementation from [64] and [290]. The MATLAB function FEM produces sparse matrices A and C from the left hippocampus surface.

```
[A, C] =FEM(hippoleft);
```

The generalized eigenvalue problem is then computed by using the sparse general eigenvalue problem routine eigs in MATLAB.

```
[V, D] = eigs(C,A,999,'sm');
```

It takes 104 seconds to compute 999 eigenfunctions in an old MacBook pro laptop while it takes 45 seconds in a 2.66Ghz Quad-Core Intel Xeon Mac computer. For larger surface meshes, you need huge memory. In MATLAB 7.9 with a Quad-Core Intel Xeon Mac computer with 32GB memory, we can construct 1000 eigenfunctions for a cortical mesh with 40962 vertices in 8 minutes. For multiple subjects, it is recommended to construct the eigenfunctions for the cortical template so the computation burden is not severe.

The computed eigenfunctions are then used in representing the heat kernel as a series expansion [319]. To smooth the left hippocampus surface with the bandwidth sigma 0.2 and 1000 eigenfunctions, we run

```
hippolefts = lb_smooth([ ],hippoleft, 0.2, 1000, V, D);
figure; figure_wire(hippolefts,'k','w')
view([140 10])
```

The first argument is where input signal should be but since the surface coordinates themselves are signal, we do not need to put input signal and leave it [ ]. Figure 13.7 shows heat kernel smoothing of the left hippocampus. We can also smooth signal on the hippocampus surface as well.

Suppose the surface signal is given by the y-coordinate. We add Gaussian white noise $N(0, 10)$ at each vertex. Heat kernel smoothing is then performed as

```
signal=hippoleft.vertices(:,2)
signal = signal + normrnd(0,10,2338,1);
figure_trimesh(hippolefts,signal,'rwb')

smoothed = lb_smooth(signal,hippoleft, 0.2, 500, V, D);
figure_trimesh(hippolefts,smoothed,'rwb');
```

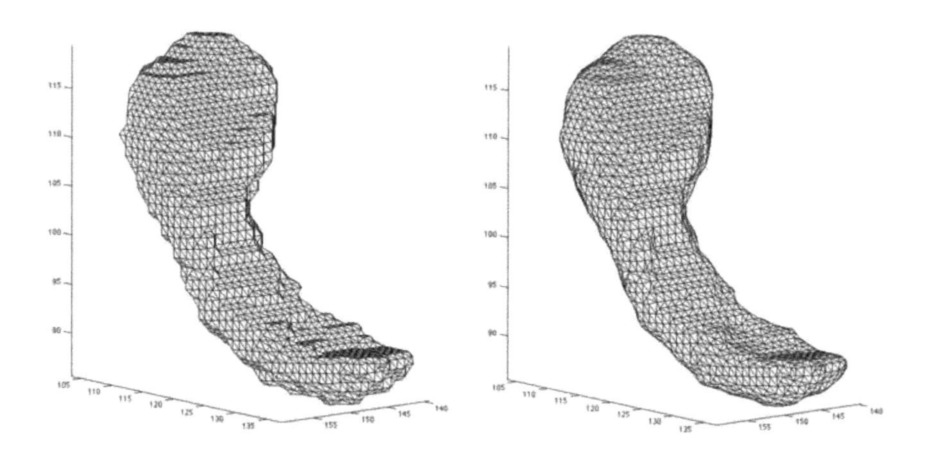

FIGURE 13.7
Heat kernel smoothing with 1000 Laplace-Beltrami eigenfunctions and bandwidth 0.2 is applied to the both surface coordinates and simulated measurement.

Before and after smoothing of the simulated surface signal is given in Figure 13.8.

13.5 Experimental Results

We applied the proposed smoothing method to mandible surfaces obtained from CT. The method is further validated against the spherical harmonics on a unit sphere.

13.5.1 Image Acquisition and Preprocessing

The CT images were obtained from several different models of GE multislice helical CT scanners. The CT scans were acquired directly in the axial plane with 1.25 mm slice thickness, matrix size of 512×512 and 15–25 cm field of view (FOV). Image resolution varied as voxel size ranged from 0.29 mm^3 to 0.49 mm^3 as determined by the ratio of FOV divided by the matrix. CT scans were converted to DICOM format and subsequently Analyze 8.1 software package (AnalyzeDirect, Inc., Overland Park, KS) was used in segmenting binary mandible structure based on histogram thresholding. Holes in the binary segmentation are resulting from differences in CT intensity between the more dense cortical bone and the interior trabecular bone [233]. By

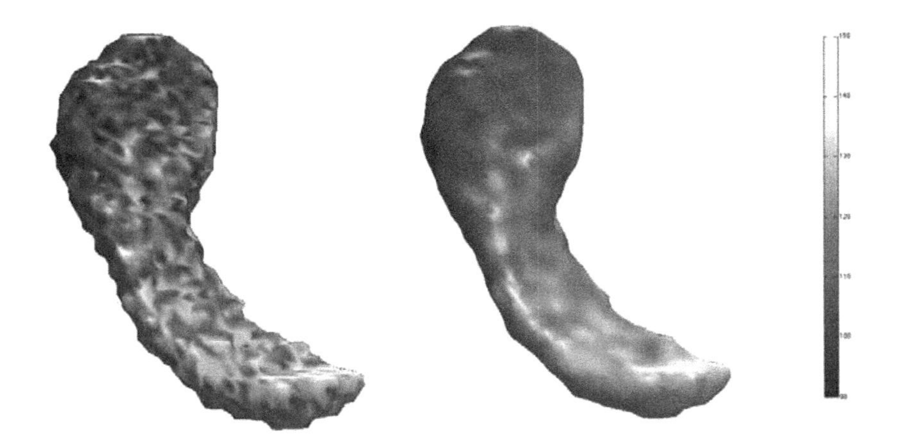

FIGURE 13.8
Left: simulated measurement on a hippocampus. Heat kernel smoothing with 1000 Laplace-Beltrami eigenfunctions and bandwidth 0.2 is applied to the both surface coordinates and simulated measurement.

checking the Euler characteristic, the holes were automatically filled up using morphological operations to make the mandible binary volume to be topologically equivalent to a solid sphere. All areas enclosed by the higher density bone that is included in the mandible definition are morphed into being included as part of the definition of the mandible object. Then, the hole-filled images were converted to surface meshes via the marching cubes algorithm.

13.5.2 Validation of Heat Kernel Smoothing

We applied the proposed method in smoothing a sample mandibular surface, treating vertex coordinates of the surface mesh as signals defined on the mesh. Figure 13.9 shows the RMSE of the smoothed coordinates for varying degrees between 5 to 200. As the degree k increases, the RMSE for each coordinate rapidly decreases and starts to flatten out.

The `MATLAB` codes used in the study are available at `http://brainimaging.waisman.wisc.edu/~chung/lb`. The numerical implementation was done with `MATLAB` 7.9 in 2×2.66 GHz Quad-Core Intel Xeon processor MAC PRO desktop with 32 GB memory. For the sample mesh with 22050 vertices, the entire process took approximately 65 seconds: 55 seconds for setting up the generalized eigenvalue problem (13.16), 10 seconds to actually solve it.

The proposed method was validated on a unit sphere, where the Laplace-Beltrami eigenfunctions are spherical harmonics. We used a spherical mesh with 40,962 uniformly sampled mesh vertices. Let Y_{lm} be the spherical har-

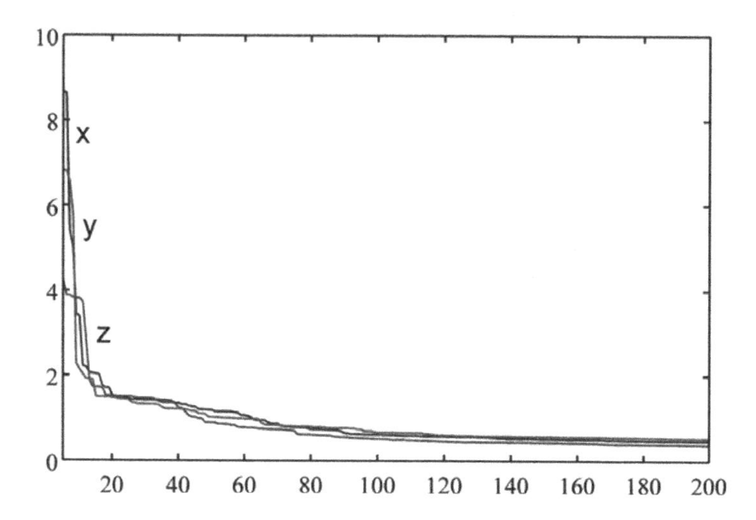

FIGURE 13.9
The plot of the root mean squared errors (RMSE) for x-, y- and z-coordinates for a sample mandibular surface, varying degree k from 5 to 200 for $\sigma = 0.5$. After degree 100, we do see much improvement in RMSE values. Most shape information is encoded in a low frequency basis.

monic of degree l and order. Due to the multiplicity, there are $2l + 1$ eigenfunctions $Y_{l,-l}, \cdots, Y_{l,l}$ corresponding to the same eigenvalue $l(l+1)$. Further, any linear combination $\sum_{m=-l}^{l} \beta_{lm} Y_{lm}$ is an eigenfunction as well. Therefore, we only checked if solving (13.16) produces the expected eigenvalues.

Here we show that the eigenvalues obtained from the FEM-discretization converges to $l * (l + 1)$ for $l = 0, 1, 2, \cdots$ on the unit sphere as the mesh resolution increases (Figure 13.12). We start with an icosahedron which has 20 triangles, 12 vertices and 30 edges. The triangle subdivision of the icosahedron increases the number of triangles by a factor of 4. At each subdivision, the vertices are projected onto the sphere. At the subsequent level of refinement, we have (42, 80), (162, 320), (642, 1280), (2562, 5120), (10242, 20480), (40962, 81920) faces and vertices. For instance, the following codes generate the spherical mesh with 10242 vertices. The first 30 eigenvalues are then computed. The codes require an icosahedron subdivision routine (`cda.psych.uiuc.edu/html/3Sphere_Tessellation`), which is a part of Brainstorm 3 package (`neuroimage.usc.edu/brainstorm`) that is mainly used for MEG/EEG analysis.

```
sphere = sphere_tri('ico',5,1);
figure;figure_wire(sphere, 'black', 'white');
[A, C]= FEM(sphere);
[V, D]= eigs(C,A,30,'sm');
```

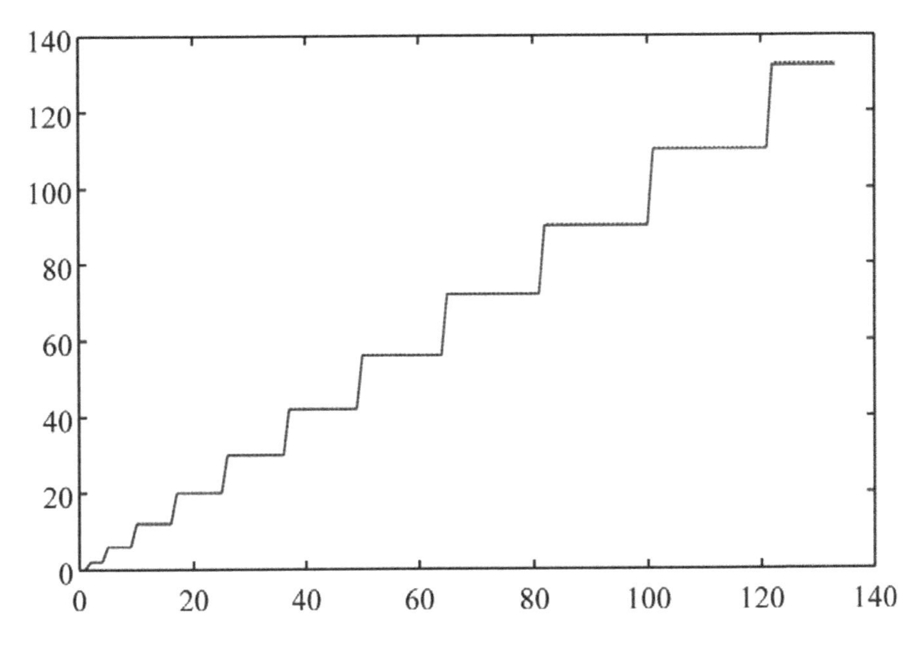

FIGURE 13.10
Comparison of eigenvalues for a unit sphere. 133 eigenvalues are numerically computed (dotted line) and compared against the ground truth (solid line) $\lambda_l = l(l+1)$ for up to degree $l = 11$. They match extremely well and the maximum possible relative error is 0.0032 (0.32%).

As expected, we should obtain the eigenvalues of spherical Laplacian given by $0, 1 \times 2, 2 \times 3, 3 \times 4, \cdots$ within the numerical accuracy:

```
>>sort(diag(D))

ans =

    0.0000
    2.0007
    2.0007
    2.0007
    6.0044
    6.0044
    6.0044
    6.0044
    6.0044
   12.0152
   12.0152
   ....
```

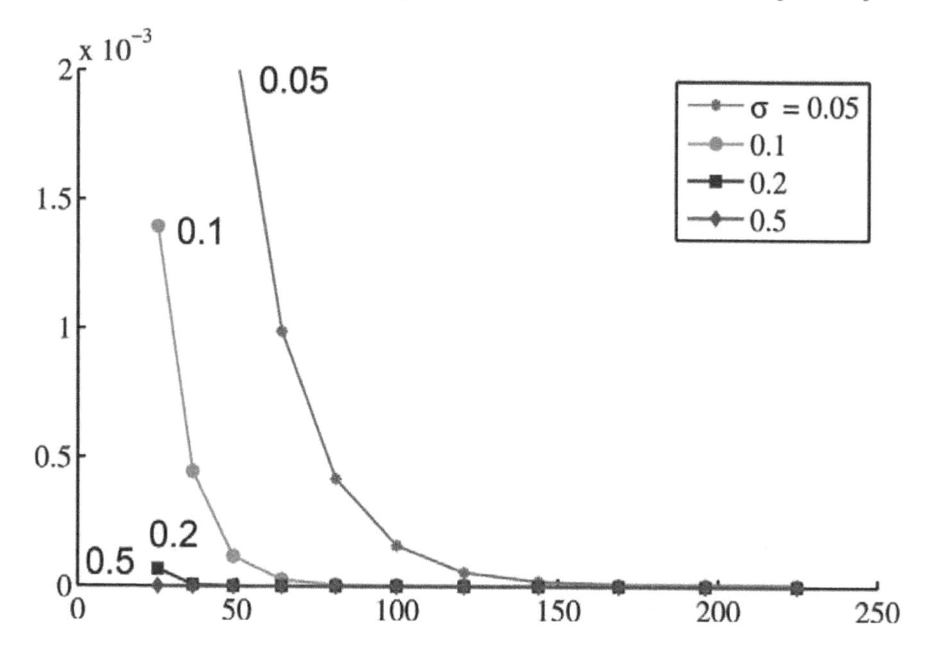

FIGURE 13.11
The plot of the root mean squared errors (RMSE) of computed heat kernel over the number of eigenfunctions used (horizontal). The ground truth is constructed using the spherical harmonics. The RMSE were computed for bandwidths 0.05, 0.1, 0.2 and 0.5. The ground truth was constructed using the spherical harmonics up to degree 85. As the number of eigenfunctions increases, our implementation converges to the ground truth.

We also checked the accuracy of the constructed heat kernel. On a unit sphere, the heat kernel is given by

$$K_\sigma(p,q) = \sum_{l=0}^{\infty} \sum_{m=-l}^{l} e^{-l(l+1)\sigma} Y_{lm}(p) Y_{lm}(q). \tag{13.21}$$

We have taken the degree $l = 85$ expansion as the ground truth and compared it to the numerically constructed heat kernel. The RMSE of heat kernel against the ground truth was computed for various bandwidths between 0.05 and 0.5 (Figure 13.11). The rate of convergence depends on the bandwidth. As the number of eigenfunctions increases, the constructed heat kernel converges to the ground truth quickly. Beyond 150 eigenfunctions, the reconstruction error is negligible. As the degrees increase, the weights $e^{-l(l+1)\sigma}$ exponentially decrease. So the accuracy of high frequency eigenfunctions will not significantly affect the accuracy of the heat kernel.

Figure 13.13 shows the result of smoothing surface coordinates with three different techniques: iterated kernel smoothing [77], the proposed heat kernel

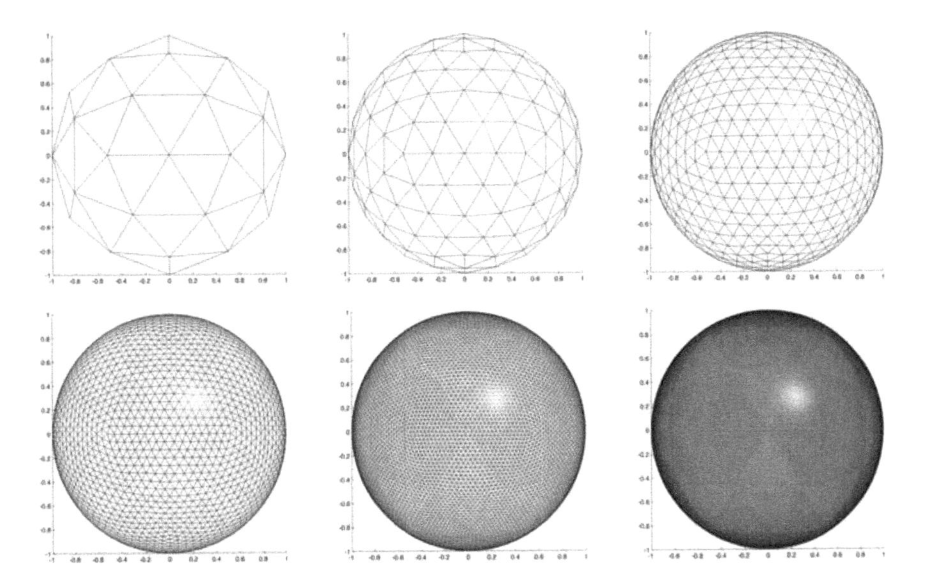

FIGURE 13.12
Icosahedron which has 20 triangles, 12 vertices and 30 edges. The triangle subdivision of the icosahedron increases the number of triangles by a factor of 4.

smoothing and diffusion smoothing [79]. Iterated kernel smoothing approximates heat kernel with Gaussian kernel and iteratively performs Gaussian kernel smoothing to diffuse the signal so it tends to compounds errors over iterations. Diffusion smoothing and heat kernel smoothing share the same FEM discretization but diffusion smoothing tends to suffer numerical instability in the finite difference scheme.

13.6 Case Study: Mandible Growth Modeling

As an illustration of the proposed technique, we performed a mandible growth analysis. The CT imaging data set consists of 77 human subjects between ages 0 and 19. Subjects are binned into three age categories: ages between 0 and 6 years (group I), between 7 and 12 years (group II), and between 13 and 19 years (group III). There are 26, 20 and 31 subjects in group I, II and III respectively. The main biological hypothesis of interest is if there is any growth spurs between these age groups. Mandible surface meshes for all subjects were constructed through the image acquisition and processing steps

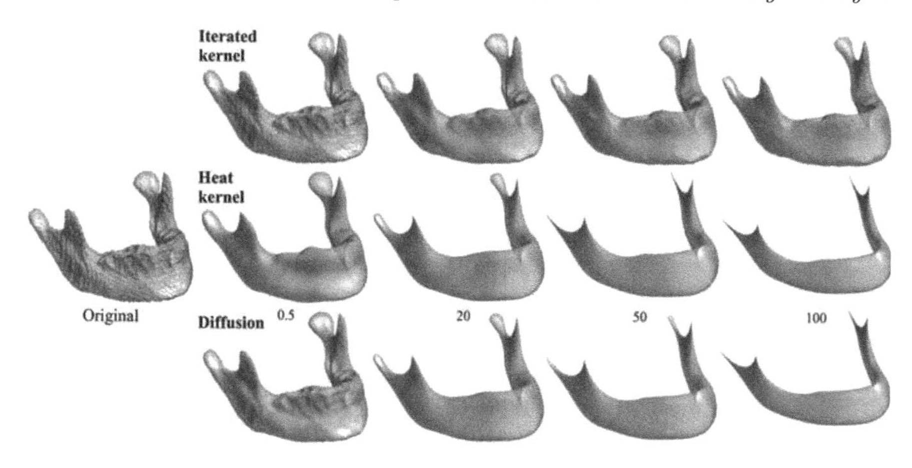

FIGURE 13.13
Smoothed mandible surfaces using three different techniques with the same bandwidths. They are all expected to be the solution of isotropic diffusion. The proposed heat kernel smoothing is done with various bandwidths, $\sigma = 0.5, 20, 50, 100$. Iterated kernel smoothing performs iterative kernel smoothing with heat kernel approximated linearly with Gaussian kernel [77]. Diffusion smoothing directly solves the diffusion equation using the same FEM discretization [79]. The diffusion smoothing and heat kernel smoothing are supposed to converge as the bandwidth increases.

explained in section 13.5.1. Subsequently diffeomorphic surface registration has been performed to align mandible surfaces across subjects.

13.6.1 Diffeomorphic Surface Registration

We have chosen a subject identified as F155-12-08 as the initial template $\mathcal{M}$ and aligned the remaining 76 mandibles to the initial template affinely (Figure 1.8). Then from the affine transformed individual mandible surfaces $\mathcal{M}_j$, we performed the additional nonlinear surface registration to the template using the large deformation diffeomorphic metric mapping (LDDMM) framework [250, 400].

LDDMM framework constructs smooth deformation fields as follows. Given the individual surface $\mathcal{M}_j$ and the initial template $\mathcal{M}_I$, we estimate the diffeomorphism d between them as a smooth streamline given by the Lagrangian evolution:

$$\frac{\partial d}{\partial t}(x,t) = v \circ d(x,t)$$

with $d(x,0) = x$, $t \in [0,1]$ for time dependent velocity field v. Note the surfaces $\mathcal{M}_j$ and $\mathcal{M}$ are the start and end points of the diffeomorphism, i.e. $\mathcal{M}_j \circ d(\cdot,0) = \mathcal{M}_j$ and $\mathcal{M}_j \circ d(\cdot,1) = \mathcal{M}_I$. Then by averaging the inverse deformation fields from the initial template $\mathcal{M}_I$ to individual subjects, we obtain yet another template $\mathcal{M}_F$. The displacement vector field d_j is then recomputed from $\mathcal{M}_F$ to individual mandible $\mathcal{M}_j$. Figure 13.14 shows the displacement differences between the groups I and II (top) and II and III (bottom). Each row shows the group differences of the displacement: group II - group I (first row) and group III - group II (second row). The arrows are the displacement differences and colors indicate their lengths in mm.

13.6.2 Random Field Theory

We are interested in determining the significance of the displacement differences. Since the length measurement provides a much easier biological interpretation, we used the length of displacement vector as a response variable. We applied the statistical parametric mapping (SPM) framework for analyzing and visualizing statistical tests performed along the template surface [10, 38, 77]. Since test statistics are constructed over all mesh vertices on the mandible, multiple comparisons need to be accounted for possibly using the random field theory [387, 396]. The random field theory assumes the measurements to be smooth Gaussian random field. Heat kernel smoothing on the length measurement will make the length measurement more smooth and Gaussian as well as increasing the signal-to-noise ratio. Therefore, heat kernel smoothing is applied with bandwidth $\sigma = 20$ using 1000 eigenfunctions. The number of eigenfunctions used is more than sufficient to guarantee relative accuracy less than 0.3%. After the displacement lengths are smoothed, we constructed the F random field testing the length difference between the age groups I and II, and II and III showing the regions of growth spur (Figure 13.15).

The multiple comparison corrected p-value computation is then given by the random field theory [2, 57, 354, 389]. For the F-field Y with α and β degrees of freedom defined on 2D manifolds $\mathcal{M}_T$, it is known that

$$P\Big(\sup_{p\in\mathcal{M}_T} Y(p) > h\Big) \approx \mu_2(\mathcal{M}_T)\rho_2(h) + \mu_0(\mathcal{M}_T)\rho_0(h) \qquad (13.22)$$

for sufficiently high threshold h. $\mu_d(\mathcal{M}_T)$ is the d-th Minkowski functional of $\mathcal{M}_T$ and ρ_d is the d-th Euler characteristic (EC) density of Y [391]. The Minkowski functionals are simply $\mu_2(\mathcal{M}_T) = 2 = \text{area}(\mathcal{M}_T)/2$ and $\mu_0(\mathcal{M}_T) = 2$. The EC-density for F-field is then given by

$$\begin{aligned}
\rho_2 &= \frac{1}{4\pi\sigma^2}\frac{\Gamma(\frac{\alpha+\beta-2}{2})}{\Gamma(\frac{\alpha}{2})\Gamma(\frac{\beta}{2})}\left(\frac{\alpha h}{\beta}\right)^{\frac{(\alpha-2)}{2}}\left(1+\frac{\alpha h}{\beta}\right)^{-\frac{(\alpha+\beta-2)}{2}}\left[(\beta-1)\frac{\alpha h}{\beta}-(\alpha-1)\right]\\
\rho_0 &= 1 - P(F_{\alpha,\beta} \le h),
\end{aligned}$$

where $P(F_{\alpha,\beta} \le h)$ is the cumulative distribution function of F-stat with α and β degrees of freedom. Note that the second order term $\mu_2(\mathcal{M}_T)\rho_2(h)$ dominates the expression (13.22) and it explicitly has the smoothing bandwidth σ.

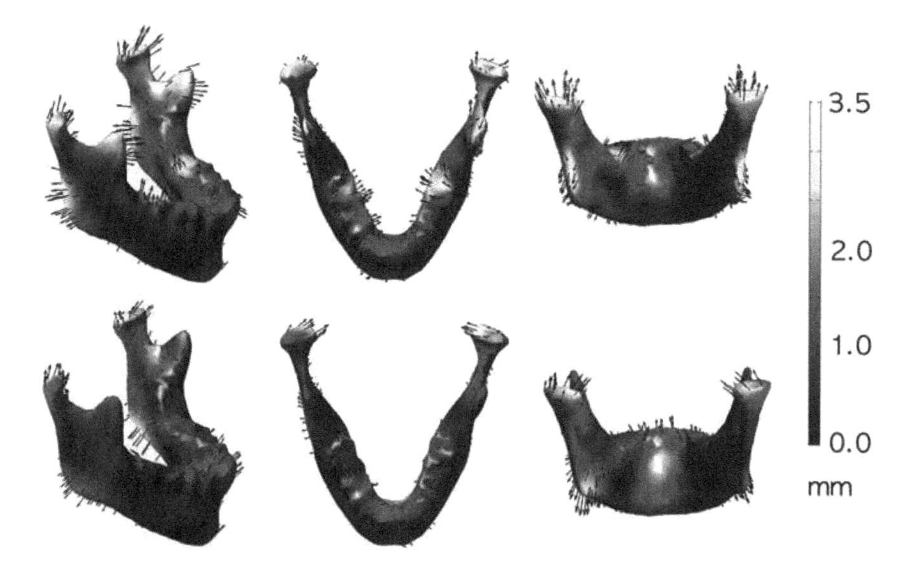

FIGURE 13.14
Mandibles are binned into three age groups: group I (ages 0 and 6), group II (ages 7 and 12) and group III (ages 13 and 19). Each row shows the mean group differences of the displacement: group II - group I (first row) and group III - group II (second row). The arrows are the mean displacement differences and colors indicate their lengths in mm.

For comparing the groups I and II, it is based on F-field with 1 and 44 degrees of freedom while for the groups I and II is based on F-field with 1 and 49 degrees of freedom. The multiple comparison corrected F-stat thresholds corresponding to $\alpha = 0.05$ and 0.01 levels are respectively 8.00 and 10.67 (group II-I) and 8.00 and 10.52 (group III- II). In the F-statistic map shown in Figure 13.15, any black and red regions are considered as exhibiting growth spurs at 0.01 and 0.05 levels respectively.

13.6.3 Numerical Implementation

Here we show the codes for performing the above statistical analysis and correcting for multiple comparisons. The full data set is stored in `mandible77subjects.mat`. The variable `id` stores gender and age information. For instance the first subject `F203-01-03` indicates the subject is female subject number 203 at age 1 year and 3 months.

```
>> id

ans =
    'F203-01-03'
```

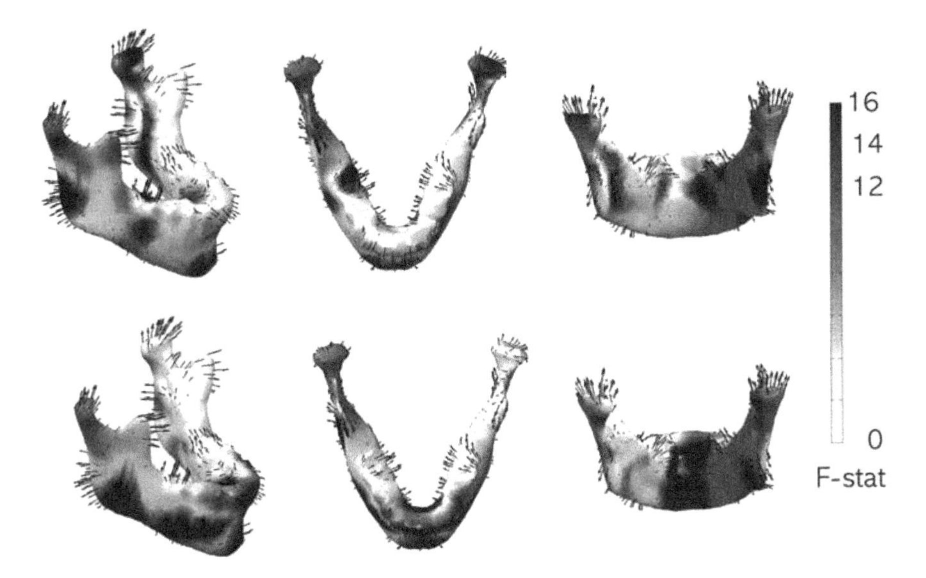

FIGURE 13.15
F-statistic map showing the regions of significant mean displacement difference shown in Figure 13.14.

```
    'F204-01-06'
    'M217-01-06'
    ...
```

We code the group variable `group` to be 1 for male and 0 for female. Months are converted to years.

```
str=cell2mat(id);
group=str(:,1)
group=[group=='M']
year = str2num(str(:,6:7));
month=str2num(str(:,9:10));
age = (12*year+month)/12;
```

The statistical analysis will be performed on the average surface which serves as the template. The average surface `template` is constructed as follows.

```
template.faces=faces;
template.vertices=voxelsize(44)*mean(vertices,3);
subplot(2,2,3); figure_patch(template,[0.74 0.71 0.61],0.7);
view([90 60]); camlight;
title('Template')
```

template consists of 43272 triangles and 21639 vertices. To speed up the computation, we reduced the size of mesh to 10% of the original size using reducepatch.

```
tt=reducepatch(template,0.1);
ind=mesh_commonvertex(template, tt);
p= vertices(ind,:,:);
```

77 subjects are grouped into three groups: age below 7 (group I), between 7 and 13 (group II), and above 13 (group III).

```
age1= age<7;
age2=(7<=age)&(age<13);
age3=13<=age;
```

We compute the displacement and its length in each group. For instance, p12 is the mean displacement between group I and group II while p12L is its length.

```
p1= p(:,:,age1);
p2= p(:,:,age2);
p3= p(:,:,age3);

p12=mean(p2,3)-mean(p1,3);
p23=mean(p3,3)-mean(p2,3);

p1L=squeeze(sqrt(sum(p1.^2,2)));
p2L=squeeze(sqrt(sum(p2.^2,2)));
p3L=squeeze(sqrt(sum(p3.^2,2)));

p12L=sqrt(sum(p12.^2,2));
p23L=sqrt(sum(p23.^2,2));
```

To visualize the group difference between group I and II, we display the mean vector difference p12 on top of its length p12L as quivers. Unfortunately, the quivers are small so we artificially enlarged them by the factor of five. The age effect is then visualized in Figure 13.14.

```
s=voxelsize(44);

figure; figure_surf(tt, s*p12L);
figure_quiver3surface(tt, s*5*p12, 5);
caxis([0 10*s]); colorbar
title('Between age group I and II')
colormap('hot'); view([90 60]); camlight;

figure; figure_surf(tt, s*p23L);
```

```
figure_quiver3surface(tt, s*5*p23, 5);
caxis([0 10*s]); colorbar
title('Between age group II and III')
colormap('hot'); view([90 60]); camlight;
```

We compute the eigenvalues `D` and eigenfunctions `V` of the Laplace-Beltrami operator along the template `tt` by the finite element method. The eigenvalues `D` are sorted in increasing order. To display the 5th eigenfunction `V(:,5)`, we run

```
[A, C]= FEM(tt);
[V, D]= eigs(C,A, 1000,'sa');

figure; figure_surf(tt,V(:,5))
view([90 60]); camlight;
```

We smooth out the length of displacement with the bandwidth 20 and 200 basis functions.

```
p1Lsmooth=lb_smooth(p1L,tt, 20, 200, V, D);
p2Lsmooth=lb_smooth(p2L,tt, 20, 200, V, D);
p3Lsmooth=lb_smooth(p3L,tt, 20, 200, V, D);
```

The two-sample t-test is then performed using the built-in function `ttest2.m`. The code below computes the t-statistic map for comparing the groups I and II. The other case is similarly done. The resulting t-static map is not shown here but its square should yield the F-statistic map in Figure 13.15.

```
tfield=[];
for i=1:length(p1Lsmooth)
[h p ci t]=ttest2(p2Lsmooth(i,:),p1Lsmooth(i,:),[],[],'equal');
tfield=[tfield t.tstat];
end;

figure; figure_surf(tt, tfield);
figure_quiver3surface(tt, s*5*p12, 4);
view([90 60]); colormap('hot'); colorbar
title('T-stat between age group I and II')
caxis([-6 6]); colorbar; camlight
```

We can also compute it slightly using Hotelling's T^2 routine `hotelT2.m`. Note that the square of t-statistic results should be the F-statistic (Figure 13.15).

```
tfield=[];
for i=1:length(p1Lsmooth)
```

```
    h= hotelT2(p2Lsmooth(i,:)', p1Lsmooth(i,:)');
    tfield=[tfield  h.t];
end;

figure; figure_surf(tt, tfield);
figure_quiver3surface(tt, s*5*p12, 4);
view([90 60]);

caxis([0 16]); colorbar
camlight; colormap('hot')

c=colormap;
colormap(c(end:-1:1,:));
title('Hotelling T-square between age group I and II')
```

The multiple comparisons corrected p-value is computed using the random field theory. It requires computing the Minkowski functional `mu0`, `mu1` and EC-density `rho0` and `rho1`. Let us compute the corrected p-value of F-statistic value of 9, 12 and 16 for the bandwidth 20.

```
t=[9 12 16]
a=1; b=sum(age1) + sum(age2) -2;
sigma=20;

mu0=2;

ss.faces = tt.faces;
ss.vertices= s*tt.vertices;
area=GETarea(ss);
mu2=sum(area.faces)/2;

rho0 = 1 - fcdf(t,a,b);
rho2 = 1/(4*sigma^2*pi)*gamma((a+b-2)/2)/ gamma(a/2)/...
       gamma(b/2)*(a*t/b).^((a-2)/2).*(1+a*t/b).^...
       (-(a+b-2)/2).*((b-1)*a*t/b-(a-1));

>>p=mu0*rho0+rho2*mu2

p =
    0.0192    0.0060    0.0014
```

This whole procedure is implemented in a single function.

```
F.tstat=[9 12 16]
F.df= [a b];
```

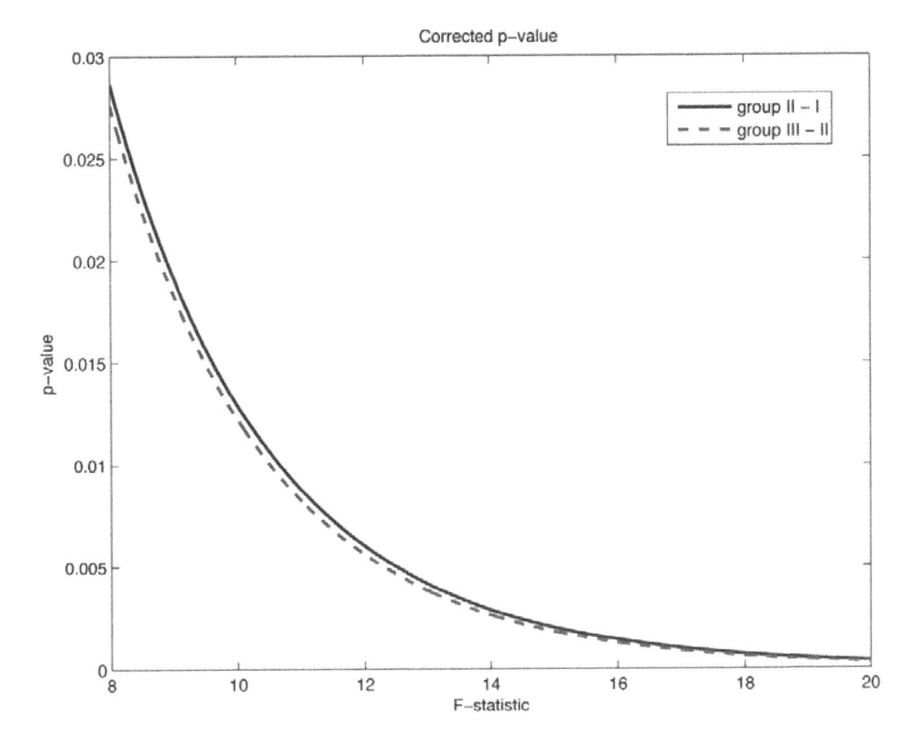

FIGURE 13.16
The plot of the corrected p-value over F-statistic for testing the group difference. The corrected p-value is computed using the random field theory.

```
>>p=stat_Fcorrected(F, sigma,ss)

p =
          tstat: [9 12 16]
             df: [1 44]
     pcorrected: [0.0192 0.0060 0.0014]
```

Between F-statistic values of 8 and 20, we plotted the corrected p-value (Figure 13.16). Since the p-value plot is monotonic, it is not that hard to identify the threshold that corresponds to the specific p-values. For instance, the F-statistic value corresponding to the corrected p-value of 0.01 is 8 for the case of comparing the groups I and II.

```
F.tstat=[8:0.01:20]
stat=stat_Fcorrected(F,sigma,ss)
figure; plot(F.tstat, stat.pcorrected, '-k', 'LIneWidth', 2)
title('Corrected p-value')
xlabel('F-statistic')
```

```
ylabel('p-value')

ind=find(stat.pcorrected<0.01);
>>F.tstat(ind(1))

ans =
     8
```

13.7 Conclusion

We presented a novel heat kernel smoothing framework where the smoothed data is expanded using the Laplace-Beltrami eigenfunctions analytically. The expansion is the solution of isotropic heat diffusion. The method was validated on a unit sphere, where heat kernel was given explicitly in terms of spherical harmonics. As demonstrated in the validation, the proposed method is highly accurate making heat kernel smoothing the possible ground truth for comparing other surface-based smoothing techniques by others.

14

Persistent Homology

We present a novel computational framework for characterizing signals in brain images *via* nonlinear pairing of critical values of the signal. Among the astronomically large number of different pairings possible, we show that representations derived from specific pairing schemes provide concise representations of the image. This procedure yields a *min-max diagram* of the image data. The representation turns out to be especially powerful in discriminating image scans obtained from different clinical populations, and directly opens the door to applications in a variety of learning and inference problems in biomedical imaging. It is noticed that this strategy significantly departs from the standard image analysis paradigm, where the mean signal is used to characterize an ensemble of images. This offers robustness to noise in subsequent statistical analyses, for example; however, the attenuation of the signal content due to averaging makes it rather difficult to identify subtle variations. The topologically oriented method introduced in this chapter seeks to address these limitations by characterizing and encoding topological features or attributes of the image. As an application, we have used this method to characterize cortical thickness measures along brain surfaces in classifying autistic subjects. Our promising experimental results provide evidence of the power of this technique. This chapter is based on [78].

14.1 Introduction

The use of critical values of measurements within classical image analysis and computer vision has been relatively limited so far, and typically appears as part of simple preprocessing tasks such as feature extraction and identification of edge pixels in an image. For example, first or second order image derivatives may be used to identify the edges of objects to serve as the contours of an anatomical shape, possibly using priors to provide additional shape context. Specific properties of critical values as a topic on its own, however, has received less attention. Part of the reason is that it is difficult to construct a streamlined linear analysis framework using critical points, or values of images. Also, the computation of critical values is a nonlinear process and almost always requires the numerical estimation of derivatives. In some applications where this is necessary, the discretization scheme must be chosen carefully, and remains an active area of research. It is noticed that in most of these applications, the interest is only in the stable estimation of these points rather than their properties, and how these properties vary as a function of images. We note that in brain imaging, on the other hand, the use of extreme values has been quite popular in other types of problems. For example, these ideas are employed in the context of multiple comparison correction using random field theory [394, 355, 203]. Recall that in random field theory, the extreme of a statistic is obtained from an ensemble of images, and is used to compute the p-value for correcting for correlated noise across neighboring voxels. Our interest in this chapter is to take a topologically oriented view of the image data. We seek to interpret the critical values in this context and assess their response as a function of brain image data. In particular, we explore specific representation schemes and evaluate the benefits they afford with respect to different applications.

The calculation of the critical values of a certain function of images such as image intensities, cortical thickness and curvature maps is the first step of our procedure. This is performed after heat kernel smoothing [71]. In this chapter, we propose to use critical points of images but our approach differs from the previous literature. Given a function of images, e.g. image intensities, cortical thickness, curvature maps etc., we first smooth it with new heat kernel smoothing [71] to obtain the analytic estimation of derivative estimation.

The obtained critical values are then paired in a nonlinear fashion following a specific pairing rule to produce so-called *min-max diagrams*, which are motivated by persistent homology. Persistent homology is a branch of computational algebraic topology that has been popularized in recent years. This field is usually referred to as topological data analysis. The main aims of the topological data analysis are to infer high dimensional structure from low dimensional representation and assemble discrete points into global structure [138]. Persistent homology is popular in computational algebraic topology with applications in protein structure analysis [311], gene expression [113], activity patterns in visual cortex [329], sensor networks [103] and complex networks [176]. However, it has not seen any applications in medical image analysis except for [68] and [78].

The main tools in persistent homology are persistence diagrams and barcodes. These diagrams visually show how the topological invariants such as Betti numbers of the sublevel sets of the signal change. The min-max diagrams are similar to the theoretical construct of persistence diagrams [114] in persistent homology, but have notable differences. The persistence diagram is a scatter plot of particular pairing that has been used to show the topological characteristics of signal. However, we wish to make it clear that our pairing rule is not that of persistence but a reduced form of the persistent diagram. Min-max diagrams resemble scatter plots, and lead to a powerful representation of the key characteristics of their corresponding images. We discuss these issues in detail, and provide a number of examples and experiments to highlight their key advantages, limitations, and possible applications to a wide variety of medical imaging problems.

In this chapter, we propose a new topologically oriented data representation framework using the min-max diagrams. We also present a new $\mathcal{O}(n \log n)$ algorithm for generating such diagrams without having to modify or adapt the complicated machinery used for constructing persistence diagrams [68, 114, 416].

14.2 Rips Filtration

Cortical surfaces already have surface meshes as the underlying topological representation. So topological computation on cortical surfaces can be easily done. However, for other type of data and images, it is crucial to obtain the most accurate topological representation that approximates them. Often an object can be easily observed and represented as an unordered collection of points in a Euclidean space [138]. Such a collection of points is often referred to as *point cloud data*. For instance, the point cloud data approximates the underlying gray shaded object in Figure 14.2. Then by connecting the point cloud data somehow, we can obtain higher dimensional topological information.

14.2.1 Topology

A high dimensional object can be approximated by the point cloud data X consisting of p number of points. If we connect points of which distance satisfy a given criterion, the connected points start to recover the topology of the object. Hence, we can represent the underlying topology as a collection of the subsets of X that consists of nodes which are connected [115, 166].

Definition 18 *Suppose $U \subset 2^X$, the collection of all possible subsets of X. Then (X, U) is a topological space on X if*

1. $\emptyset, X \subset U$,
2. $u_1, u_2 \subset U$ *implies* $u_1 \cup u_2 \subset U$ *and*
3. $u_1 \cap u_2 \subset U$.

Note that every metric space is a topological space. In general, given a point cloud data set X with a rule for connections, the topological space is a simplicial complex and its element is a simplex [415]. For point cloud data, the Delaunay triangulation is probably the most widely used method for connecting points. The Delaunay triangulation represents the collection of points in space as a graph whose face consists of triangles.

Point cloud data can be obtained manually using a mouse as follows.

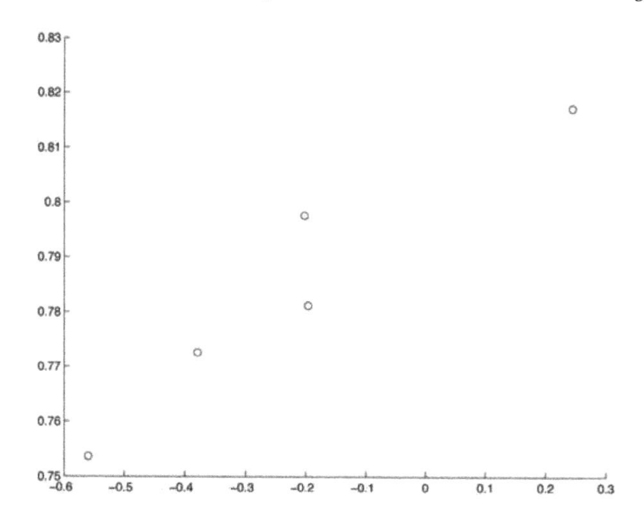

FIGURE 14.1
A sample point cloud data constructed using `ginput.m`.

```
figure; axis on
hold on
xy = [];
n = 0;
but = 1;
while but == 1
    [xi,yi, but] = ginput(1);
    plot(xi,yi,'ko')
    n = n+1;
    xy(:,n) = [xi;yi];
end

>> xy

xy =

    0.2431   -0.5593   -0.3791   -0.1967   -0.2029
    0.8173    0.7537    0.7726    0.7812    0.7976
```

The left mouse button puts a point in a figure and the right mouse button terminates `ginput.m`. For instance, cloud point data consisting of five points is given in Figure 14.1. The list of coordinates `xy` will be used in constructing the Rips complex later.

14.2.2 Simplex

Definition 19 *The k-simplex σ is the convex hull of $(k+1)$ independent points $x_0, \cdots, x_k$.*

A point is a 0-simplex, an edge is a 1-simplex, and a triangle is a 2-simplex. A complete graph with k nodes is a $(k-1)$-simplex.

Definition 20 *A* simplicial complex *K is a finite collection of simplices satisfying [115]*

1. *any face of $\sigma \in K$ is also in K, and*
2. *for $\sigma_1, \sigma_2 \in K$, $\sigma_1 \cap \sigma_2$ is a face of both σ_1 and σ_2.*

Hence a graph is a simplicial complex consisting of 0-simplices (nodes) and 1-simplices (edges). There are various simplicial complexes. One of them is the Rips complex.

14.2.3 Rips Complex

The Rips complex has been the main building block for persistent homology [138] and defined on top of the point cloud data.

Definition 21 *The* Rips complex *is a graph constructed by connecting two data points if they are within specific distance ϵ.*

Figure 14.2 shows an example of the Rips complex that approximates the gray object with 36 nodes and 56 edges. Given a point cloud data X, the Rips complex $R_X(\epsilon)$ is a simplicial complex whose k-simplices correspond to unordered $(k+1)$-tuples of points which are pairwise within distance ϵ [138]. While a graph has at most 1-simplices, the Rips complex has at most k-simplices. One major problem of the Rips complex is that given n points, it exactly produces a graph with n nodes so the resulting graph becomes very complicated when n becomes large.

Figure 14.2 is simulated as follows. The mouse clicking routines are now incorporated into a single function `mouse2points.m` on the image `toy-key.tif`.

```
endp = mouse2points(500,500, 'toy-key.tif');
imshow('toy-key.tif')
set(gcf,'Color','white','InvertHardcopy','off');

[adj, prob, vertices] = points2graph(endp, 70, 'rips');
figure_graph_color(adj,vertices, ones(size(adj,1),1));
```

The `points2graph.m` inputs the point cloud data `endp` and constructs the Rips complex using 70mm. Since the Rips complex is a graph, we can represent it as using an adjacency matrix `adj` and display the planner graph (Figure 14.2).

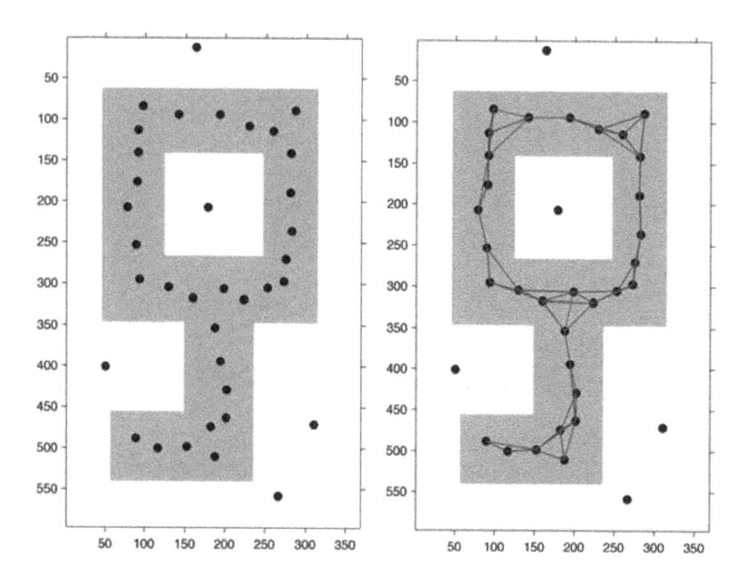

FIGURE 14.2
Left: A point cloud that approximates the underlying object (gray). Right: The rips complex of the point cloud data. $\epsilon = 70$ is used for construction. If two points are within the ϵ radius, we connect them with a link. There are 36 nodes and 56 edges. Three outlying data are not connected to the largest connected component.

14.2.4 Constructing Rips Filtration

The Rips complex has the property that

$$R_X(\epsilon_0) \subset R_X(\epsilon_1) \subset R_X(\epsilon_2) \subset \cdots$$

for $0 = \epsilon_0 \leq \epsilon_1 \leq \epsilon_2 \leq \cdots$. When $\epsilon = 0$, the Rips complex is simply the point cloud V. By increasing the ϵ-value, we are connecting more nodes so the size of the edge set increases. Such the nested sequence of the Rips complexes is called a Rips filtration, the main object of interest in the persistent homology [114]. The increasing ϵ values are called the filtration values. The Rips filtration can be constructed by iteratively applying `points2graph` routine with increasing radius. Figure 14.3 shows the Rips filtration between 0 and 100mm radius at 20mm increment.

```
figure;
subplot(2,3,1);
imshow('toy-key.tif')
set(gcf,'Color','white','InvertHardcopy','off');
```

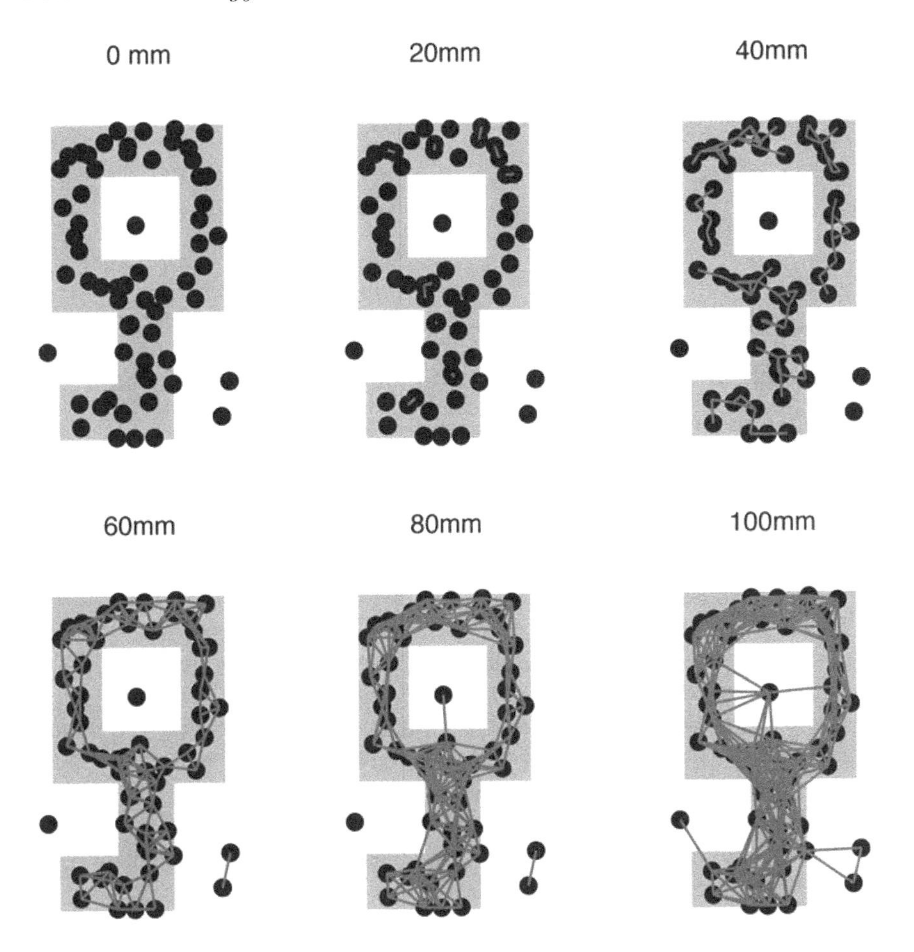

FIGURE 14.3
The Rips filtration is the collection of Rips complexes with increasing radius. The Rips complex with a smaller radius is enclosed in the Rips complex with a larger radius.

```
[adj, prob, vertices] = points2graph(endp,0, 'rips');
figure_graph_color(adj,vertices, ones(size(adj,1),1));
str= ['0 mm']
title(str)

for i=1:5
    subplot(2,3,i+1);
    imshow('toy-key.tif')
    set(gcf,'Color','white','InvertHardcopy','off');
```

```
    [adj, prob, vertices] = points2graph(endp,20*i, 'rips');
    figure_graph_color(adj,vertices, ones(size(adj,1),1));
    str= [num2str(20*i), 'mm']
    title(str)
end;
```

14.3 Heat Kernel Smoothing of Functional Signal

Consider measurements f from images given as [248]

$$f(t) = \mu(t) + \epsilon(t),\ t \in \mathcal{M} \subset \mathbb{R}^d, \tag{14.1}$$

where μ is the unknown mean signal (to be estimated) and ϵ is noise. The unknown mean signal is estimated *via* image smoothing over $\mathcal{M}$, and denoted as $\widehat{\mu}$. Traditionally, the estimate for the residual $f - \widehat{\mu}$ is used to construct a test statistic corresponding to a hypothesis about the signal. The mean signal may not be able to fully characterize complex imaging data, and as a result, may have limitations in the context of inference. Hence, we propose to use a new topologically motivated framework called the *min-max diagram*, which is the scatter plot of specific pairing of critical values. Intuitively, the collection of critical values of μ can approximately characterize the shape of the continuous signal μ. By pairing critical values in a nonlinear fashion and plotting them, we construct the min-max diagram. We will provide additional details shortly.

In order to generate the min-max diagram, we need to find the critical values of μ. It requires estimating the unknown signal smoothly so that derivatives can be computed. We avoid the diffusion equation based implicit smoothing techniques [53, 10, 53, 83] since the approach tends to result in unstable derivative estimation. Instead, we present a more flexible spectral approach called *heat kernel smoothing* that explicitly represents the solution to the diffusion equation analytically [71, 76]. Our approach offers at least two advantages over the traditional implicit smoothing: the ability to analytically differentiate smoothed images and resample at any scale without additional interpolation.

Heat kernel smoothing analytically solves the following equation

$$\frac{\partial F}{\partial \sigma} = \Delta F,\ F(t, \sigma = 0) = f(t).$$

The solution is given in terms of eigenfunctions ψ_k (and the corresponding eigenvalues λ_k) of the Laplace-Beltrami operator, i.e., $\Delta f + \lambda f = 0$. Define the heat kernel K_σ as

$$K_\sigma(t, s) = \sum_{k=0}^{\infty} e^{-\lambda_k \sigma} \psi_k(t) \psi_k(s).$$

The heat kernel smoothing estimate of μ is then given by

$$\widehat{\mu} = \int_{\mathcal{M}} K_\sigma(t,s) f(s)\, d\eta(s) = \sum_{i=0}^{\infty} e^{-\lambda_k \sigma} f_k \psi_k(t). \tag{14.2}$$

where $d\eta$ is the Lebesgue measure that makes the basis ψ_i orthonormal. We can rewrite (14.2) in a Fourier series expansion as

$$K_\sigma * f(t) = \sum_{i=0}^{\infty} e^{-\lambda_i \sigma} f_i \psi_i(t), \tag{14.3}$$

The Fourier coefficients $f_k = \int_{\mathcal{M}} f \psi_k \, d\eta$ are estimated using least squares [344]. For a given bandwidth, the expansion is truncated at the degree where adding more terms will not increase the goodness of fit [70]. The details on the iterative estimation procedure are given in [71].

Example 9 *For $\mathcal{M} = [0,1]$, with the additional constraints $f(t+2) = f(t)$ and $f(t) = f(-t)$, the eigenfunctions are $\psi_0(t) = 1, \psi_k(t) = \sqrt{2}\cos(k\pi t)$ with the corresponding eigenvalues $\lambda_k = k^2\pi^2$. For simulation in Figure 14.4, we used $\sigma = 0.0001$ and truncated the series at the 100-th degree.*

For $\mathcal{M} = S^2$, the eigenfunctions are the spherical harmonics $Y_{lm}(\theta, \varphi)$ and the corresponding eigenvalues are $\lambda_l = l(l+1)$. The bandwidth $\sigma = 0.001$ and degree $k = 42$ were used for cortical thickness example in Figure 14.5. We found that bandwidths larger than 0.001 smooth out relevant anatomical detail.

The advantage of using the series expansion to smooth out signals is that smoothing is given as an explicit functional representation so various manipulation such as differentiation and resampling can be done analytically. The explicit analytic derivative of the expansion (14.2) is simply given by

$$\mathcal{D}\widehat{\mu} = \sum_{i=0}^{\infty} e^{-\lambda_i \sigma} f_i \mathcal{D}\psi_i(t)$$

where $\mathcal{D}$ is $\frac{\partial}{\partial t}$ for $[0,1]$ and $(\frac{\partial}{\partial \theta}, \frac{\partial}{\partial \varphi})$ for S^2. For the unit interval, the derivatives are $\mathcal{D}\psi_l(t) = -\sqrt{2} l\pi \sin(l\pi t)$. For S^2, the partial derivatives with respect to θ can be given in slow iterative formulas [70]. To speed up the computation, the convexity of the first order neighbor of a vertex in a cortical mesh is used in determining a critical point. Figure 14.5 shows the result of minimum and maximum detection after heat kernel smoothing.

14.4 Min-max Diagram

A function is called a Morse function if all critical values are distinct and non-degenerate, i.e., the Hessian does not vanish [252]. For images (where

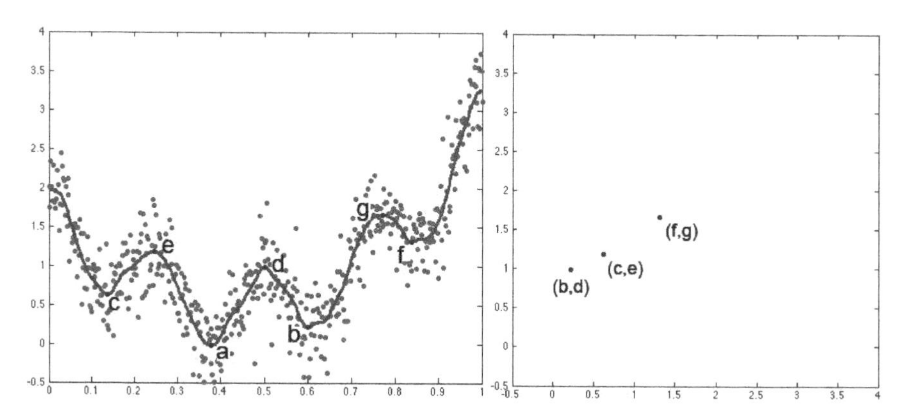

FIGURE 14.4
The birth and death process of sublevel sets. Here $a < b < c < f$ are minimums and $d < e < g$ are maximums. At $y = b$, we add a new component to the sublevel set. When we increase the level to $y = d$, we have the death of the component so we pair them. In this simulation, we pair (f, g), (c, e) and (b, d) in the order of parings generated in Algorithm 1.

intensities are given as integers), critical values of intensity may not all be distinct; however, the underlying continuous signal μ in (14.1) can be assumed to be a Morse function. For a Morse function $\widehat{\mu}$, define a sublevel set as $R(y) = \widehat{\mu}^{-1}(-\infty, y]$. The sublevel set is the subset of $\mathcal{M}$ satisfying $\widehat{\mu}(t) \leq y$. As we increase y from $-\infty$, the number of connected components of $R(y)$ changes as we pass through critical values. Hence the sequence of the sublevel sets form a so called Morse filtration. Let $\#R(y)$ be the number of connected components in the sublevel set.

Let us denote the local minimums as $g_1, \cdots, g_m$ and the local maximums as $h_1, \cdots, h_n$. Since the critical values of a Morse function are all distinct, we can strictly order the local minimums from the smallest to the largest as $g_{(1)} < g_{(2)} < \cdots < g_{(m)}$ and similarly for the local maximums as $h_{(1)} < h_{(2)} < \cdots < h_{(n)}$ by sorting them. We further order all critical values together and let

$$z_{(1)} < z_{(2)} < \cdots < z_{(m+n)},$$

where z_i is either h_i or g_i. At each minimum, the sublevel set adds a new component while at a local maximum, two components merge into one. By keeping track of the birth and death of components, it is possible to compute topological invariants of sublevel sets such as Euler characteristics and Betti numbers [114], i.e.

$$\#R(g_i - \epsilon) = \#R(g_i) + 1$$

for sufficiently small ϵ. The new component is identified with the local minimum g_i. Similarly for at each maximum, we have the death of a component, i.e.

$$\#R(h_i - \epsilon) = \#R(h_i) - 1,$$

and two components will merge as one. The number of connected components will only change if we pass through critical points and we can iteratively compute $\#R$ at each critical value as

$$\#R(z_{(i+1)}) = \#R(z_{(i)}) \pm 1.$$

The sign depends on if $z_{(i)}$ is maximum (-1) or minimum $(+1)$. This is the basis of the Morse theory [252] that says the topological characteristics of a Morse function such as the number of connected components, Euler characteristic and Betti numbers are completely characterized by critical points.

Example 10 *The birth and death processes are illustrated in Figure 14.4, where the gray dots are simulated with Gaussian noise with mean 0 and variance* 0.2^2 *as*

$$f(t) = t + 7(t - 1/2)^2 + \cos(8\pi t)/2 + N(0, 0.2^2).$$

The signal is estimated using the 1D heat kernel smoothing. Let us increase y *from* $-\infty$ *to* ∞*. When we hit the first critical value* $y = a$*, the sublevel set consists of a single point, i.e.* $\#R(a) = 1$*. When we hit the minimum at* $y = b$*, we have the birth of another component at* b*, i.e.* $\#R(b) = 2$*. This process continues until we exhaust all critical values. At* $y = b$*, we add a new component to the sublevel set* $R(y)$*. When we increase the level to* $y = d$*, we have the death of the component so we pair* b *and* d*. In this simulation, we need to pair* (b, d)*,* (c, e) *and* (f, g)*.*

Example 10 is implemented in `MATLAB` as follows.

```
x=[0:0.002:1]';
s= x + 7*(x - 0.5).^2 + cos(8*pi*x)/2;
e=normrnd(0,0.2,length(x),1);
Y=s+e;
figure;
plot(x,Y,'ko','MarkerEdgeColor',[0.5 0.5 0.5], ...
'MarkerFaceColor',[0.7 0.7 0.7], 'MarkerSize',4)
```

We used heat kernel smoothing in estimating the underlying functional signal.

```
k=100; sigma=0.0001;
[wfs, beta]=WFS_COS(Y,x,k,sigma);
hold on; plot(x,wfs,'k','LineWidth',5);
```

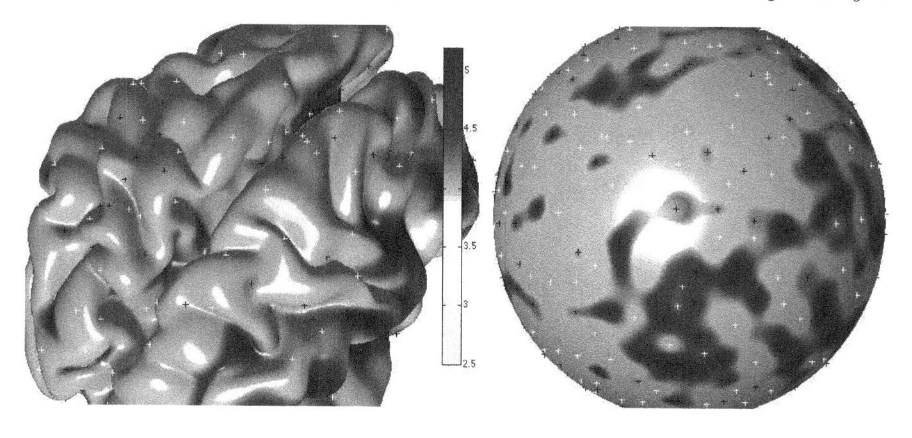

FIGURE 14.5
Heat kernel smoothing of cortical thickness and surface coordinates with $\sigma = 0.001$ and degree $k = 42$. For better visualization, it has been flattened onto the unit sphere. The white (black) crosses are local minimums (maximums). They will be paired in a specific manner to obtain the min-max diagram. The min-max diagram is invariant to whether it is constructed from the cortical surface or from the unit sphere.

14.4.1 Pairing Rule

When we pass a maximum and merge two components, we pair the maximum with the higher of the minimums of the two components [114]. Doing so we are pairing the birth of a component to its death. Note that the paired critical values may not be adjacent to each other. The min-max diagram is then defined as the scatter plot of these pairings.

For higher dimensional Morse functions, saddle points can also create or merge sublevel sets so we also have to be concerned with them. However, we will not consider the saddle points in pairing since the saddle points do not yield a clear statistical interpretation compared to local minimums and maximums. In fact, there is no statistical methodology developed for saddle points so far. If we include saddle points in the pairing rule, we obtain *persistence diagrams* [68, 114] instead of min-max diagrams. In one dimension, the two diagrams are identical since there are no saddle points in 1D Morse functions. For higher dimensions, persistence diagrams will have more pairs than min-max diagrams. The addition of the saddle points makes the construction of the persistence diagrams much more complex. We note that [416] presents an algorithm for generating persistence diagrams based on filtration of Morse complexes.

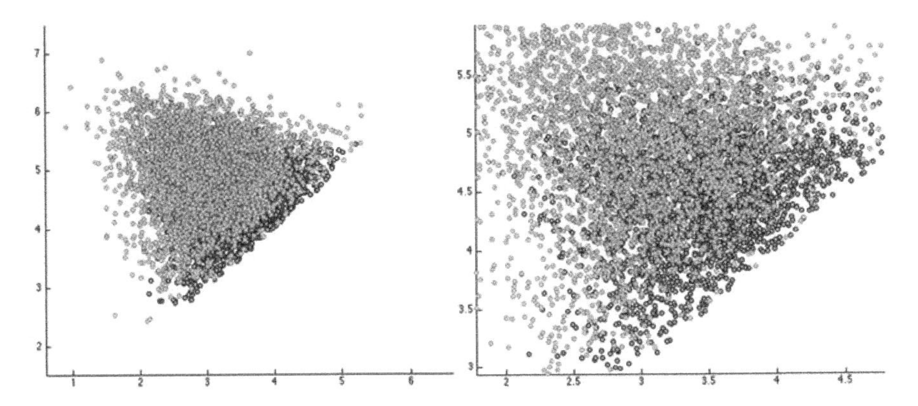

FIGURE 14.6
Min-max diagram for 11 control (gray) and 16 autistic (black) subjects. The pairings for autism often occur closer to $y = x$ line indicating there is greater high frequency noise in autism. This observation is consistent with the autism literature where it has been found that there is greater anatomical variability in autism subjects than the control subjects. This figure suggests that the min-max diagram may indeed be useful for discriminating populations.

14.4.2 Algorithm

We have developed a new simpler algorithm for pairing critical values. Our algorithm generates min-max diagrams as well as persistence diagrams for 1D Morse functions. At first glance, the nonlinear nature of pairing does not seem to yield a straightforward algorithm. The trick is to start with the maximum of minimums and go down to the next largest minimum in an iterative fashion. The algorithm starts with $g_{(m)}$ (step 3). We only need to consider maximums above $g_{(m)}$ for pairing. We check if maximums h_j are in a neighborhood of $g_{(m)}$, i.e. $h_j \sim g_{(m)}$. The only possible scenario of not having any larger maximum is when the function is unimodal and obtains the global minimum $g_{(m)}$. In this situation we have to pair $(g_{(m)}, \infty)$. Since ∞ falls outside our 'plot', we leave out $g_{(m)}$ without pairing. Other than this special case, there exists at least one smallest maximum h^*_m in a neighborhood of $g_{(m)}$ (intuitively, if there is a valley, there must be mountains nearby). Once we paired them (step 4), we delete the pair from the set of extreme values (step 5) and go to the next maximum of minimums $g_{(m-1)}$ and proceed until we exhaust the set of all critical values (step 6). Due to the sorting of minimums and maximums, the running time is $\mathcal{O}(n \log n)$. The Iterative Pairing and Deletion Algorithm is as follows.

1. $H \leftarrow \{h_1, \cdots, h_n\}$.
2. $i \leftarrow m$.
3. $h^*_i = \arg\min_{h_j \in H} \{h_j | h_j > g_{(i)}, h_j \sim g_{(i)}\}$.

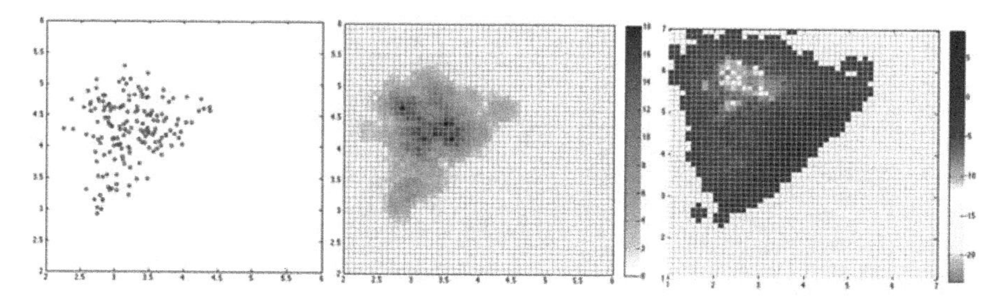

FIGURE 14.7
Left: Min-max diagram of an autistic subject from Figure 14.5. Middle: The concentration map of the min-max diagram is constructed by discretizing the square $[1, 7]^2$ into 50^2 uniform pixels and evaluating the number of pairs within a circle ($r = 0.2$) centered on the pixel. Right: The t-test statistic (autism - control) shows significant group differences in black regions ($t \geq 3.61$) *vs* white ($t \leq -4.05$) regions at level 0.05 (corrected).

4. If $h_i^* \neq \emptyset$, pair $(g_{(i)}, h_i^*)$
5. $H \leftarrow H - h_i^*$.
6. If $i > 1$, $i \leftarrow i - 1$ and go to Step 3.

Higher dimensional implementation is identical to the 1D version except how we define neighbors of a critical point. The neighborhood relationship $\sim$ is established by constructing the Delaunay triangulation on all critical points.

The persistence diagram in Figure 14.4 is then generated by

```
pairs=pairing_1D(x,wfs);
set(gcf,'Color','w')
figure; plot(pairs(1,:),pairs(2,:),'.k')
```

`pairs` produce the coordinates for the persistent diagram like

```
>> pairs

pairs =

    1.3126    1.6816
    0.5948    1.1636
    0.2130    0.8886
```

Since we are putting Gaussian noise in Example 10, the coordinates will vary a bit from simulation to simulation.

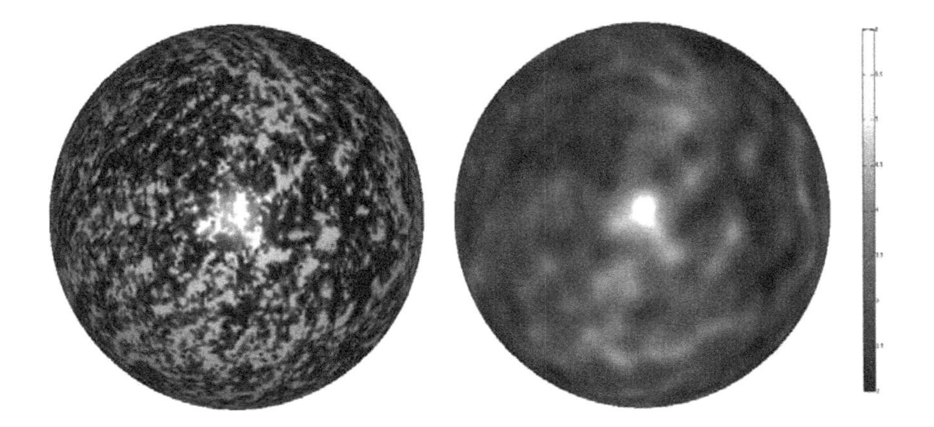

FIGURE 14.8
Cortical thickness of the first autistic subject mapped onto a sphere. Weighted spherical harmonic representation of cortical thickness with 42 degrees and bandwidth 0.001. Smoothness of cortical thickness is necessary to be able to identify critical values.

14.5 Case Study: Cortical Thickness Analysis

Among various cortical measures, we used cortical thickness [120] as a measure for characterizing autistic subjects. Cortical thickness has been used as a main cortical measure for characterizing various clinical populations [249, 402, 76]. We used an MRI dataset of 16 highly functional autistic subjects and 11 normal control subjects (aged-matched right-handed males). These images were obtained from a 3-Tesla GE SIGNA scanner, and went through intensity nonuniformity correction [330], spatially normalized into the MNI stereotaxic space *via* a global affine transformation, and tissue segmentation [85]. Subsequently a supervised neural network classifier was used for tissue segmentation. Brain substructures such as the brain stem were removed to make both the outer and the inner surfaces to be topologically equivalent to a sphere. A deformable surface algorithm [237] was used to obtain the inner cortical surface by deforming from a spherical mesh (Figure 14.5). The outer surface $\mathcal{M}$ was obtained by deforming the inner surface further. The cortical thickness f is then defined as the distance between the two surfaces, this measure is known to be relevant for autism.

Since the deformable surface algorithm starts with a spherical mesh, there is no need to use other available surface flattening algorithms such as [152, 180, 12, 367] for mapping thickness to the unit sphere S^2. Let $\zeta : \mathcal{M} \to S^2$ be a sufficiently smooth surface flattening obtained from the deformable surface algorithm. Then the pullback $(\zeta^{-1})^* \widehat{\mu} = \widehat{\mu} \circ \zeta^{-1}$ projects the cortical thickness from the cortical surface $\mathcal{M}$ to the unit sphere.

Since the critical values do not change even if we geometrically change the underlying manifold from $\mathcal{M}$ to S^2, the min-max diagram must be topologically invariant as well. Therefore, the min-max diagram is constructed on the unit sphere by projecting the cortical data on to the sphere. Figure 14.6 shows the superimposed min-max diagram for 11 control (blue) and 16 autistic (red) subjects. A single subject example is shown in Figure 14.7. Pairings for autistic subjects are more clustered near $y = x$ indicating higher frequency noise in autism. More pairing occurs at high and low thickness values in the controls showing additional topological structures not present in autism.

14.5.1 Numerical Implementation

We will use the cortical surface data set used in [78]. There are total 16 autistic and 11 control subjects in `AUTISM.coordinates.mat`. Cortical thickness is computed as the L_2 distance between the two surfaces. The cortical thickness is projected onto the unit sphere that has the identical mesh topology.

```
load AUTISM.coordinates.mat
thick_au=squeeze(sqrt(sum((autism_coord-autism_coordw).^2,2)));

load unitsphere.mat
figure_trimesh(sphere,thick_au(1,:)')
caxis([2 6]); colormap('hot')
```

The cortical thickness of the first autistic subject is displayed on a unit sphere (Figure 14.8). Since the cortical thickness data is fairly noisy, it is necessary to smooth the data using the weighted spherical harmonic representation with degree 42 and bandwidth 0.001 [71, 70]. The smoothed cortical thickness is given in Figure 14.8. `SPHARMsmooth2` is used to smooth the cortical thickness.

```
directory='/basis/';
SPHARMconstruct(directory,42);

L=42;
sigma=0.001;

surf.faces=sphere.faces;
surf.vertices=zeros(40962,3);
surf.vertices(:,1)=thick_au(1,:)';

[surf_smooth, fourier]=SPHARMsmooth2(surf,sphere,L,sigma);
figure; figure_trimesh(sphere,surf_smooth.vertices(:,1))
```

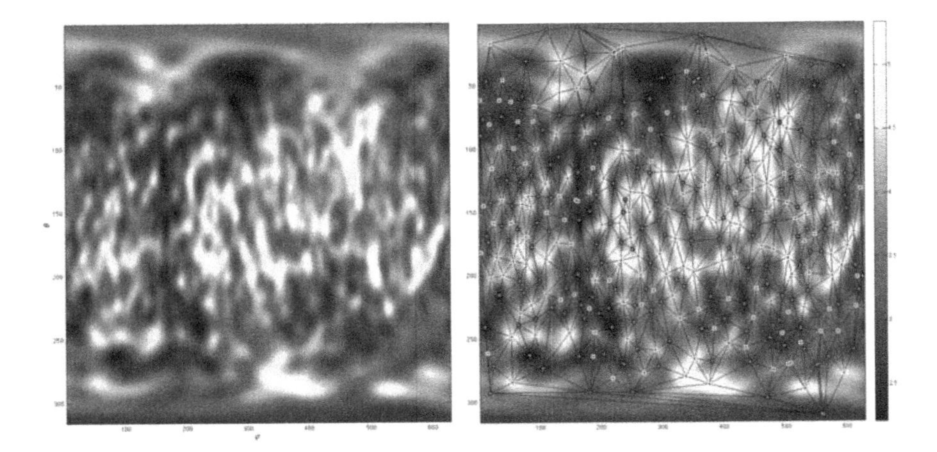

FIGURE 14.9
Left: the flat map of smoothed cortical thickness. Right: maximums and minimums are identified.

```
caxis([2 6]); colormap('hot')
```

The min-max diagram is constructed by pairing minimums and maximums using the iterative pairing and deletion algorithm . To simplify the problem, we first identify critical values in 2D flat map. The computation for extreme values is done by the code `extrema.m` and `extrema2.m` written by Carlos Adrian Vargas Aguilera of Universidad de Guadalajara (`www.mathworks.com/matlabcentral/fileexchange/12275-extrema-m-extrema2-m`). This is just one way of identifying extreme values and there are many other methods. Let us map cortical thickness given in Figure 14.8 onto a plane for better visualization. Figure 14.9 shows the identified extreme values.

```
square=SPHARM2square(fourier.x(:,:,1),42,0.001);
[lmax, lmin] = figure_extrema(square);
```

Subsequently, the minimum and maximums in the cortical thickness data are paired. The pairing for all critical values identified in Figure 14.9 can be done by

```
value=surf_smooth.vertices(:,1);
pairs= pairing_mesh(sphere, value, L, sigma);
figure; plot(pairs(:,1),pairs(:,2),'ko','MarkerEdgeColor',...
    'k', 'MarkerFaceColor',[0.7 0.7 0.7], 'MarkerSize',5)
xlim([2.5 5]); ylim([2.5 5])
```

The code used `FINDnbr.m` that finds neighboring vertices of a given mesh vertex. The resulting pairing is given in Figure 14.10.

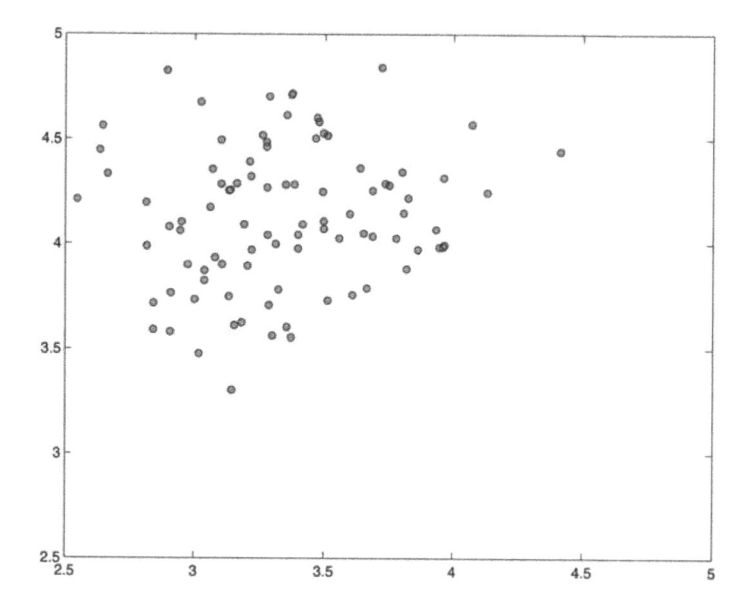

FIGURE 14.10
The min-max diagram of pairing extreme values in Figure 14.9.

14.5.2 Statistical Inference

We have formally tested our hypothesis of different topological structures between the groups. Given a min-max diagram in the square $[1,7]^2$, we have discretized the square with the uniform grid such that there are a total of 50^2 pixels (Figure 14.7-b). A concentration map of the pairings was obtained by counting the number of pairs in a circle of radius 0.2 centered at each pixel. Notice that this approach is somewhat similar to the voxel-based morphometry [14], where brain tissue density maps are used as a shapeless metric for characterizing concentration of the amount of tissue. The inference at the 0.05 level (corrected for multiple comparison) was done by performing 5000 random permutations on the maximum of t-statistic of concentration maps (Figure 14.7-c).

If data is white noise, pairings occur close to $y = x$ line. The deviation from $y = x$ indicates signal. In the t-test result, we detected two main clusters of pairing difference. High number of pairings occurs around (2,6) for controls and (4,4) for autism. This is only possible if surfaces have more geometric features and signals in the controls. On the other hand, the autism shows a noisier characteristic.

The concentration maps are constructed for all min-max diagrams and the two sample t-statistic map $T(t)$ (with equal variance) is constructed (Figure 14.7). Unfortunately, the concentration maps do not follow Gaussian distributional assumptions. Note that this is the usual multiple comparison problem

[394] due to correlated t statistic values across neighboring pixels. So we have performed a permutation test with 5000 random permutations to empirically estimate the distribution of $\sup_{t\in[1,7]^2} T(t)$ and $\inf_{t\in[1,7]^2} T(t)$ to determine the statistical significance. By thresholding the tails of the empirical distribution at 0.05, we obtain the quantile points 3.61 and -4.05. We have detected two main clusters of pairing concentration difference. The blue cluster below -4.05 is the region of high pairing concentration for controls while the red cluster above 3.61 is the region of high pairing concentration for autism. The implication is that more pairing occurs at high and low thickness values in the control subjects.

14.6 Discussion

We have presented a unified framework of the min-max diagram based signal characterization in images. While unconventional, we believe that this representation is very powerful and holds considerable promise for a variety of learning and inference problems in neuroimaging. To demonstrate these ideas, we applied the methods to characterize cortical thickness data in a dataset of autistic and control subjects, *via* the use of a new Iterative Pairing and Deletion algorithm (to generate the min-max diagram). Our results indicate that significant improvements in classification accuracy are possible (relative to existing methods) merely by representing the input data as a set of min-max diagrams. Finally, we note that this chapter only scratches the surface, and future research will clearly bring up other applications where these ideas might be useful.

15

Sparse Networks

In this chapter, we introduce a whole brain multimodal structural connectivity study in characterizing white matter abnormalities using MRI and DTI. The Jacobian determinant (JD) from MRI and the fractional anisotropy (FA) from DTI over predefined nodes are taken as response vectors in a multivariate general linear model (MGLM). However, when the number of nodes is larger than the number of samples, the covariance matrix estimation is ill-conditioned. So it is necessary to regularize the covariance matrix using various sparse regressions such as compressed sensing, LASSO and sparse likelihood. These sparse regressions are known to cause huge computational bottlenecks. By exploiting hidden persistent homological structure embedded in the sparse regressions, we show how to completely bypass the time consuming L1-optimization and still be able to construct sparse networks and do a network inference. This chapter is based on [75].

15.1 Introduction

We present novel multivariate tensor-based morphometry (TBM) for characterizing white matter abnormalities. Traditionally TBM is used in quantifying tissue volume changes in a massive univariate fashion. At each voxel, the Jacobian determinant obtained from TBM is used as the response variable in a general linear model (GLM) and a test statistic is constructed. However, this obvious approach cannot be used in testing, for instance, if the change in one voxel is related to other voxels. To address this limitation of univariate-TBM, we propose a novel multivariate framework for more complex relational hypotheses across brain regions. To develop multivariate-TBM, it is necessary to regularize ill-conditioned covariance matrix by incorporating sparse penalty. Unfortunately, most sparse models like compressed sensing, sparse likelihood and LASSO cause a serious computational bottleneck. The computational bottleneck can be bypassed by exploiting hidden persistent structures in the sparse models. The proposed methods are applied to quantify abnormal white matter in maltreated children to show multivariate-TBM combined with persistent homology can extract additional information that cannot be obtained in univariate-TBM.

15.2 Massive Univariate Methods

Tensor-based morphometry (TBM) uses the spatial derivatives of deformation fields obtained during nonlinear image registration [14, 16, 361, 81]. The morphological tensor maps are subsequently computed and used to quantify variations in high order morphometric changes at each voxel. From these tensor maps, statistical parametric maps are created for a group of subjects in the 3D whole brain volume, on the 2D cortical surface [360, 10, 83, 225, 377] or on the surface of the brain substructures such as the hippocampus and amygdala [375, 376].

Previous TBM analyses have been massively univariate in that response variables are fitted using a linear model at each voxel producing massive number of test statistics (Figure 15.1). However, univariate-TBM is ill-suited for testing more complex hypotheses about multiple anatomical regions. For example, the univariate-TBM cannot answer how the volume increase in one voxel is related to other voxels. To address this type of more complex relational hypothesis across different brain regions, multivariate-TBM is needed.

Motivated by the limitation of traditional univariate-TBM, we present a novel multivariate framework for testing more complex relational hypotheses for multiple brain regions. We propose to correlate the Jacobian determinant across different voxels and quantify how the volume change in one voxel is correlated to the volume changes in other voxels. However, the direct application of existing multivariate statistical methods exhibits serious defects in applying them to the whole brain regions due to the *small-n large-p problem* [129, 314, 369]. Specifically, the number of voxels p are substantially larger than the number of subjects n so the often used maximum likelihood estimation of the covariance matrix shows the rank deficiency and it is no longer positive definite. In turn, the estimated correlation matrix is not considered as a good approximation to the true correlation matrix. The small-n large-p problem can be solved by regularizing the ill-conditioned covariance matrix by sparse regularization terms. Unfortunately, many sparse regression frameworks such as compressed sensing, sparse likelihood and LASSO (least absolute shrinkage and selection operator) cause a serious computational bottleneck when trying to apply the methods to the whole brain.

Sparse model $\mathcal{A}$ is usually parameterized by a tuning parameter λ that controls the sparsity of the representation. So the sparse model $\mathcal{A}(\lambda)$ can be viewed as a function of λ. Increasing the sparse parameter λ makes the representation more sparse. To overcome the computational bottleneck in obtaining sparse solutions, we propose to identify the persistent homological structures in $\mathcal{A}(\lambda)$ for reducing computational complexity. Within the persistent homology framework, $\mathcal{A}(\lambda)$ is *persistent* if it has the nested subset structure under changes in λ value [219, 116, 138]. Then by exploiting the hidden persistent homology in $\mathcal{A}(\lambda)$, we will show that it is possible to completely bypass the

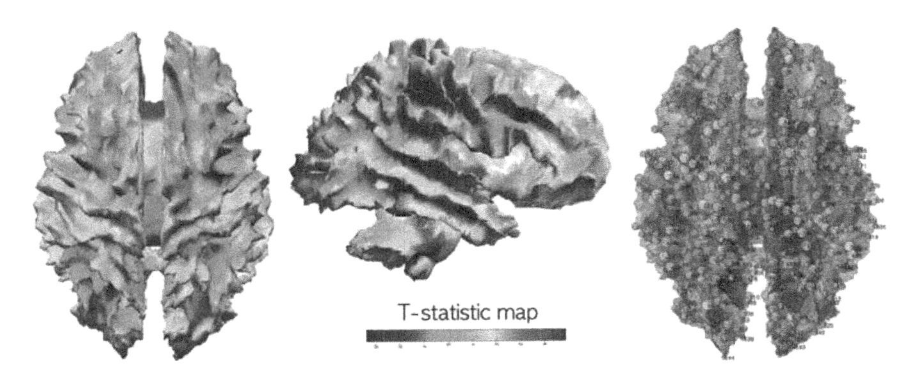

FIGURE 15.1
Left, Middle: T-statistic map of group differences (PI-controls) on Jacobian determinants. Red regions above 4.86 are considered as statistically significant at 0.05 (corrected). Right: 548 uniformly sampled nodes where multivariate-TBM will be performed. The nodes are sparsely sampled in the template to guarantee there is no spurious high correlation due to proximity between nodes. (See color insert.)

FIGURE 15.2
548 uniformly sampled nodes along the white matter surface where multivariate-TBM will be performed. Exactly the same nodes used for Jacobian determinants are also selected. (See color insert.)

computational bottlenecks and speed up the computation by the factor of more than ten-thousand times.

The proposed framework is applied in characterizing abnormal white matter alterations in children who experienced maltreatment while living in post-institutional (PI) settings in Eastern Europe and China before being adopted by families in the US. These children will be compared to age-matched children who did not experience maltreatment.

The main contributions of the proposed framework are (1) the introduction

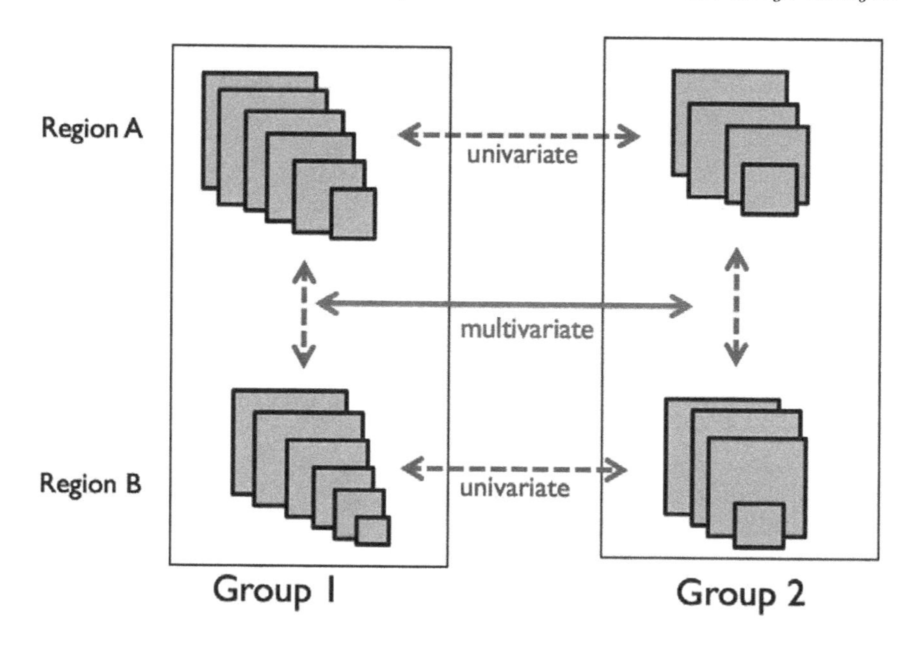

FIGURE 15.3
Schematic showing the difference between univariate- and multivariate-TBM. Each square represents the Jacobian determinant computed at a specific voxel. In univariate-TBM, the Jacobian determinants are compared between the groups at each voxel. In the proposed multivariate-TBM, the change of Jacobian determinants across different voxels is compared between the groups.

of new multivariate-TBM *via* sparse regression, (2) the identification of hidden persistent homological structures in sparse regression such as LASSO and sparse likelihood, (3) the utilization of persistent structures in reducing computational complexity, and (4) its novel application to clinical brain imaging data.

15.3 Why Are Sparse Models Needed?

Let $\mathbf{J}_{n\times p} = (J_{ij})$ be the data matrix of Jacobian determinant for subject i at voxel position j. The subscripts denote the dimension of the matrices. Assume there are p voxels of interest and n subjects. Through the chapter, we will use the following notations. The Jacobian determinants of all subjects at the j-th voxel are denoted as $\mathbf{x}_j = (J_{1j}, \cdots, J_{nj})'$. The Jacobian determinants of all voxels for the i-th subject are denoted as $\mathbf{y}_i = (J_{i1}, \cdots, J_{ip})'$. $\mathbf{x}_j$ is the j-th column and $\mathbf{y}_i$ is the i-th row of the data matrix $\mathbf{J}$.

If we are interested in quantifying the relationship among Jacobian determinants in every voxel simultaneously, the standard procedure is to set up a multivariate general linear model (MGLM). MGLM generalizes widely used univariate general linear models (GLM) by incorporating vector valued response and explanatory variables [9, 129, 394, 396, 355, 80]. MGLM assumes $\mathbf{y}_i$ are independent and identically distributed multivariate normal with mean vector μ and covariance $\mathbf{\Sigma}$. When $p = 1$, MGLM collapses to GLM and the resulting test statistics becomes Hotelling's T^2 statistic often used for inference on vector data [359, 194, 56, 131, 81]. Note that the covariance matrix of $\mathbf{y}_i$ is given by

$$\text{Cov}\ (\mathbf{y}_i) = \mathbf{\Sigma}_{p\times p} = (\sigma_{kl}).$$

For a notational convenience, suppose we center the Jacobian determinant such that

$$\mathbf{y}_i \leftarrow \mathbf{y}_i - \mathbb{E}\mathbf{y}_i.$$

Basically we are subtracting the group mean from individual Jacobian maps to make $\mathbb{E}\mathbf{y}_i = 0$. Then neglecting constant terms, the log-likelihood function L of the data matrix is given by

$$L(\mathbf{\Sigma}) \quad = \quad \log\det\mathbf{\Sigma}^{-1} - \frac{1}{n}\sum_{i=1}^{n}\mathbf{y}_i'\mathbf{\Sigma}^{-1}\mathbf{y}_i \tag{15.1}$$

The maximum likelihood estimate (MLE) of $\mathbf{\Sigma}$ is trivially given as

$$S = \frac{1}{n}\sum_{i=}^{n}\mathbf{y}_i\mathbf{y}_i' = \frac{1}{n}\mathbf{J}_{p\times n}'\mathbf{J}_{n\times p}, \tag{15.2}$$

which is the sample covariance. However, there is a serious defect with MLE (15.2), namely the estimated covariance matrix S is ill-conditioned for $n < p$, which is true for almost all neuroimaging studies. Note that there are more voxels (p) than the number of subject (n) in most studies. Hence, MLE does not yield a good estimation in estimating the covariance matrix [129, 314]. This is the main reason why MGLM was rarely employed over the whole brain and instead massive univariate approaches are still used in most neuroanatomical studies.

To remedy this small$-n$ and large-p problem, we propose to regularize the likelihood term with L_1-penalty and maximize the sparse likelihood:

$$L(\mathbf{\Sigma}^{-1}) = \log\det\mathbf{\Sigma}^{-1} - \ \text{tr}\Big(\mathbf{\Sigma}^{-1}S\Big) - \lambda\|\mathbf{\Sigma}^{-1}\|, \tag{15.3}$$

where $\|\cdot\|$ is the sum of the absolute values of the elements. The sparse-likelihood is given as a function of $\mathbf{\Sigma}^{-1}$ to emphasize that we are actually interested in estimating the inverse covariance. The tuning parameter $\lambda > 0$ controls the sparsity of the off-diagonal elements of the covariance matrix. Then we maximize L over the space of all possible symmetric positive definite matrices. (15.3) is a convex problem and we solve it using the graphical-lasso (GLASSO) algorithm [22, 127, 179]. By increasing λ, the estimated covariance matrix becomes more sparse.

Since the different choice of sparse parameter λ will produce different results, we propose to use the collection of $\mathbf{\Sigma}(\lambda)$ for every possible value of λ for the subsequent statistical inference. This avoids the problem of identifying the optimal sparse parameter that may not be optimal in practice. Unfortunately, GLASSO is a fairly time consuming algorithm [127, 179]. For instance, solving GLASSO for 548 nodes takes about 6 minutes on a desktop computer. To reduce the computational burden, we propose to identify hidden persistent homological structures in sparse regression and exploit their features for reducing the computational complexity [138, 219].

Sparse representation $\mathcal{A}$ is usually parameterized by a tuning parameter λ which controls the sparsity of the representation. So the representation $\mathcal{A}(\lambda)$ can be viewed as a function of λ. Since $\mathcal{A}(\lambda)$ gets more sparse as λ increases, it might be possible to construct a nested subset structure called *filtration* on the tuning parameter such that

$$\mathcal{A}(\lambda_1) \supset \mathcal{A}(\lambda_2) \supset \mathcal{A}(\lambda_3) \supset \cdots \tag{15.4}$$

for $\lambda_1 \leq \lambda_2 \leq \cdots$. In this section, we will explicitly construct such persistent structures for the first time.

15.4 Persistent Structures for Sparse Correlations

To simplify the argument, we assume the measurement vector $\mathbf{x}_j$ at the j-th node is centered with zero mean and unit variance. These conditions are achieved by centering and normalizing data such that $\mathbf{x}_i'\mathbf{x}_i = 1$ and $\sum_{i=1}^{n} x_{ij} = 0$. Let $\mathbf{\Gamma} = (\gamma_{jk})$ be the correlation matrix, where γ_{jk} is the correlation between the nodes j and k. The sample correlation $\widehat{\gamma}_{jk} = \mathbf{x}_j'\mathbf{x}_k$ is shown to satisfy

$$\widehat{\gamma}_{jk} = \arg\min_{\gamma_{jk}} \sum_{j=1}^{p} \sum_{k \neq j} \| \mathbf{x}_j - \gamma_{jk}\mathbf{x}_k \|_2^2 . \tag{15.5}$$

The sparse version of (15.5) is the minimization of

$$F(\gamma_{jk}) = \frac{1}{2} \sum_{j=1}^{p} \sum_{k \neq j} \| \mathbf{x}_j - \gamma_{jk}\mathbf{x}_k \|_2^2 + \lambda \sum_{j,k=1}^{p} |\gamma_{jk}|. \tag{15.6}$$

This is the compressed sensing or LASSO type of sparse regression. By increasing $\lambda \geq 0$, the estimated correlation matrix $\widehat{\mathbf{\Gamma}}(\lambda)$ becomes more sparse. The minimum of F is then achieved when

$$0 = \frac{\partial F}{\partial \gamma_{jk}} = \gamma_{jk} - \mathbf{x}_j'\mathbf{x}_k \pm \lambda.$$

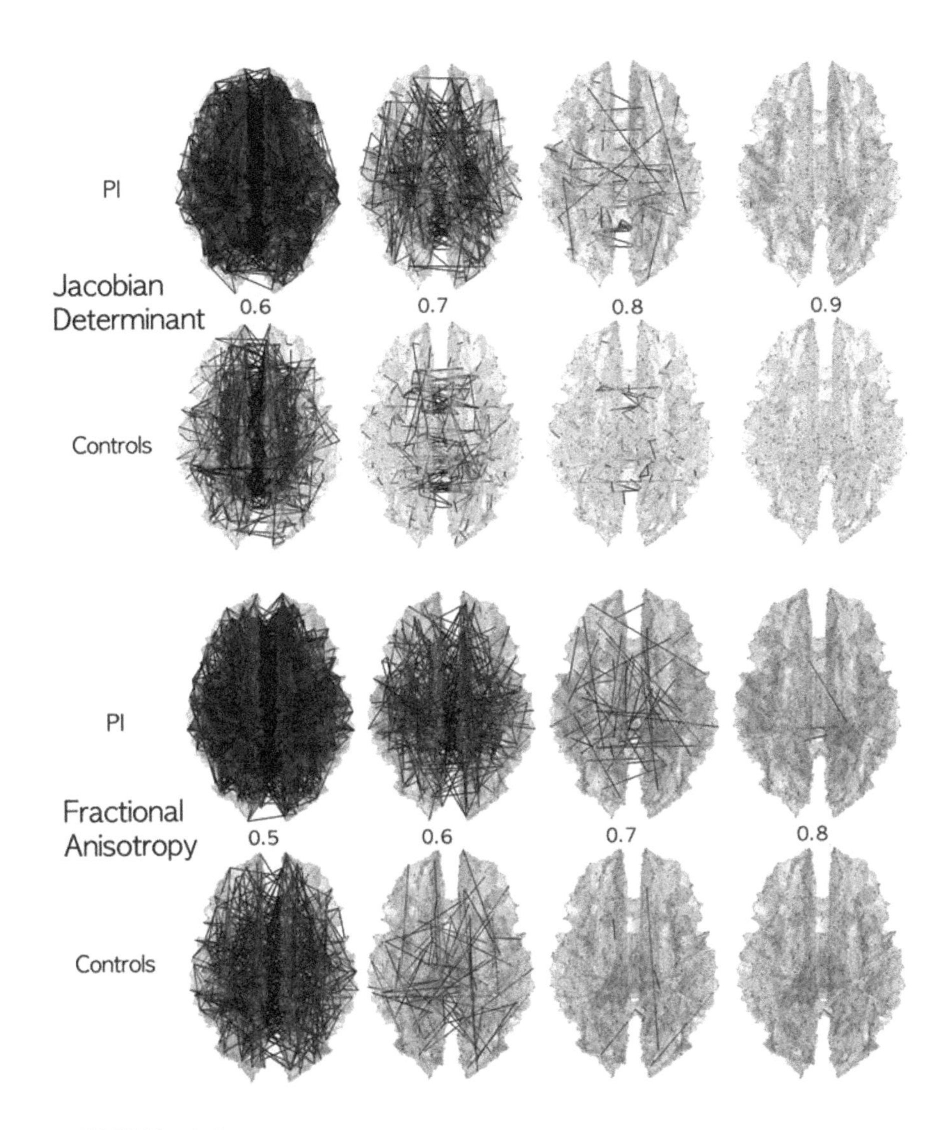

FIGURE 15.4
Graph $\mathcal{G}(\lambda)$ obtained from sparse correlation for the Jacobian determinant from MRI and fractional anisotropy from DTI at different λ values. $\mathcal{G}(\lambda)$ forms a filtration over increasing λ. PI shows more dense network at a given filtration value. Since PI is more homogenous in the white matter region, there are more dense high correlations between nodes. The filtration over the correlation can be also visualized using a dendrogram [58, 219], which also shows more dense connections for PI.

The sign of λ depends on the sign of γ_{jk}. Due to this simple expression, there is no need to optimize (15.6) numerically using the coordinate descent learning or the active-set algorithm often used in compressed sensing [274, 127]. Then for $\lambda \geq 0$, the sparse correlation estimation is given by

$$\widehat{\gamma}_{jk}(\lambda) = \begin{cases} \mathbf{x}_j'\mathbf{x}_k - \lambda & \text{if } \mathbf{x}_j'\mathbf{x}_k > \lambda \\ \mathbf{x}_j'\mathbf{x}_k + \lambda & \text{if } \mathbf{x}_j'\mathbf{x}_k < -\lambda \ . \\ 0 & \text{otherwise} \end{cases} \tag{15.7}$$

Using the sparse solution (15.7), we implicitly construct a persistent homological structure. We will basically build a graph $\mathcal{G}$ using spare correlations. Let $A(\lambda) = (a_{ij})$ be the adjacency matrix defined as

$$a_{jk}(\lambda) = \begin{cases} 1 & \text{if } \widehat{\gamma}_{jk} \neq 0; \\ 0 & \text{otherwise.} \end{cases}$$

This is equivalent to the adjacency matrix $B = (b_{jk})$ defined as

$$b_{jk}(\lambda) = \begin{cases} 1 & \text{if } |\mathbf{x}_j'\mathbf{x}_k| > \lambda; \\ 0 & \text{otherwise.} \end{cases} \tag{15.8}$$

The adjacency matrix B is simply obtained by thresholding the sample correlations. Then the adjacency matrices A and B induce a identical graph $\mathcal{G}(\lambda)$ consisting of $\kappa(\lambda)$ number of partitioned subgraphs

$$\mathcal{G}(\lambda) = \bigcup_{l=1}^{\kappa(\lambda)} G_l(\lambda) \ \text{ with } G_l = \{V_l(\lambda), E_l(\lambda)\},$$

where V_l and E_l are node and edge sets respectively. Note

$$G_l \bigcap G_m = \emptyset \ \text{ for any } \ l \neq m.$$

and no two nodes between the different partitions are connected. The node and edge sets are denoted as $\mathcal{V}(\lambda) = \bigcup_{l=1}^{\kappa} V_l$ and $\mathcal{E}(\lambda) = \bigcup_{l=1}^{\kappa} E_l$ respectively. Then we have the following theorem:

Theorem 10 *The induced graph from the spare correlation forms a filtration:*

$$\mathcal{G}(\lambda_1) \supset \mathcal{G}(\lambda_2) \supset \mathcal{G}(\lambda_3) \supset \cdots \tag{15.9}$$

for $\lambda_1 \leq \lambda_2 \leq \lambda_3$. Equivalently, the node and edge sets also form filtrations as well:

$$\mathcal{V}(\lambda_1) \supset \mathcal{V}(\lambda_2) \supset \mathcal{V}(\lambda_3) \supset \cdots, \quad \mathcal{E}(\lambda_1) \supset \mathcal{E}(\lambda_2) \supset \mathcal{E}(\lambda_3). \tag{15.10}$$

The proof can be easily obtained from the definition of adjacency matrix (15.8).

Hence we have the persistent homological structure induced from the compressed sensing type of the form (15.6). Figure 15.4 shows filtrations obtained from sparse correlations between Jacobian determinants on preselected 548 nodes in the two groups showing group difference.

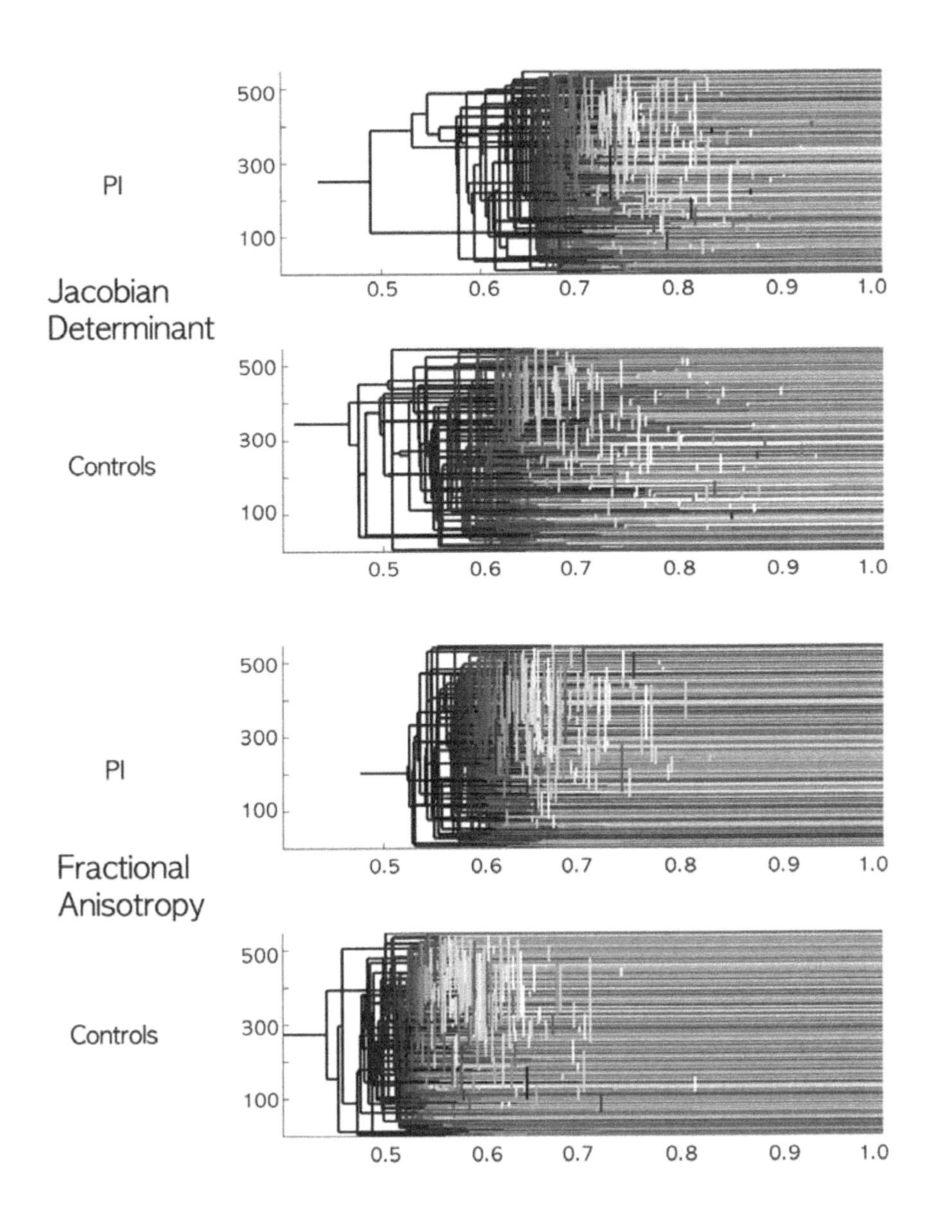

FIGURE 15.5
The filtration over the correlation in Figure 15.4 can be also visualized using a dendrogram [58, 219], which also shows more dense connections for PI.

15.4.1 Numerical Implementation

In this section, we show how to perform various LASSO type of sparse regression in MATLAB. Unlike the persistent homological approach, this requires numerical optimization and can be fairly time consuming for large data sets. For L_1-regularization, we will base our examples on the widely used tool l_1-MAGIC written by Emmanuel Candès and Justine Romberg of the California Institute of Technology (users.ece.gatech.edu/~justin/l1magic). It is necessary to download the package from the website and set up the proper path first:

```
path(path, './l1magic-1.1');
path(path, './l1magic-1.1/Optimization');
```

Let us solve an underdetermined system $b = Ax$, where the matrix dimension is as follows:

$$b_{50\times 1} = A_{50\times 100} x_{100\times 1}.$$

We assume there are $\mathtt{p} = 100$ number of nodes while the number of subjects are $\mathtt{n} = 50$. We put $\mathtt{T} = 4$ spikes as unknown sparse signal.

```
p = 100;
T = 4;
n = 50;
```

A and b are given and it is required to estimate x sparsely by minimizing

$$\min_{Ax=b} \|x\|_1,$$

where $\|x\|_1 = \sum_{i=1}^{n} |x_i|$. This is called the *basis pursuit*. We let all the node values be zero except four positions where we randomly assign values $+1$ or -1. This is illustrated in Figure 15.6.

```
x = zeros(p,1);
q = randperm(p);
x(q(1:T)) = sign(randn(T,1));
figure; subplot(2,2,1); plot(x)
title('Unknown parameter')
```

We let A be a random matrix given by randn(n,p). The least squares estimation is simply computed using matrix inversion and given in the bottom left in Figure 15.6. The least squares estimation provides a really bad fit for sparse data.

```
A = randn(n,p);
b = A*x;
subplot(2,2,2); plot(b);
title('Observation')
```

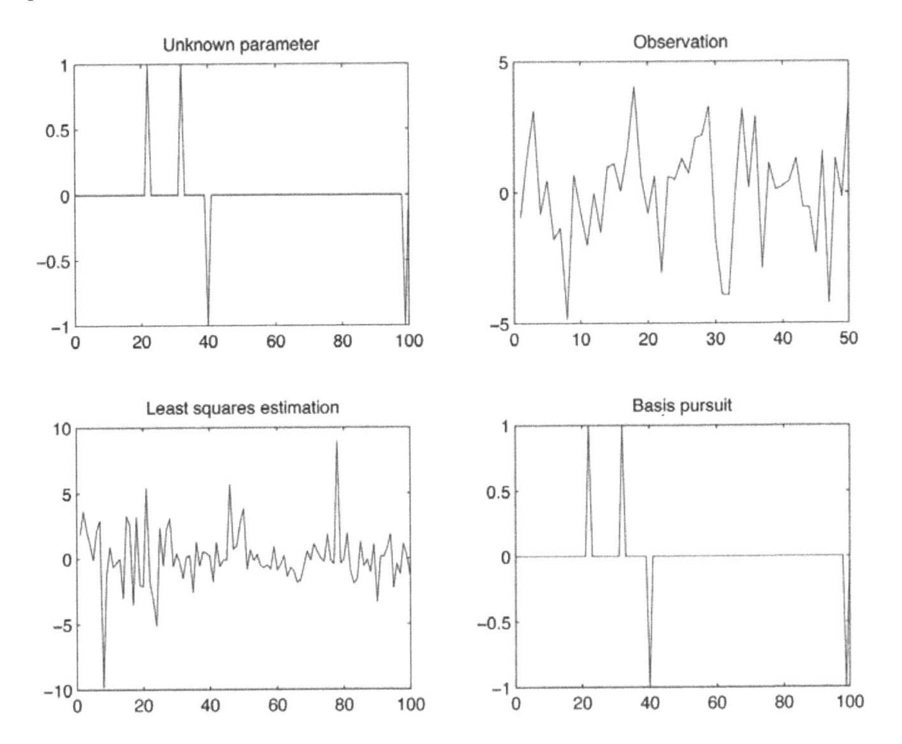

FIGURE 15.6
Unknown parameter vector x is estimated based on the observation b. The least squares estimation gives a really bad estimate while the basis pursuit gives an almost perfect estimation.

```
x_lse=inv(A'*A)*A'*b;
subplot(2,2,3); plot(x_lse)
title('Least squares estimation')
```

The basis pursuit estimates x by minimizing the L_1-norm of x (Bottom right in Figure 15.6). The least squares estimation is taken as the initial guess x0 in the basis pursuit.

```
x0 = inv(A'*A)*A'*b;
x_bp = l1eq_pd(x0, A, [], b, 1e-3);
subplot(2,2,4);  plot(x_bp)
title('Basis pursuit')
```

The basis pursuit basically solves the system $b = Ax$, which is unrealistic due to the lack of noise in practice. A more realistic model is to add noise and solve for

$$\min_{\|b-Ax\|_2 \leq \epsilon} \|x\|_1. \quad (15.11)$$

(15.11) is shown to be equivalent for solving

$$\min_x \|b - Ax\|_2^2 + \lambda\|x\|_1,$$

which is more well known as LASSO. For this example, we will use the same x used in the previous basis pursuit method. For this example, we assume the observations are contaminated by noise `0.7*randn(n,1)`.

```
figure; subplot(2,2,1); plot(x)
title('Unknown parameter')

e = 0.7*randn(n,1);
subplot(2,2,2); plot(e);
title('Noise')

b1 = A*x + e;
subplot(2,2,2);
plot(b,'--r');
hold on; plot(b1);
title('Observation')
```

If we estimate the parameters x with the basis pursuit, we end up with a bad fit. One the other hand, the LASSO gives a better estimation (Figure 15.7). Note that LASSO is a shrinkage estimation so even though it will precisely estimate the node positions, it will shrink the size of the underlying signal.

```
x0 = A'*b1;
x_sp = l1eq_pd(x0, A, [], b1, 1e-3);
subplot(2,2,3);  plot(x_sp)
title('Basis pursuit')

epsilon = 5
x_lasso = l1qc_logbarrier(x0, A, [], b1, epsilon, 1e-3);
subplot(2,2,4); plot(x_lasso)
title('LASSO')
```

15.5 Persistent Structures for Sparse Likelihood

The identification of a persistent homological structure out of the inverse covariance $\widehat{\boldsymbol{\Sigma}}^{-1}(\lambda)$ for sparse-likelihood (15.3) is similar. Let $A(\lambda) = (a_{ij})$ be

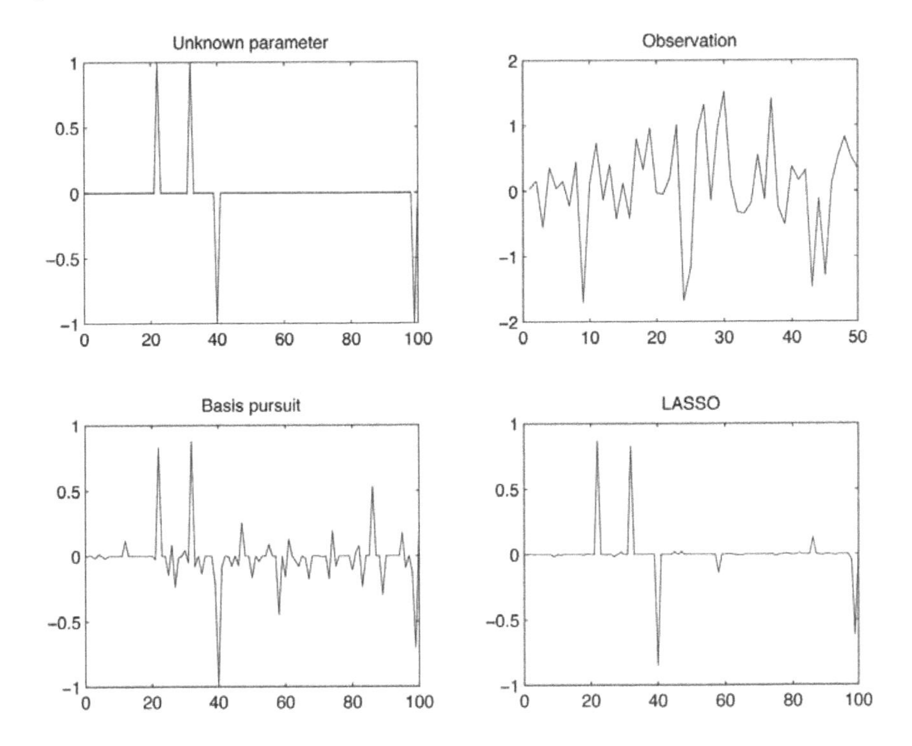

FIGURE 15.7
Unknown parameter vector x is estimated based on the observation b. The basis pursuit gives a really bad estimate while the LASSO gives a better fit.

the adjacency matrix given by

$$a_{ij}(\lambda) = \begin{cases} 1 & \text{if } \widehat{\sigma}^{ij} \neq 0; \\ 0 & \text{otherwise.} \end{cases} \tag{15.12}$$

The adjacency matrix A induces a graph $\mathcal{G}(\lambda)$ consisting of $\kappa(\lambda)$ number of partitioned subgraphs

$$\mathcal{G}(\lambda) = \bigcup_{l=1}^{\kappa(\lambda)} G_l(\lambda) \text{ with } G_l = \{V_l(\lambda), A_l(\lambda)\}.$$

Motivated by the sparse correlation (15.8), let us similarly define a corresponding adjacency matrix $B(\lambda) = (b_{ij})$ as

$$b_{ij}(\lambda) = \begin{cases} 1 & \text{if } |\widehat{s}_{ij}| > \lambda; \\ 0 & \text{otherwise.} \end{cases} \tag{15.13}$$

The adjacency matrix B similarly induces a graph with $\tau(\lambda)$ disjoint sub-

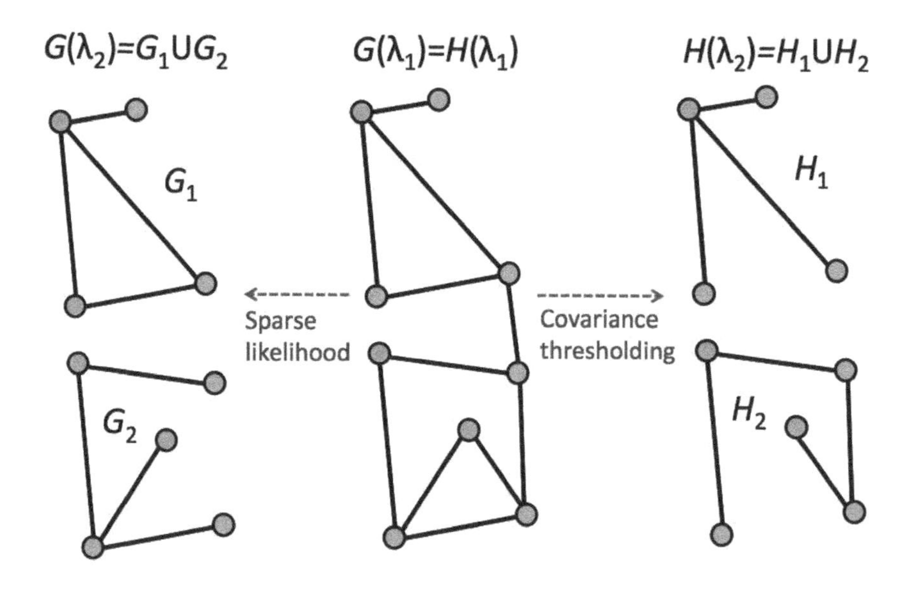

FIGURE 15.8
Schematic of graph filtrations obtained by sparse-likelihood (15.12) and sample covariance thresholding (15.13). At filtration λ_1, we start with $\mathcal{G}(\lambda_1) = \mathcal{H}(\lambda_1)$. For the next filtration value $\lambda_2 \geq \lambda_1$, $\mathcal{G}(\lambda_2) \neq \mathcal{H}(\lambda_2)$. However, the partitioned vertex sets of $\mathcal{G}(\lambda_2)$ and $\mathcal{H}(\lambda_2)$ exactly match (Theorem 11). Exploiting this hidden topological structure, we can drastically speed up network construction and topological computation.

graphs:

$$\mathcal{H}(\lambda) = \bigcup_{l=1}^{\tau(\lambda)} H_l(\lambda) \text{ with } H_l = \{W_l(\lambda), B_l(\lambda)\}.$$

Unlike the sparse correlation case, $\mathcal{G} \neq \mathcal{H}$ and we do not have full persistency on both the node and edge sets. However, the partitioned graphs are shown to be partially nested in a sense that the node sets exhibits persistency.

Theorem 11 *For any $\lambda > 0$, the adjacency matrices (15.12) and (15.13) induce the identical vertex partition so that $\kappa(\lambda) = \tau(\lambda)$ and $V_l(\lambda) = W_l(\lambda)$. Further, the node sets V_l and W_l form a filtration over the sparse parameter:*

$$V_l(\lambda_1) \supset V_l(\lambda_2) \supset V_l(\lambda_3) \supset \cdots \tag{15.14}$$
$$W_l(\lambda_1) \supset W_l(\lambda_2) \supset W_l(\lambda_3) \supset \cdots \tag{15.15}$$

for $\lambda_1 \leq \lambda_2 \leq \lambda_3$.

From (15.13), it is trivial to see the filtration holds for W_l. The filtration for

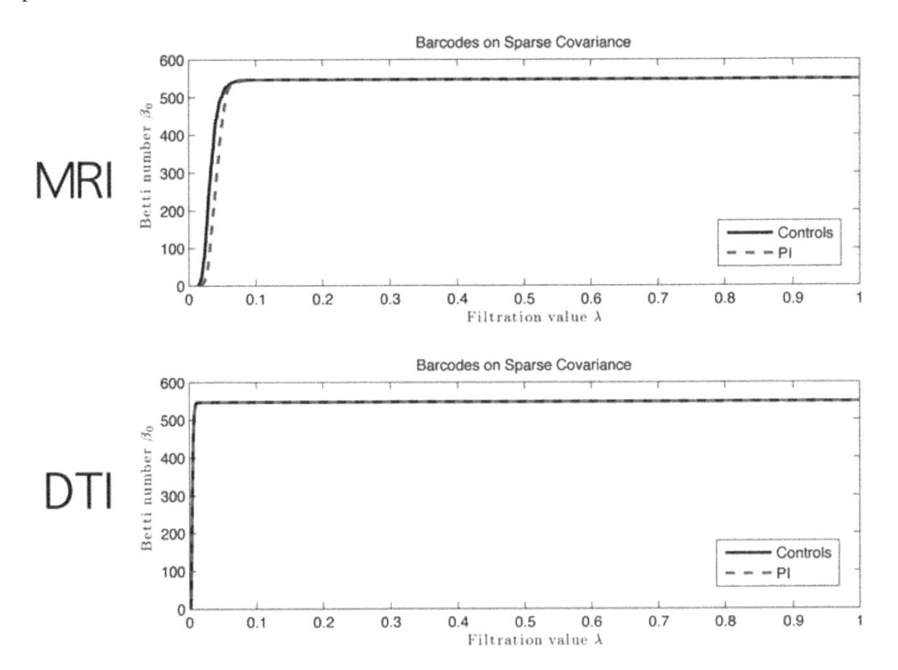

FIGURE 15.9
The barcodes on the sparse inverse covariance for Jacobian determinant from MRI and FA from DTI. The inverse covariance does not show huge group separation between normal controls and post-institutionalized (PI) children.

V_l is proved in [179]. The equivalence of the node sets $V_l = W_l$ is proved in [245]. Note that the edge sets may not form a filtration. The construction of the filtration on the node sets V_l (15.14) is very time consuming since we have to solve the sequence of GLASSO. For instance, for 548 node sets and 547 different filtration values, the whole filtration takes more than 54 hours in a desktop. Theorem 11 is illustrated in Figure 15.8 with two levels of filtration.

15.6 Case Study: Application to Persistent Homology

15.6.1 MRI Data and Univariate-TBM

T1-weighted MRI were collected using a 3T GE SIGNA scanner for 23 children who experienced maltreatment while living in post-institutional (PI) settings in Eastern Europe and China before being adopted by families in the US, and age-matched 31 normal control subjects. The average age for PI is 11.26 $\pm$

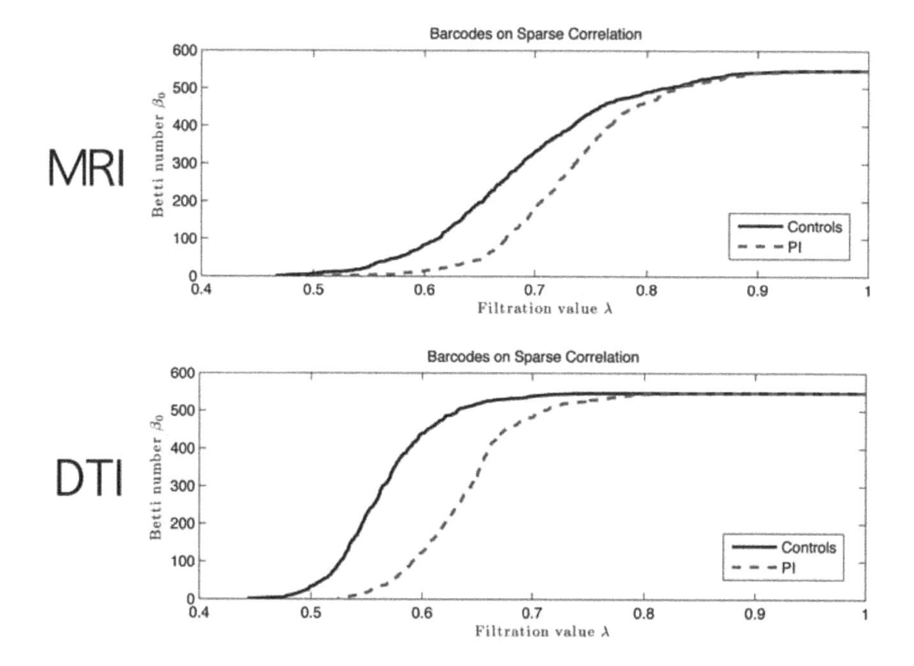

FIGURE 15.10
The barcodes on the sparse correlation for Jacobian determinant from MRI and FA from DTI. Unlike inverse covariance, the correlation shows huge group separation between normal controls and post-institutionalized (PI) children (p-value < 0.001).

1.71 years while that of controls is 11.58 $\pm$ 1.61 years. There are 10 boys and 13 girls in PI, and 18 boys and 13 girls in the control subjects.

A study specific template was constructed using the diffeomorphic shape and intensity averaging technique through Advanced Normalization Tools (ANTS) [19]. Image normalization of each individual image to the template was done using symmetric normalization with cross-correlation as the similarity metric. The deformation fields are then smoothed out using Gaussian kernel with bandwidth $\sigma = 4$mm, which is equivalent to the full width at half maximum (FWHM) of 4mm. Then the Jacobian determinant of the inverse deformation was computed at each voxel.

The computed Jacobian maps were feed into univariate-GLM at each voxel for testing the group effect while accounting for nuisance covariates such as age and gender. Figure 15.1 shows the significant group difference between PI and controls. Any region above 4.86 or below -4.86 is considered significant at 0.05 (corrected) [396]. However, what the univariate-TBM can not test is the dependency of Jacobian determinants at two different positions. It is possible that structural abnormality at one region of the brain might be related to the other regions due to interregional dependency. For this type of more complex hypothesis, we need the proposed multivariate approach.

15.6.2 Multivariate-TBM via Barcodes

Since Jacobian determinants at neighboring voxels are highly correlated, we uniformly subsampled 548 nodes along the white matter boundary in order not to have spurious high correlation between two adjacent nodes (Figure 15.1). The number of predefined regions is still larger than most regions of interest (ROI) approaches in MRI and DTI [407]. Subsequently we applied the proposed multivariate framework and obtained sparse correlations and inverse covariance, and constructed the filtrations on them. Without solving the optimization problem (15.8), the sample correlation and covariance are simply thresholded to obtain the filtrations. For 547 levels of filtration, the sequence of GLASSO would take more than 54 hours in a desktop (6 min. per GLASSO). But it took less than 1 min. using the simple thresholding method to construct the filtrations.

A filtration is usually quantified by the barcode which plots the change of Betti numbers [116, 138, 219]. The first Betti number $\beta_0(\lambda)$ counts the number of connected components at the filtration value λ. Given barcode $\beta_0^i(\lambda)$ for group i, we are interested in testing the null hypothesis

$$H_0 : \beta_0^1(\lambda) = \beta_0^2(\lambda) \text{ for all } \lambda \in [0, 1]$$

against the alternative hypothesis

$$H_1 : \beta_0^1(\lambda) \neq \beta_0^2(\lambda) \text{ for some } \lambda \in [0, 1].$$

Since barcodes are similar to the shape of cumulative probability distribution functions, Kolmogorov-Smirnov (KS) like test statistic can be used:

$$T = \sup_{\lambda \in [0,1]} \left| \beta_0^1(\lambda) - \beta_0^2(\lambda) \right|.$$

Since each group produces one barcode, we used the jackknife resampling technique for inference. For a group with n subjects, one subject is removed and the remaining $n - 1$ subjects are used in computing the sparse inverse covariance and correlations. This process is repeated for each subject to produce n covariance and correlations. Then the filtration is performed on jackknife resampled covariance and correlations (Figures 15.9 and 15.10). The jackknife resampling produces 23 and 31 barcodes respectively for PI and controls. Then KS-like test statistic K is constructed between 23×31 pairs of barcodes. Under the null, K is expected to be zero. One-sample student T-test is then subsequently performed to show almost perfect group separation (p-value < 0.001).

15.6.3 Connection to DTI Study

Severe stress and maltreatment during early development is found to be related to structural abnormality in various brain regions [368, 181, 164, 163].

Thus we expect white matter differences not only in the Jacobian determinants but the fractional anisotropy (FA) values obtained from DTI as well. The MRI data in this study have the corresponding DTI. The DTI acquisition is done in the same 3T GE SIGNA scanner and the acquisition parameters can be found in [163]. We applied the proposed methods in obtaining the sparse correlation and inverse covariance maps in the same 548 nodes. The resulting filtration patterns also show a similar pattern of rapid increase in disconnected components (Figure 15.4 and 15.10). The jackknife-based one-sample T-test also shows significant group difference (p-value < 0.001). These results are due to consistent abnormalities observed in both MRI and DTI modalities. The PI group exhibited stronger white matter homogeneity and less spatial variability compared to normal controls in both MRI and DTI measurements.

15.7 Sparse Partial Correlations

So far we explored how to construct sparse networks using sparse correlations and the sparse likelihood method. In this section, we introduce the sparse partial correlation based method. This section is based on [75].

15.7.1 Partial Correlation Network

Let p be the number of nodes in the network. In most applications, the number of nodes is expected to be larger than the number of observations n, which gives an underdetermined system. The i-th measurement f_i is then discretely sampled at p nodes, which we will simply index by integers. To simplify the notation, we denote $x_{ij} = f_i(j)$. At node j, we have the random variable x_j, which is realized by the random sample $x_{1j}, \cdots, x_{nj}$. We will denote this realization as $\mathbf{x}_j = (x_{1j}, \cdots, x_{nj})'$. The collection of random variables x_j are assumed to be distributed with mean zero and covariance $\Sigma = (\sigma_{jj'})$ i.e.

$$\mathbb{E}x_j = 0, \ \mathbb{E}(x_j x_{j'}) = \sigma_{jj'}.$$

If $\mathbb{E}x_j \neq 0$, we can always center the data by translation. The correlation $\gamma_{jj'}$ between the two nodes j and j' is given by

$$\gamma_{jj'} = \frac{\sigma_{jj'}}{\sqrt{\sigma_{jj}\sigma_{j'j'}}}.$$

By thresholding the correlation, we can establish a link between two nodes. However, there is a problem with this simplistic approach in that it fails to explicitly factor out the confounding effect of other nodes. To remedy this problem, partial correlations can be used in factoring out the dependency of other nodes [170, 242, 178, 179, 274].

If we denote the inverse covariance matrix as $\Sigma^{-1} = (\sigma^{jj'})$, the *partial correlation* between the nodes j and j' while factoring out the effect of all other nodes is given by

$$\rho_{jj'} = -\frac{\sigma^{jj'}}{\sqrt{\sigma^{jj}\sigma^{j'j'}}}. \tag{15.16}$$

Equivalently, we can compute the partial correlation *via* a linear model as follows. Consider a linear model of correlating measurement at node j to all other nodes:

$$x_j = \sum_{k \neq j} \beta_{jk} x_k + \epsilon_k. \tag{15.17}$$

The parameters β_{jk} are estimated by minimizing the sum of squared residual of (15.17)

$$L(\beta) = \sum_{j=1}^{p} \|\mathbf{x}_j - \sum_{k \neq j} \beta_{jk} \mathbf{x}_k\|^2 \tag{15.18}$$

in a least squares fashion. If we denote the least squares estimator by $\widehat{\beta_{jk}}$, the residuals are given by

$$r_j = x_j - \sum_{k \neq j} \widehat{\beta_{jk}} x_k. \tag{15.19}$$

The partial correlation is then obtained by computing the correlation between the residuals of the model fit (15.17) [170, 227, 274]:

$$\rho_{jj'} = \text{corr}\ (r_j, r_{j'}).$$

The minimization of (15.18) is exactly given by solving the normal equation:

$$\mathbf{x}_j = \sum_{k \neq j} \beta_{jk} \mathbf{x}_k, \tag{15.20}$$

which can be turned into standard linear form $y = A\beta$ [221]. Note that (15.20) can be written as

$$\mathbf{x}_j = \underbrace{[\mathbf{x}_1, \cdots, \mathbf{x}_{j-1}, \mathbf{0}, \mathbf{x}_{j+1}, \cdots, \mathbf{x}_p]}_{\mathbf{X}_{-j}} \underbrace{\begin{pmatrix} \beta_{j1} \\ \beta_{j2} \\ \vdots \\ \beta_{jp} \end{pmatrix}}_{\beta_j},$$

where $\mathbf{0}_{n\times 1}$ is a column vector of all zero entries. Then we have

$$\underbrace{\begin{pmatrix} \mathbf{x}_1 \\ \mathbf{x}_2 \\ \vdots \\ \mathbf{x}_p \end{pmatrix}}_{y_{np\times 1}} = \underbrace{\begin{pmatrix} \mathbf{X}_{-1} & \mathbf{0} & \cdots & \mathbf{0} \\ \mathbf{0} & \mathbf{X}_{-2} & \cdots & \mathbf{0} \\ \vdots & \vdots & \ddots & \vdots \\ \mathbf{0} & \mathbf{0} & \cdots & \mathbf{X}_{-p} \end{pmatrix}}_{A_{np\times p^2}} \underbrace{\begin{pmatrix} \beta_1 \\ \beta_2 \\ \vdots \\ \beta_p \end{pmatrix}}_{\beta_{p^2\times 1}}, \tag{15.21}$$

where A is a block diagonal matrix and $\mathbf{0}_{n\times p}$ is a matrix of all zero entries.

15.7.2 Sparse Network Recovery

There is a serious problem with the least squares estimation framework discussed in the previous section. Since $n \ll p$, this is a significantly underdetermined system. This is also related to the covariance matrix Σ being singular so we cannot just invert the covariance matrix in (15.16). So we need to regularize (15.21) by incorporating l_1 LASSO-penalty J [365, 274, 221]:

$$J = \sum_{j,j'} |\beta_{jj'}|.$$

The sparse estimation of $\beta_{jj'}$ is then given by minimizing $L + \lambda J$. Since there is dependency between y and A, (15.21) is not exactly a standard compressed sensing problem. Nevertheless, as an exploratory data analysis, we will model as if they are independent as has been done in others [274, 221]. It should be intuitively understood that sparsity makes the linear equation (15.20) less underdetermined. The larger the value of λ, the more sparse the underlying topological structure gets. Since

$$\rho_{jj'} = \beta_{jj'} \sqrt{\frac{\sigma^{jj}}{\sigma^{j'j'}}},$$

the sparsity of $\beta_{jj'}$ directly corresponds to the sparsity of $\rho_{jj'}$, which is the strength of the link between nodes j and j' [274, 221]. Once the sparse partial correlation matrix ρ is obtained, we can simply link nodes j and j', if $\rho_{jj'} > 0$ and assign the weight $\rho_{jj'}$ to the edge. This way, we obtain the weighted graph. To simplify the problem, we will only consider positive partial correlations ρ^+ (Figure 15.11). Since the partial correlation matrix is likely to be very sparse, the resulting weighted graph will have an easily interpretable topological structure.

15.7.3 Sparse Network Modeling

The majority of functional and structural connectivity studies in brain imaging are usually performed following the standard analysis framework

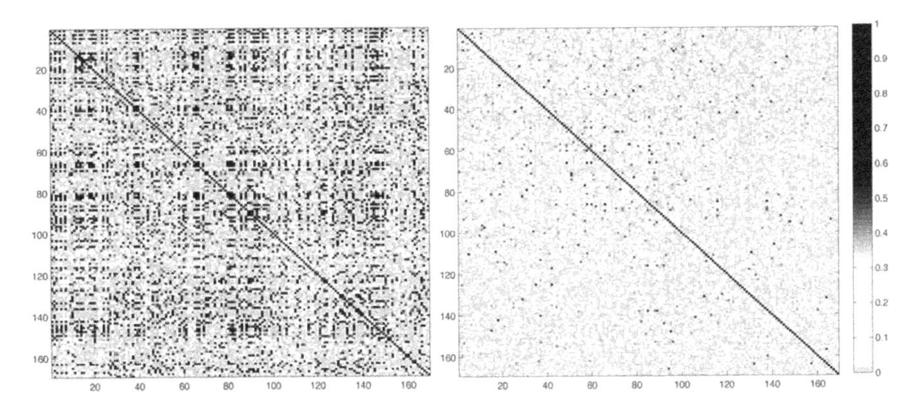

FIGURE 15.11
Partial correlation estimation using least squares estimation (left) and LASSO with $\lambda = 100$ (right). Only positive correlations are shown. The LASSO penalty reduces the number of links in the network by forcing sparsity on partial correlations.

[145, 158, 123, 407]. From 3D whole brain images, n regions of interest (ROI) are identified and serve as the nodes of the brain network. Measurements at ROIs are then correlated in a pair-wise fashion to produce the connectivity matrix of size $n \times n$. The connectivity matrix is then thresholded to produce the adjacency matrix consisting of zeros and ones that define the link between two nodes. The binarized adjacency matrix is then used to construct the brain network. Then various graph complexity measures such as degree, clustering coefficients, entropy, path length, hub centrality and modularity are defined on the graph and the subsequent statistical inference is performed on these complexity measures.

For a large number of nodes, simple thresholding of correlation will produce a large number of links which makes the interpretation difficult. For example, for 3×10^5 voxels in an image, we can possibly have a total of 9×10^{10} links in the graph. For this reason we used the sparse data recovery framework in obtaining a far smaller number of significant links.

15.7.4 Application to Jacobian Determinant

We have applied the method to a group of 33 normal control (NC) subjects. T1-weighted MRIs were collected using a 3T GE SIGNA scanner. Details on the image preprocessing pipelines are explained in [164]. Using symmetric normalization (SyN) algorithm [19], MRIs were registered to a template. The nonlinear transformations resulting from the SyN algorithm characterize the voxelwise shape changes from the template to each subjectŠs brain. Afterward

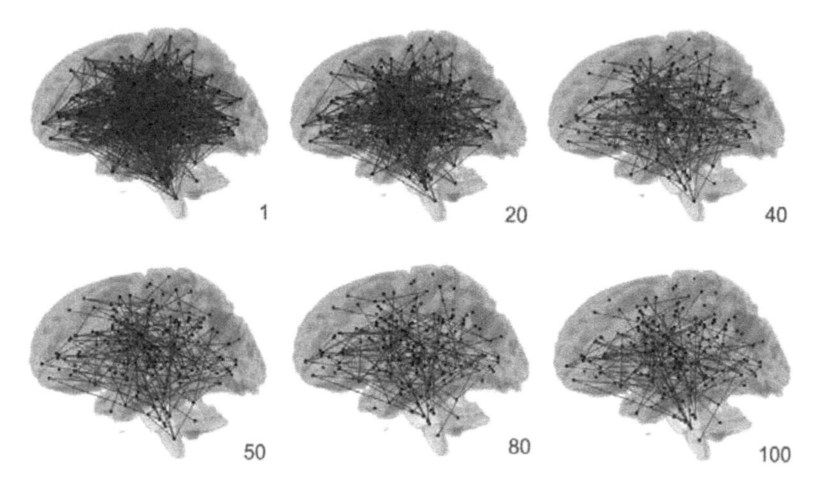

FIGURE 15.12
Structural brain network of 33 control subjects. The LASSO-penalty is used for various λ values from 1 to 100. Increasing the λ value increases the sparsity of connections and, in turn, simplifies the topological structure of the network. Persistent topological features such as Betti numbers over increasing λ can be used in the characterization of brain networks [220]. The sparsity parameter λ can be automatically determined depending on if the change of the amount of sparsity is significant.

the Jacobian determinant of deformation from individual MRI to a template is computed at each voxel. The Jacobian determinant quantifies the fractional volume expansion and contraction at each voxel with respect to the template. Taking the Jacobian determinant for the i-th subject as the functional measurement f_i, we performed the topological network modeling. 169 voxels in the white matter are uniformly selected to form nodes of the network. We have sparsely estimated the partial correlation of the Jacobian determinants across nodes (Figure 15.11). This is a huge l_1-minimization problem and requires to estimate a total of 169^2 parameters. For instance, the matrix A is of size 5577×28561. We have used the interior-point method with various λ values in sparsely estimating ρ^+ [208]. Using $\rho^+_{jj'}$ as link weights, we can construct the weighted graphs (Figure 15.12).

15.7.5 Limitations of Sparse Partial Correlations

We have presented a novel framework that formulates signal detection as an underlying topological structure recovery. Functional measurements are represented as simplicial complexes or networks. The strength of the links between nodes is measured using partial correlations. The sparse data recovery framework is used to sparsely estimate the partial correlations and obtain sparse topological structures. The sparse topological structures are easier to interpret as demonstrated in brain network modeling.

However, this new framework has a serious computational bottleneck. For n measurements over p nodes, it is required that we solve a linear system with an extremely large A matrix of size $np \times p^2$, so that the complexity of the problem increases by a factor of p^3! Consequently, for a large number of nodes, the problem immediately becomes almost intractable for a small computer. For example, for 1 million nodes, we have to compute 1 trillion possible pairwise relationships between nodes. One practical solution is to modify (15.17) so that the measurement at node i is represented more sparsely over some possible index set S_i:

$$x_i = \sum_{S_i} \beta_{ij} x_j + \epsilon_i.$$

making the problem substantially smaller.

An alternate approach is to simply follow the *homotopy path*, which adds network links one by one with a very limited increase of computational complexity so there is no need to compute β repeatedly from scratch [108, 286, 269]. The trajectory of the optimal solution β in LASSO follows a piecewise linear path as we change λ. By tracing the linear path, we can substantially reduce the computational burden of reestimating β when λ changes.

15.8 Summary

This chapter explored three major sparse network modeling framework: sparse correlations, sparse partial correlations and sparse likelihood. For sparse correlations and sparse likelihood, we showed how to speed up network construction and computations by exploiting hidden topological structures using persistent homology. For sparse partial correlations, we believe there is a similar relationship but it is unclear what exactly is the relationship and we leave it for a future study.

16

Sparse Shape Models

We present a new sparse shape modeling framework on the Laplace-Beltrami (LB) eigenfunctions. Traditionally, the LB-eigenfunctions are used as a basis for intrinsically representing surface shapes as a form of Fourier descriptors. To reduce high frequency noise, only the first few terms are used in the expansion and higher frequency terms are simply thrown away. However, some lower frequency terms may not necessarily contribute significantly in reconstructing the surfaces. Motivated by this simple idea, we present a LB-based method to filter out only the significant eigenfunctions by imposing an L_1-norm penalty. The sparse surface shape model is applied in investigating the influence of age and gender on amygdala and hippocampus shapes in the normal population. This chapter is based on [205].

16.1 Introduction

The influence of age on the changes of subcortical structures has been somewhat controversial [346, 373]. The effect of aging on amygdalar and hippocampal structures has drawn much attention [41, 112, 147, 295, 345, 373]. While many cross sectional and longitudinal studies reported significant reduction in regional volume of amygdala and hippocampus due to aging [41, 112, 295], others failed to confirm such relationship [147, 345, 373]. Gender may be another factor that affects these structures. One study reported significant gender effect in amygdala and hippocampus volume [146] whereas others failed to replicate the finding [154].

In these traditional volumetric studies, the total volume of the amygdala and hippocampus were typically estimated by tracing the region of interest (ROI) manually and counting the number of voxels within the ROI. The limitation of the ROI-based volumetry is that it cannot determine if the volume difference is diffuse over the whole ROI or localized within specific regions of the ROI [81]. The proposed sparse shape representation can localize the volume difference up to the mesh resolution at each surface mesh vertex.

Starting with the 3D deformation field derived from the spatial normalization of MRI, we can model how the surfaces of subcortical structures are different from each other at the vertex level. Since the deformation field is noisy, it is necessary to smooth out the field along the surface to increase the signal-to-noise ratio (SNR). Further, smoothness of data is desirable in satisfying the assumptions of the random field theory, which is used in correcting for multiple comparisons [4, 394]. With these motivations, we present a framework that sparsely filters out significant coefficients in the LB-eigenfunction expansion using the L_1-norm penalty, which is often used in compressed sensing and sparse regression. The proposed framework is then used in examining the effect of age and gender on amygdala and hippocampus shapes contrasting with the traditional volumetric analysis. We further show how to model the emotional response on the subcortical structure shapes.

The proposed pipeline for sparse shape modeling is as follows:

(1) Obtain a mean volume of a subcortical structure by averaging the spatially normalized binary masks and extract a template surface from the averaged binary volume.

(2) Interpolate the 3D displacement vector field onto the vertices of the surface meshes.

(3) Estimate a sparse representation of Fourier coefficients with L_1-norm penalty for the displacement length along the template surface to reduce noise.

(4) Apply a general linear model (GLM) testing the effect of age and gender on the displacement.

16.2 Amygdala and Hippocampus Shape Models

Due to the limitations of the current MRI acquisition technique such as insufficient resolutions and low contrasts, fully-automated segmentation of subcortical structures is not yet satisfactory [291]. Instead, full or partial use of manual segmentation has been utilized with surface modeling frameworks for the subcortical structures to sensitize the power of detection. In [362], hippocampus and temporal horn were manually delineated then the surfaces are parametrized. The distance between the medial axis and boundary of the surface is measured as a local feature of the subcortical structure in order to characterize local expansion and contraction. In [292, 291], hippocampus masks are first automatically segmented using FreeSurfer then were corrected by injecting a subcortical template based on manual segmentations. Then the atrophy due to normal aging is quantified by normal surface momentum while the group difference was inferred on the Laplace-Beltrami eigenvalues. In [80], manually segmented amygdalar surfaces were modeled using spherical harmonics. The displacement vector field between the template and the indi-

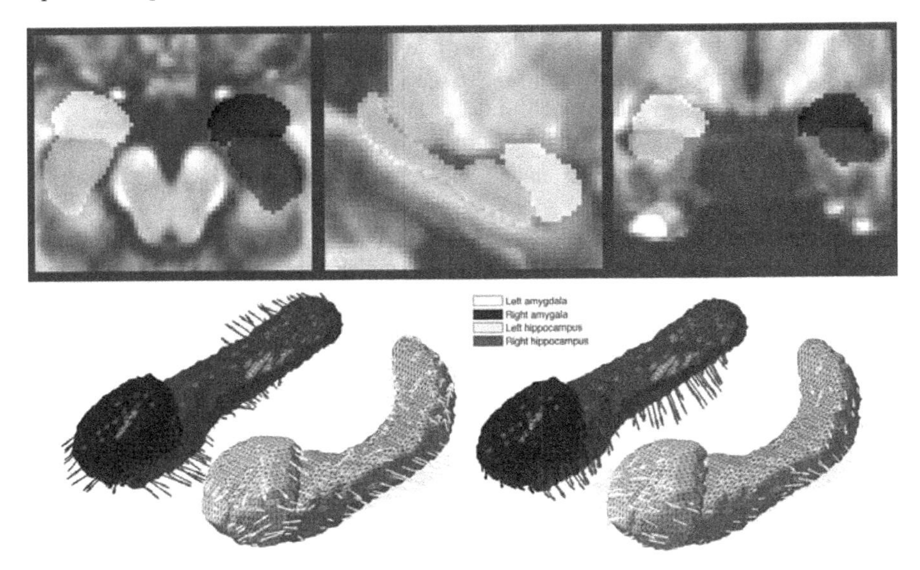

FIGURE 16.1
Manual segmentation of hippocampus and amygdala. Template surfaces are constructed by averaging the binary segmentation in the normalized space. The arrows are the displacement of warping from the template to individual surfaces.

vidual surfaces was used as a multivariate response variable in multivariate general linear models.

16.3 Data Set

We have high-resolution T1-weighted inverse recovery fast gradient echo MRI, collected in 124 contiguous 1.2-mm axial slices (TE=1.8 ms; TR=8.9 ms; flip angle = 10°; FOV = 240 mm; 256 × 256 data acquisition matrix) of 52 middle-age and elderly adults ranging between 37 to 74 years (mean age = 55.52 ± 10.40 years). There are 16 men and 36 women in the study. Trained raters manually segmented the amygdala and hippocampus structures. Brain tissues in the MRI scans were automatically segmented using Brain Extraction Tool (BET) [332]. Then we performed a nonlinear image registration using the diffeomorphic shape and intensity averaging technique with the cross-correlation as the similarity metric through Advanced Normalization Tools (ANTS) [19]. Then a study-specific template was constructed from a random subsample of 10 subjects.

Using the deformation field obtained from warping the individual image to the template, we aligned the amygdala and hippocampus binary masks to the template space. The normalized masks were then averaged to produce the subcortical structure template. The isosurfaces of the subcortical structure template were extracted using the marching cube algorithm [232].

The displacement vector field is defined on each voxel, while the vertices of mesh are located within a voxel. So we linearly interpolated the vector field on mesh vertices from the voxels (Fig. 16.1). The length of the displacement vector at each vertex is computed and used as a response variable measuring the local shape variation with respect to the template space.

16.4 Sparse Shape Representation

Since the lengths of displacement defined on mesh vertices are expected to be noisy due to errors associated with image acquisition and image preprocessing, it is necessary to smooth out the noise and increase the signal-to-noise ratio [80, 83]. Many previous surface data smoothing approaches have used heat diffusion type of smoothing to reduce surface noise [10, 80, 239, 275, 333, 351, 353].

Instead, we propose to use the Laplace-Beltrami (LB) eigenfunctions in parametrically representing the surface data. In previous LB-eigenfunction and similar spherical harmonic (SPHARM) expansion approaches only the first few terms are used in the expansion and higher frequency terms are simply thrown away [290, 319, 318, 344] to reduce the high frequency noise. However, some lower frequency terms may not necessarily contribute significantly in reconstructing the surfaces. Motivated by this idea, we propose to sparsely filter out insignificant eigenfunctions by imposing the L_1-penalty.

Consider a real-valued functional measurement $Y(p)$ on a manifold $\mathcal{M} \subset \mathbb{R}^3$. We assume the following additive model:

$$Y(p) = \theta(p) + \epsilon(p), \tag{16.1}$$

where $\theta(p)$ is the unknown mean signal to be estimated and $\epsilon(p)$ is a zero-mean Gaussian random field. We may further assume $Y \in L^2(\mathcal{M})$, the space of square integrable functions on $\mathcal{M}$ with the inner product

$$\langle f, g \rangle = \int_{\mathcal{M}} f(p)g(p)d\mu(p),$$

where μ is the Lebesgue measure such that $\mu(\mathcal{M})$ is the volume of $\mathcal{M}$.

Solving

$$\Delta \psi_j = \lambda_j \psi_j, \tag{16.2}$$

on $\mathcal{M}$, we find the eigenvalues λ_j and eigenfunctions ψ_j. The eigenfunctions ψ_j

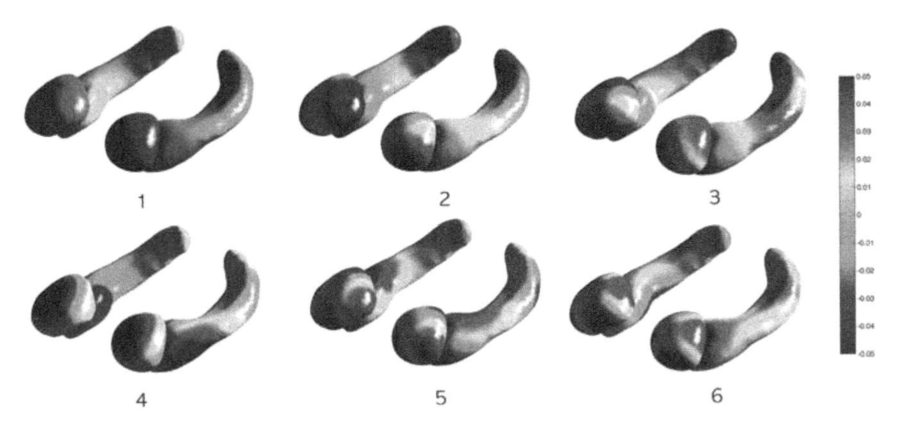

FIGURE 16.2
First six LB-eigenfunctions on amygdala and hippocampus surfaces. (See color insert.)

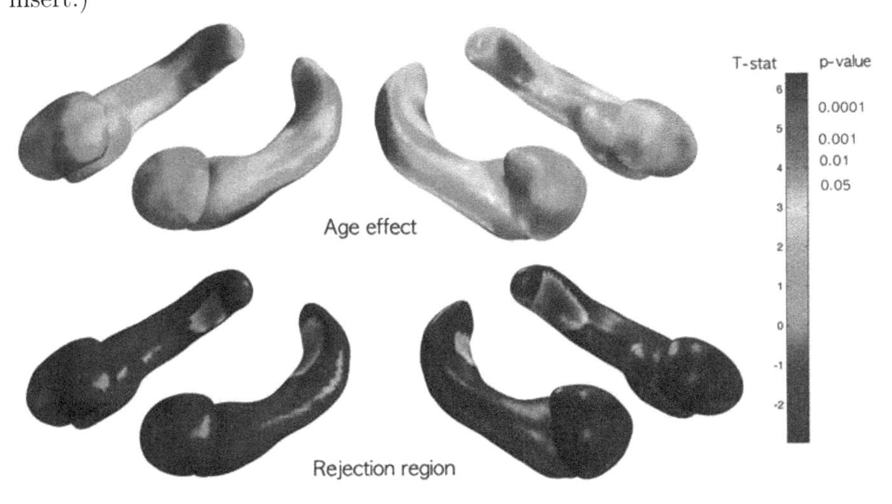

FIGURE 16.3
Age effect on hippocampus shape. The T-stat. and the corrected p-value are also shown. There is no age effect on amygdala. Rejection regions $\mathcal{M}_1$ corresponding to 0.05 level are also shown. (See color insert.)

form an orthonormal basis in $L^2(\mathcal{M})$, the space of square integrable functions on $\mathcal{M}$ [229]. We may order eigenvalues as

$$0 = \lambda_0 < \lambda_1 \leq \lambda_2 \cdots$$

and corresponding eigenfunctions as $\psi_0, \psi_1, \psi_2, \cdots$.

Since the closed form expression for the eigenfunctions of the LB-operator on an arbitrary curved surface is unknown, the eigenfunctions are numerically estimated by discretizing the LB-operator. Using the cotan discretization [79, 290], (16.2) is linearized as the generalized eigenvalue problem:

$$\mathbf{C}\boldsymbol{\psi} = \lambda \mathbf{A}\boldsymbol{\psi}, \tag{16.3}$$

where $\mathbf{C}$ is the stiffness matrix, $\mathbf{A}$ is the mass matrix and $\boldsymbol{\psi} = (\psi(p_1), \cdots, \psi(p_n))'$ is the unknown eigenfunction evaluated at n mesh vertices [319]. The first six LB-eigenfunctions are shown in Figure 16.2.

Once we obtained the eigenfunctions ψ_j, we can parametrically estimate the unknown mean signal $\theta(p)$ as the Fourier expansion as

$$\theta(p) = \sum_{i=0}^{k} \beta_j \psi_j,$$

where β_j is the Fourier coefficients to be estimated. The truncation degree k is usually low. In Styer et al., 12 and 15 degree SPHARM expansions were used for hippocampus and caudate respectively [344]. The Fourier coefficients can be obtained by the usual least squares estimation (LSE) by solving

$$\mathbf{Y} = \boldsymbol{\psi}\boldsymbol{\beta}, \tag{16.4}$$

where $\mathbf{Y} = (Y(p_1), \cdots, Y(p_n))'$, $\boldsymbol{\beta} = (\beta_1, \cdots, \beta_k)'$ and $\boldsymbol{\psi} = (\psi_i(p_j))$ is an $n \times k$ matrix of eigenfunctions evaluated at mesh vertices [408]. The LSE is then given as

$$\widehat{\boldsymbol{\beta}} = (\boldsymbol{\psi}'\boldsymbol{\psi})^{-1}\boldsymbol{\psi}'\mathbf{Y}. \tag{16.5}$$

The estimation, however, may include low degree coefficients that do not contribute significantly. Therefore, instead of using LSE, we introduce the additional L_1-norm penalty to sparsely filter out insignificant low degree coefficients by minimizing [207, 365]:

$$||\mathbf{Y} - \boldsymbol{\psi}\beta||_2^2 + \lambda||\beta||_1, \tag{16.6}$$

where the parameter $\lambda > 0$ controls the amount of sparsity.

16.5 Case Study: Subcortical Structure Modeling

We have applied our sparse shape modeling framework in determining the effects of age and gender on the shape of amygdala and hippocampus. We demonstrate that the proposed L_1-penalty approach can detect the localized effects within the cortical substructures while the traditional method cannot.

16.5.1 Traditional Volumetric Analysis

In the traditional approach, the volume of a structure is simply computed by counting the number of voxels within the binary mask. In order to account for the effect of inter-subject variability in brain size, the brain volume except cerebellum was estimated and covariated in general linear models (GLM).

The brain volume is significantly correlated with the amygdala (left: $r = 0.57$, left: $p < 0.0001$; right: $r = 0.55$, $p < 0.0001$) and the hippocampus volumes (left: $r = 0.61$, $p < 0.0001$; right: $r = 0.63$, $p < 0.0001$). Here p indicates the p-value. Since amygdala and hippocampus volumes are dependent on the whole brain volume, we need to factor out the brain volume in the subsequent statistical analysis.

We model the `volume` of amygdala and hippocampus as

$$\texttt{volume} = \beta_1 + \beta_2 \cdot \texttt{brain} + \beta_3 \cdot \texttt{age} + \beta_4 \cdot \texttt{gender} + \epsilon, \tag{16.7}$$

where ϵ is zero mean Gaussian noise and `brain` is the total brain volume. The age and gender effects were determined by testing the significance of parameters β_3 and β_4 at $\alpha = 0.05$.

We did not find a significant age effect on the amygdala (left $p = 0.40$; right $p = 0.23$; combined $p = 0.29$) nor the hippocampus (left $p = 0.25$; right $p = 0.53$; combined $p = 0.34$). We only found a significant gender effect on the left hippocampus ($p = 0.04$), but no others (left amygdala $p = 0.26$; right amygdala $p = 0.47$; amygdalae combined $p = 0.34$; right hippocampus $p = 0.12$; hippocampi combined $p = 0.05$). Since the results are based on the whole volume of the amygdala and hippocampus, it is still unclear if there are any localized shape differences within the parts of the subcortical structures.

16.5.2 Sparse Shape Analysis

The length of displacement vector field along the template surface was estimated using the sparse framework with $\lambda = 1$ and$k = 1000$ eigenfunctions. This is sufficient number of basis functions. Only the 50 largest coefficients among 1000 estimated coefficients are used in the sparse representation. The age effect on the displacement length `length` is regressed over the total brain volume and other variables:

$$\texttt{length} = \beta_1 + \beta_2 \cdot \texttt{brain} + \beta_3 \cdot \texttt{age} + \beta_4 \cdot \texttt{gender} + \epsilon, \tag{16.8}$$

where ϵ is a zero mean Gaussian field. The age effect was determined by performing t-tests on the parameter β_3. We used the `SurfStat` `MATLAB` toolbox for fitting the parameters and correcting multiple comparisons [80]. The results are displayed in Figure 16.3.

We found the regions of significant effect of age on the posterior part of hippocampi at $\alpha = 0.05$. Particularly, on the caudal regions of the left and right hippocampi, we found highly localized signals. It is consistent with other shape modeling studies on hippocampus [292, 398]. We did not find any age effects on the amygdala surface at $\alpha = 0.05$.

16.6 Statistical Power

The effect of the proposed L_1-penalty framework is quantified using the statistical power. We can show that our sparse selection of LB coefficients can boost the statistical power. The statistical power computation for random fields is fairly complicated due to multiple comparisons.

16.6.1 Type-II Error

The power computation relies on type-II error, which is somewhat involving under the multiple comparisons setting [168].

Given the null hypothesis H_0 and the alternate hypothesis H_1, let us define what is the power of a test statistic.

Definition 22 *The probabilities of type-I (α) and type-II (β) errors are defined as:*

$$\begin{aligned}
\alpha &= P(\textit{Type I error}) \\
&= P(\textit{reject } H_0 \mid H_0 \textit{ true}). \\
\beta &= P(\textit{Type II error}) \\
&= P(\textit{not reject } H_0 \mid H_0 \textit{ false }) \\
&= 1 - P(\textit{reject } H_0 \mid H_1 \textit{ true}).
\end{aligned}$$

Definition 23 *The* power *of the test is defined to be* $1 - \beta$.

$$\textit{Power} = P(\textit{reject } H_0 \mid H_1 \textit{ ture}).$$

The power of a statistical test is given as the probability of rejecting the null hypothesis that there is no signal when there is an actual signal. When the test procedure has the power of 0.9, it implies that we can correctly reject the null hypothesis H_0 90% of the time when the alternate hypothesis H_1 is true. The sample size computation is then based on power. So the power is usually given in terms of the sample size and α level. In designing an imaging experiment, we are interested in the minimum number of samples in achieving a specific power that is usually set at 90%.

16.6.2 Statistical Power for t-Test

Let us start with the power computation for scalar measurement at a voxel. Consider two samples

$$X_1, \cdots, X_{n_1} \sim N(\mu_1, \sigma^2),$$

$$Y_1, \cdots, Y_{n_2} \sim N(\mu_2, \sigma^2).$$

We are interested in testing

$$H_0 : \mu_1 - \mu_2 = 0 \text{ vs. } H_1 : \mu_1 - \mu_2 = c\sigma \neq 0.$$

The constant c represents the mean difference with respect to the standard deviation and is called the *effect size*. The effect size c is usually estimated from the sample mean and standard deviation.

For a test statistic, we use the t-statistic with the equal variance assumption:

$$T = \frac{\bar{X} - \bar{Y} - (\mu_1 - \mu_2)}{S_p\sqrt{1/n_1 + 1/n_2}}, \tag{16.9}$$

where $\bar{X}$ and $\bar{Y}$ are the sample means and S_p^2 is the pooled sample variance. If the sample variance of the i-th group is denoted by S_i^2, the pooled sample variance is given by

$$S_p^2 = \frac{(n_1 - 1)S_1^2 + (n_2 - 1)S_2^2}{n_1 + n_2 - 2}.$$

For computing the power, the α-level of the test has to be specified first. Under H_0, the test statistic T follows

$$T \sim t_{n_1+n_2-2},$$

the student t-distribution with $n_1 + n_2 - 2$ degrees of freedom. The rejection region corresponding to the α-level is then given by

$$\frac{|\bar{X} - \bar{Y}|}{S_p\sqrt{1/n_1 + 1/n_2}} > t^*_{\alpha/2},$$

where $t^*_{\alpha/2}$ is the quantile satisfying

$$P(T \geq t^*_{\alpha/2}) = \alpha/2.$$

Here we assumed $\sigma \approx S_p$.

Under H_1, $X_i \sim N(\mu_1, \sigma^2)$ and $Y_i \sim N(\mu_1 + c\sigma, \sigma^2)$. So it follows

$$T = \frac{\bar{X} - \bar{Y} - c\sigma}{S_p\sqrt{1/n_1 + 1/n_2}} \sim t_{n_1+n_2-2}.$$

Then the power is given by

$$\begin{aligned}\text{Power} &= P\Big(-t^*_{\alpha/2} < \frac{\bar{X}-\bar{Y}}{S_p\sqrt{1/n_1+1/n_2}} < t^*_{\alpha/2}\Big|H_1\Big)\\ &= P\Big(-t^*_{\alpha/2} - \frac{c}{\sqrt{1/n_1+1/n_2}} < T < t^*_{\alpha/2} - \frac{c}{\sqrt{1/n_1+1/n_2}}\Big).\end{aligned}$$

For sufficiently large n_1 and n_2, we may assume $T \sim N(0,1)$. Let Φ be the cumulative distribution for the standard normal distribution. Note that

$$\Phi(x) = \frac{1}{\sqrt{2\pi}}\int_{-\infty}^{x} e^{-t^2/2}\,dt.$$

In MATLAB, Φ is implemented using `normcdf`. In other nonstatistical computing environments, the error function is more often used. The error function `erf` is defined as

$$\texttt{erf}(x) = \frac{2}{\sqrt{\pi}}\int_{0}^{x} e^{-t^2/2}\,dt.$$

The relationship between `normcdf` and `erf` is

$$\texttt{normcdf}(x) = \frac{1}{2} + \frac{1}{2}\texttt{erf}(x/\sqrt{2}).$$

Then using Φ, the power is approximated as

$$\text{Power}(n_1,n_2) = 1+\Phi\Big(-z^*_{\alpha/2} - \frac{c}{\sqrt{1/n_1+1/n_2}}\Big) - \Phi\Big(z^*_{\alpha/2} - \frac{c}{\sqrt{1/n_1+1/n_2}}\Big),$$

where $z^*_{\alpha/2}$ is the quantile corresponding to

$$\Phi(z^*_{\alpha/2}) = 1 - \frac{\alpha}{2}.$$

Assuming $n = n_1 = n_2$, we can plot the power as a function of the sample size. For example, in order to obtain power of 0.8 for a $\alpha = 0.05$ test in differentiating the effect size $c = 0.2$, we need $n = n_1 = n_2 = 393$. For given sample size `n` and effect size `c`, the power is computed using the following function:

```
function power=power_ttest(n,c,alpha)
z=norminv(1-alpha);
power=1 + normcdf(-z-c./sqrt(2./n))-normcdf(z-c./sqrt(2./n));
```

Figure 16.4 shows the power plot for various `n` and effect size `c`. In general, we should design a method that achieves at least 80% power.

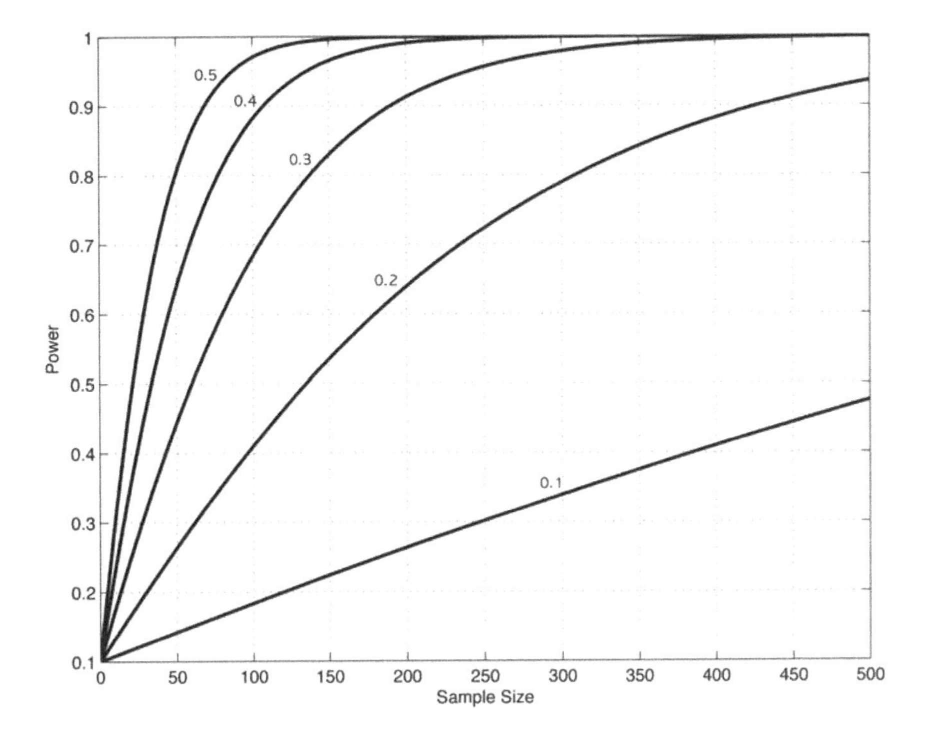

FIGURE 16.4
Power plot for the two sample t-test for the effect size between 0.1 and 0.5. For the effect size of 0.5, we need at least 50 samples in each group to achieve 80% power.

16.7 Power Under Multiple Comparisons

So far, the power computation is based on scalar measurements. Assume we have two functional measurements $X_1(t), \cdots, X_{n_1}(t)$ and $Y_1(t), \cdots, Y_{n_2}(t)$ over continuous index $t \in \mathcal{M}$. X_i and X_j can be then modeled as random fields over $\mathcal{M}$. At each fixed t, we have the same test statistic $T(t)$ given in (16.9). Note that $T(t)$ is a random field defined over the whole brain $\mathcal{M}$.

The usual point-wise hypotheses are given by

$$H_0(t) : \mu_1(t) - \mu_2(t) = 0 \text{ vs. } H_1(t) : \mu_1(t) - \mu_2(t) = c\sigma > 0$$

for each fixed t. Instead of the point-wise inference, what we need is a global inference for the whole parameter space $\mathcal{M}$. The usual global hypotheses accounting for multiple comparisons are then given by

$$J_0 : \mu_1(t) - \mu_2(t) = 0 \text{ for all } t \in \mathcal{M}$$
$$\text{vs. } J_1 : \mu_1(t) - \mu_2(t) = c\sigma > 0 \text{ for some } t \in \mathcal{M}.$$

The relationship between the point-wise hypotheses $H_0(t), H_1(t)$ and the global hypotheses J_0, J_1 are

$$J_0 = \bigcap_{t \in \mathcal{M}} H_0(t), \; J_1 = \bigcup_{t \in \mathcal{M}} H_1(t).$$

16.7.1 Type-I Error Under Multiple Comparisons

In order to compute the power over $\mathcal{M}$, it is necessary to determine the type-I error first. Note that we reject J_0 if $T(t) > h$ for some thresholding h for all $t \in \mathcal{M}$. This is equivalent to the event $\sup_{t \in \mathcal{M}} T(t) > h$. Hence, the type-I error computation requires knowing the distribution of the random variable $\sup_{t \in \mathcal{M}} T(t)$ which can be very involving. The type-I error over $\mathcal{M}$ is given by

$$\alpha = P\Big(\sup_{t \in \mathcal{M}} T(t) > t^*_\alpha \Big),$$

where t^*_α is the quantile corresponding to the random variable $\sup_{t \in \mathcal{M}} T(t)$. If necessary, the quantile can be determined numerically using the permutation test.

16.7.2 Type-II Error Under Multiple Comparisons

Let

$$T(t) = \frac{\bar{X}(t) - \bar{Y}(t)}{S_p(t)\sqrt{1/n_1 + 1/n_2}}.$$

Under J_0, $T(t)$ this is a t-random field with $n_1 + n - 2$ degrees of freedom. The rejection region of J_0 corresponding to the α level is given by $\sup_{t \in \mathcal{M}} T(t) > t^*_\alpha$. Under J_1, we have

$$X_i(t) \sim N(\mu_1, \sigma^2) \text{ and } Y_i(t) \sim N(\mu_1 + c\sigma, \sigma^2)$$

at $t \in \mathcal{M}_1$. In the other region $\mathcal{M}/\mathcal{M}_1$, we have

$$X_i(t), Y_i(t) \sim N(\mu_1, \sigma^2).$$

Figure 16.5 shows the schematic view of when J_1 is true. Therefore, in $\mathcal{M}_1$, we have

$$T'(t) = T(t) - \frac{c}{\sqrt{1/n_1 + 1/n_2}} \sim t_{n_1+n_2-2}$$

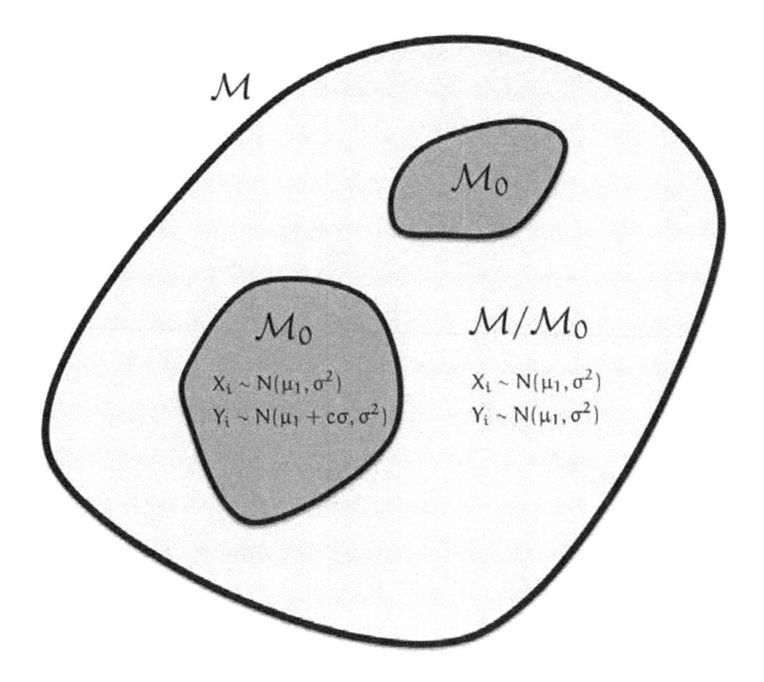

FIGURE 16.5
Under J_1, there exist nonempty regions $\mathcal{M}_0$ where the means of the two groups are different.

pointwisely. Note that T' is not t-field in $\mathcal{M}/\mathcal{M}_1$. Then the over all power over $\mathcal{M}$ is given as

$$\begin{aligned} \text{Power} &= P\Big(\sup_{t\in\mathcal{M}} T(t) > t^*_\alpha \Big| J_1\Big) \\ &= P\Big(\sup_{t\in\mathcal{M}_1} T'(t) > t^*_\alpha - \frac{c}{\sqrt{1/n_1 + 1/n_2}}\Big). \end{aligned}$$

Since the analytic derivation of the exact probability is intractable, we will approximate the power using the first order term in the Euler characteristic method. For sufficiently large n_1 and n_2, we approximate the t-field T' using the Gaussian field Z with zero mean and unit variance.

Let Ψ be the tail distribution of the supremum of Z field in $\mathcal{M}$:

$$\Psi(h, \mathcal{M}) = P\Big(\sup_{t\in\mathcal{M}} Z(t) > h\Big).$$

If $\mathcal{M}$ is a 2D surface, Ψ is dominated by

$$\Psi(h, \mathcal{M}) \approx \frac{\mu(\mathcal{M})}{\text{FWHM}^2} \frac{4\ln 2}{(2\pi)^{3/2}} h \exp(-h^2/2), \tag{16.10}$$

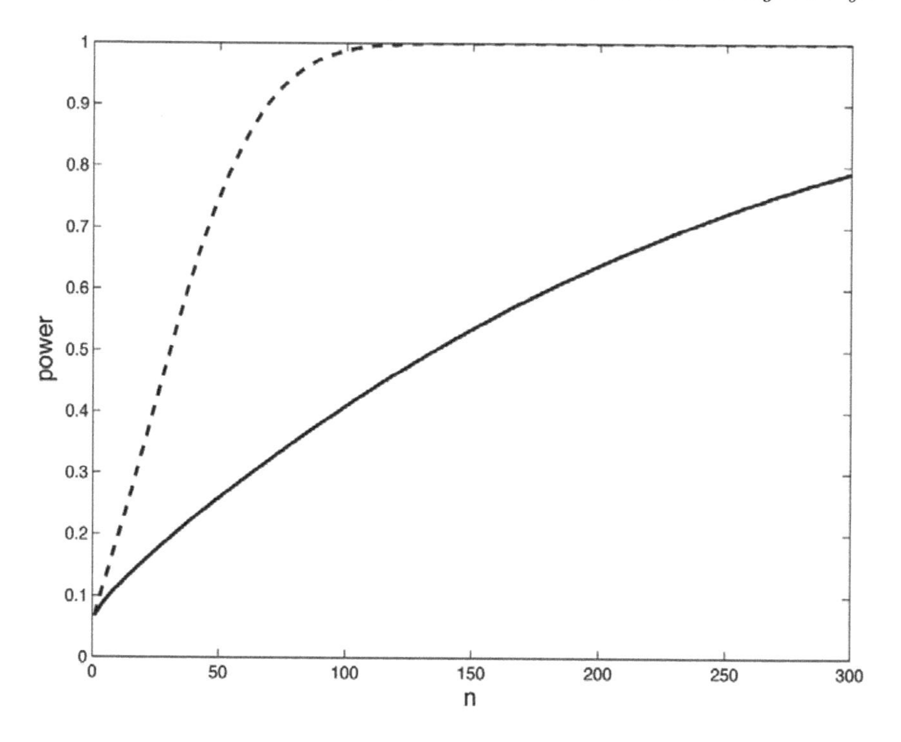

FIGURE 16.6
Power vs. sample size for the two sample t-test. Solid line is for at each voxel and the dotted line is for whole brain surface accounting under multiple comparisons. To obtain power of 0.8 at significance $\alpha = 0.05$, in differentiating the group difference 0.2σ, we need a significantly smaller sample size under multiple comparisons. This is the reason why we usually need smaller sample sizes in imaging studies.

where $\mu(\mathcal{M})$ is the surface area of $\mathcal{M}$ and FWHM is the full-width-at-half-maximum of signal or smoothing kernel [394]. Since Ψ is a really small number for high threshold h, the Taylor expansion

$$\exp\big[-\Psi(h,\mathcal{M})\big] \approx 1 - \Psi(h,\mathcal{M})$$

is a very good approximation for high h. Then we can write

$$P\Big(\sup_{t\in\mathcal{M}} T'(t) > h\Big) \approx 1 - \exp\big[-\Psi(h,\mathcal{M})\big].$$

This transformation guarantees the power estimation to be bound between 0 and 1 for any h value [168]. Without the transformation, we can possibly get the power estimate larger than 1 for small h value. Subsequently, the power is given by

$$\text{Power} = 1 - \exp\Big[- \Psi\big(t^*_\alpha - \frac{c}{\sqrt{1/n_1 + 1/n_2}}, \mathcal{M}_1\big)\Big].$$

Suppose we use 5mm FWHM of smoothing kernel and the outer cortical surface area of 302180mm^2 in cortical thickness analysis [83]. Assuming $n = n_1 = n_2$, we can plot the power as a function of the sample size (Figure 16.6). In order to obtain the power of 0.8 for a $\alpha = 0.05$ (corrected) test in differentiating $\mu_1 - \mu_2 = 0.2\sigma$, we need significantly smaller n compared to the power computation at each voxel.

16.7.3 Statistical Power of Sparse Representation

Since the power computation under multiple comparisons is fairly involving, a statistical resampling technique can be used in determining the power numerically. For this, we first need to identify the threshold h corresponding to a specific α-level.

For a T-random field $T(x)$, the threshold h corresponding to the type-I error at 0.05 is given by

$$P\big(\sup_{x \in \mathcal{M}} T(x) > h\big) = 0.05,$$

where $\mathcal{M}$ is the surface area. For the power computation, it is needed to identify the rejection region $\mathcal{M}_0$ as well (Figure 16.5). The rejection region is taken as the region

$$\mathcal{M}_0 = \{x \in \mathcal{M} | T(x) > h\}.$$

Then the statistical power is computed as

$$\texttt{Power} = 1 - P\big(\sup_{x \in \mathcal{M}_0} T(x) > h\big),$$

where the maximum is restricted to the rejection region $\mathcal{M}_1$ (Figure 16.3). The distribution of the supremum of T field is empirically determined using 5000 permutations.

As seen in the overall power curves (Figure 16.7), L_1-norm minimization gives higher power for a given sample size. The proposed sparse regression requires smaller sample size to achieve a given power level demonstrating the advantage of the proposed method.

16.8 Conclusion

We have presented a new subcortical structure shape modeling framework based on the sparse representation of Fourier coefficients constructed with the

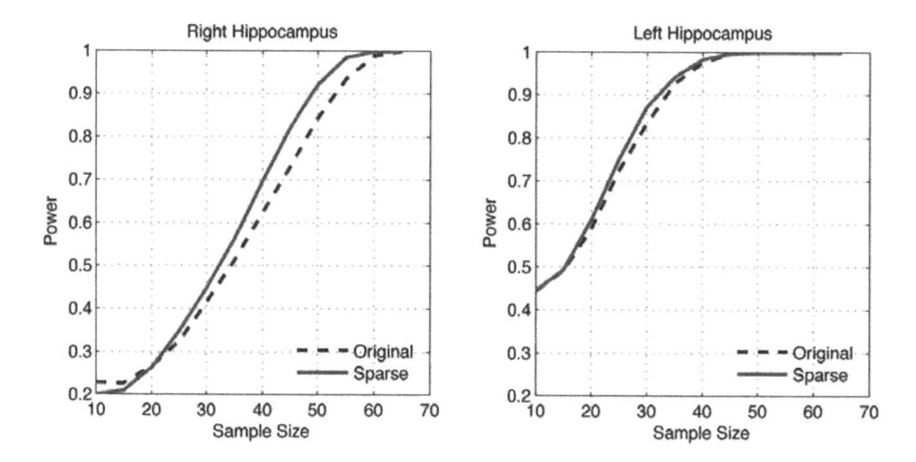

FIGURE 16.7
Statistical power over sample size computed under multiple comparisons.

LB-eigenfunctions. The proposed framework demonstrated a higher sensitivity in modeling shape variations compared to the traditional volumetric analysis.

We found significant structural changes on hippocampi due to normal aging in the present dataset using the proposed framework, consistent with the previous shape analysis methods [362, 292]. The age effect was not found in the traditional volumetric analysis. It demonstrates the higher sensitivity of the proposed method in detecting shape variation. The ability to localize precise morphological difference may provide anatomical evidence for the functional organization within human subcortical structures. It is shown that the higher statistical power can be achieved by the proposed method.

17

Modeling Structural Brain Networks

DTI offers a unique opportunity to characterize the structural connectivity of the human brain non-invasively by tracing white matter fiber tracts. Whole brain tractography studies routinely generate up to half a million tracts per brain, which serve as edges in an extremely large 3D graph with up to half a million edges. Currently there is no agreed-upon method for constructing the brain structural network graphs out of a large number of white matter tracts. In this paper, we present a scalable iterative framework called the ϵ-neighbor method for building a network graph and apply it to testing abnormal connectivity in autism. This chapter is based on [66].

17.1 Introduction

Structural brain connectivity has been usually modeled as a network graph using white matter fiber tracts in DTI. The whole gray matter has been traditionally parcellated into n disjoint regions. White matter fibers provide information on how one gray matter region is connected to another via a $n \times n$ connectivity matrix. The connectivity matrix is then thresholded to produce a binarized adjacency matrix, which is further used in constructing a graph with n nodes [123, 407, 145, 158]. However, there is no gold standard for gray matter parcellation which makes the identification of nodes depend on the choice of parcellation. Depending on the scale of parcellation, the parameters of the graph, which characterize graph topology, vary considerably up to 95% [123, 407]. Another problem of the parcellation is the arbitrariness of thresholding connectivity matrices. The topological parameters such as sparsity and clustering coefficients change substantially depending on the level of threshold [145].

The problems of parcellation and the subsequent arbitrary thresholding can be avoided if we do not use any parcellation in building the network. So the question is whether if it is possible to construct a network graph without the usual parcellation scheme. This paper presents a novel network graph modeling technique called the ϵ-*neighbor construction* that avoids parcellation and the subsequent thresholding of the connectivity matrix. Instead of using the pre-specified parcellation, we propose to use the two end points of fibers as network nodes while the fibers themselves serve as the edges connecting nodes.

The ϵ-neighbor construction is motivated by the Rips complex of point cloud data [138], which has been used to characterize the topology of the point cloud data. The Rips complex is a graph constructed by connecting two data points if they are within specific distance ϵ. The problem of the Rips complex is that given n data points, it exactly produces a graph with n nodes so the resulting graph becomes very complicated when n becomes large. Unlike the Rips complex, the ϵ-neighbor method does not use all data points in constructing a graph so it significantly reduces the complexity of data. Further, while the point cloud data does not have any hidden topological constraint, the two end points of white matter fibers are connected so we are actually dealing with paired point cloud data. So the ϵ-neighbor construction is different from building the Rips complex and offers a substantial computational advantage.

17.2 DTI Acquisition and Preprocessing

The DTI data from 31 subjects were used in this study: (i) 17 subjects with high functioning autism spectrum disorders (ii) 14 control subjects matched for age, handedness, IQ, and head size. The Autism Diagnostic Interview-Revised [231] was used for diagnoses by trained researchers [98]. Diffusion weighted images were acquired in 12 non-collinear diffusion encoding directions with diffusion weighting factor of b=1000 s/mm2 in addition to a single (b=0) reference image. Eddy current related distortion and head motion of each data set were corrected using AIR [386] and distortions from field inhomogeneities were corrected using custom software algorithms based on [185]. The six tensor elements were calculated using non-linear fitting methods [8, 90, 188].

Spatial normalization of DTI plays a key role in constructing brain network graphs that are spatially compatible across different subjects. The quality of spatial normalization determines the extent to which white matter tracts are aligned. It has direct impact on the successful removal of shape confounds and consequently on the validity, specificity, and sensitivity of the subsequent statistical inferences of group differences. Inadequate normalization with coarse registration algorithms can result in insufficient removal of shape differences that is necessary for obtaining a topologically invariant network graph.

We have used the nonlinear tensor image registration algorithm given in [408] for spatial normalization. This approach combines full tensor co-registration and high-dimensional diffeomorphic spatial normalization. The registration is based on an iterative strategy [408, 192], where the initial template was computed as the average of original DTI. Then DTI was first affinely aligned to the template. The tensor images after the affine alignment were pro-

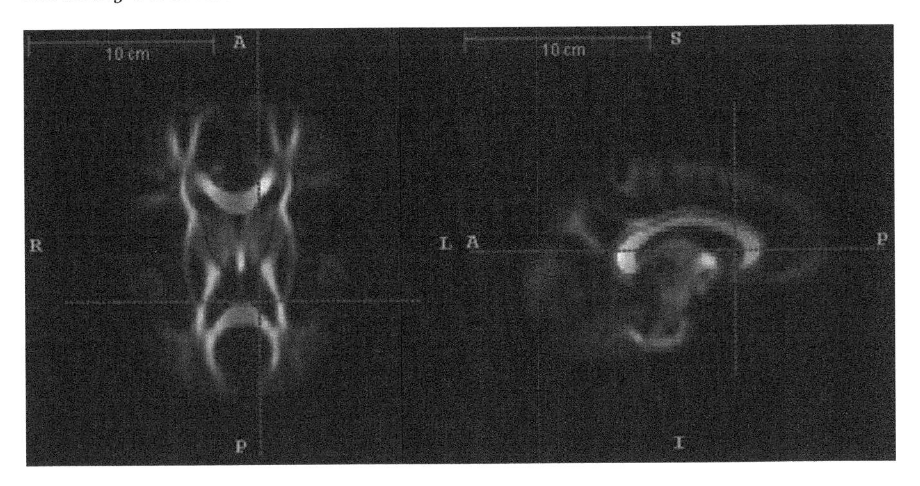

FIGURE 17.1
FA map of the population specific template obtained after 6 iterations of diffeomorphic registration. L: left, R: right, A: anterior, P: posterior, I: inferior, S: superior. This figure was generated by Nagesh Adluru of University of Wisconsin-Madison.

vided as the input to the registration algorithm. The algorithm leverages full tensor-based similarity metrics while optimizing tensor orientation explicitly. We have used the L_2-distance between the anisotropic parts of diffusion profiles associated with the diffusion tensors [410]. The algorithm then approximates smooth transformations using a dense piecewise affine parameterization which is sufficient when the required deformations are not large. Compute a refined template as an average of the normalized images. If the change between templates from consecutive iterations is sufficiently small, we stop the iteration otherwise we continue the iterative process of getting a new template and refitting. The FA-map of the template derived from the tensor template is illustrated in Figure 17.1. After DTI were alined to a template space, we perform streamline based tractography using the TENsor Deflection (TEND) algorithm [218, 90], Figure 1.11 shows the subsampled tractography result for a single subject.

17.3 ϵ-Neighbor Construction

Suppose a whole brain tractography result yields k number of tracts. The i-th tract consists of two end points e_{i1} and e_{i2}. In constructing the network graph, we only need to worry about the two end points of a tract since all other points along the tract are connected to these two end points. The end points of tracts

FIGURE 17.2
Stream tube representation of white matter fiber tracts. (See color insert.)

are considered as nodes of a graph while the tracts are considered as edges of the graph. The TEND algorithm can generate upward of half a million tracts but it is not practical to construct a massive graph with half a million edges. So we have developed a scalable iterative network graph construction technique called the *ϵ-neighbor construction.*

Let $\mathcal{G}_k = \{\mathcal{V}_k, \mathcal{E}_k\}$ be a 3D graph with vertex set $\mathcal{V}_k$ and edge set $\mathcal{E}_k$ at the k-th iteration.

Definition 24 *The distance $d(p, \mathcal{G}_k)$ of a point p to the graph $\mathcal{G}_k$ is the shortest Euclidean distance between p and all points in $\mathcal{V}_k$*

$$d(p, \mathcal{G}_k) = \min_{q \in \mathcal{V}_k} \|p - q\|.$$

Definition 25 *p is the* ϵ-neighbor *of graph $\mathcal{G}_k$ if $d(p, \mathcal{G}_k) \leq \epsilon$. The threshold ϵ will be called the resolution of the graph and it determines the scale at which we construct the graph.*

The larger the value of ϵ, the cruder the constructed graph becomes with the less number of nodes. Consider a tract with two end points e_{11} and e_{12}. The algorithm then starts with the graph $\mathcal{G}_1 = \{\mathcal{V}_1, \mathcal{E}_1\}$ with $\mathcal{V}_1 = \{e_{11}, e_{12}\}, \mathcal{E}_1 = \{e_{11}e_{12}\}$. For the edge set $\mathcal{E}_1$, we simply denote the edge connecting two nodes e_{11} and e_{12} as $e_{11}e_{12}$. In the next iteration, we add the second tract with two end points e_{21}, e_{22} to the existing graph $\mathcal{G}_1$. There are four possibilities:

(1) e_{21} and e_{22} are all ϵ-neighbors of $\mathcal{G}_1$. Since the end points e_{21} and e_{22} are close to the already existing graph $\mathcal{G}_1$, we do not change the vertex set, i.e. $\mathcal{V}_2 = \mathcal{V}_1$. Now check if the edge $e_{21}e_{22}$ is in the edge set $\mathcal{E}_1$ and add them if it is not found in the edge set. In this case we have $\mathcal{E}_2 = \mathcal{E}_1 \cup \{e_{21}e_{22}\}$.

(2) Only e_{21} is an ϵ-neighbor. We only to add e_{22} to $\mathcal{V}_1$ and let

$$\mathcal{V}_2 = \mathcal{V}_1 \cup \{e_{21}\},\ \mathcal{E}_2 = \mathcal{E}_1 \cup \{e_{21}e_{22}\}.$$

(3) Only e_{22} is an ϵ-neighbor. We add e_{21} to $\mathcal{V}_1$ and let

$$\mathcal{V}_2 = \mathcal{V}_1 \cup \{e_{22}\},\ \mathcal{E}_2 = \mathcal{E}_1 \cup \{e_{21}e_{22}\}.$$

(4) e_{21} and e_{22} are not ϵ-neighbors. We add the end points to the vertex set and add the edge to the edge set. In this case, $e_{21}e_{22}$ forms a disjoint edge and we have

$$\mathcal{V}_2 = \mathcal{V}_1 \cup \{e_{21}, e_{22}\},\ \mathcal{E}_2 = \mathcal{E}_1 \cup \{e_{21}e_{22}\}.$$

The procedure is iteratively performed to every tract until we exhaust all the tracts. The `MATLAB` code for performing ϵ-neighbor construction is given in `brainimaging.waisman.wisc.edu/~chung/graph`.

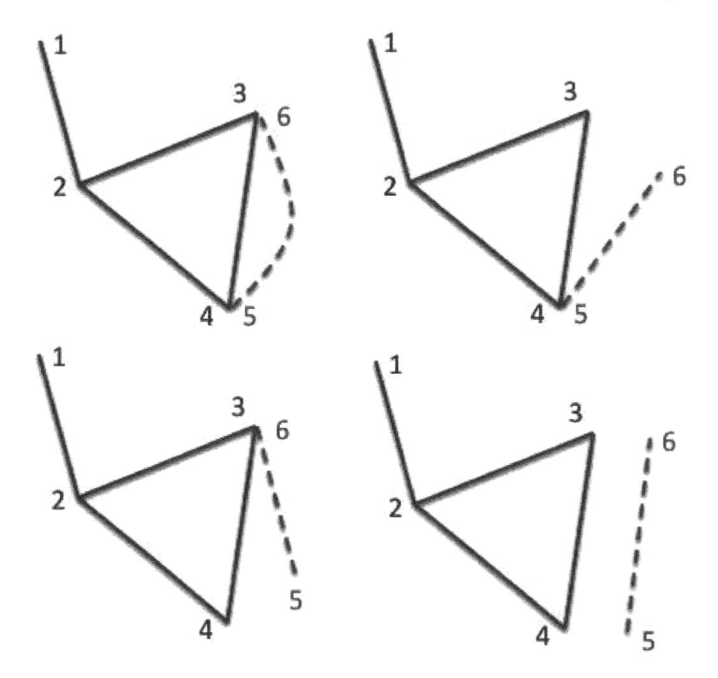

FIGURE 17.3
Example of constructing a 3D network graph by adding one tract at a time to the existing graph. There are four possible cases of connecting the two end points (indexed 5 and 6) of the tract to the existing graph with nodes $\{1, 2, 3, 4\}$. Depending on the proximity of $\{5, 6\}$ to nodes $\{1, 2, 3, 4\}$, we either do nothing, or add one node (either 5 or 6) or add two nodes (both 5 and 6) to the graph.

17.4 Node Degrees

The constructed 3D networks graph can be uniquely parameterized by transforming the graph into adjacent matrices. The adjacency matrix $A = (a_{ij})$ of a graph is constructed on the fly at each iteration by checking if we are adding a new edge to the existing edge set. If nodes i and j are connected, we let $a_{ij} = 1$ and $a_{ij} = 0$ otherwise. The adjacency matrix is symmetric. Depending on how we define connectivity, the diagonal terms a_{ii} can be either zero or one. In this study, we simply let $a_{ii} = 0$. The adjacency matrix contains sufficient information to reconstruct the graph. The resulting 6, 10 and 20 mm-neighbor graphs and the corresponding adjacency matrices are given in Figure 17.4.

The degree of connectivity of a node is obtained by summing up the corresponding rows in the adjacency matrix. The distribution of nodes is computed

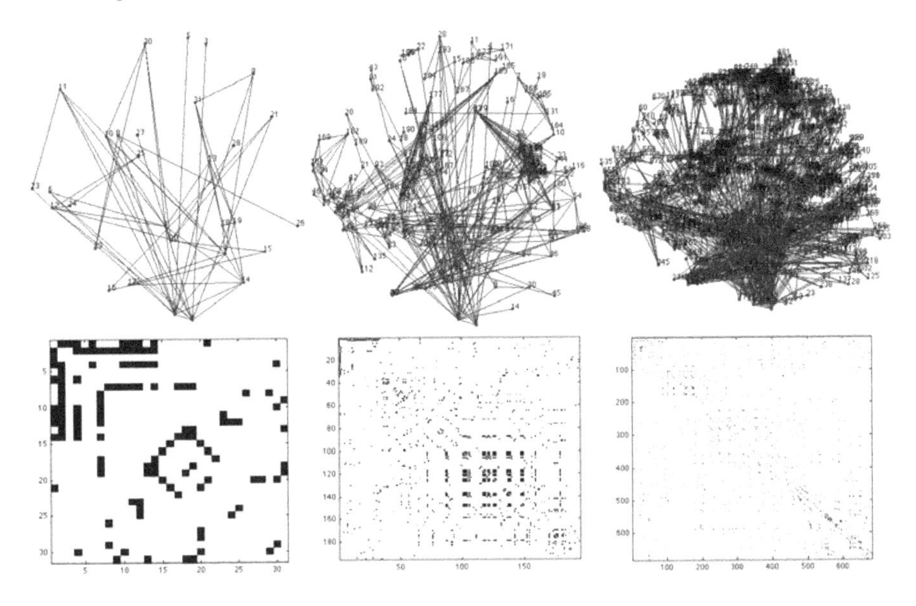

FIGURE 17.4
Scalable 3D connectivity graphs constructed from the proposed algorithm and the corresponding adjacency matrices. The nodes are indexed by numbers. From the left to right, graphs are at 20, 10 and 6mm resolution.

from degree 1 up to 25 and renormalized (Figure 17.5). We did not go beyond degree 25 since there are not many nodes with a degree larger than 25 so the tail region is fairly noisy. The autistic subjects show significantly more low degree nodes (degree 1 to 3) compared to the control subjects (p-values $= 0.024, 0.015, 0.080$). Therefore, an autistic subject is characterized by the under-connectivity of nodes indicating the abnormal brain network connectivity.

The degree distributions show the global network difference; however, it does not tell us local graph characteristics. To characterize local network difference, we need to incorporate spatial coordinate information into the statistical inference. One way of doing this is to superimpose individual degree maps on the template and perform statistical analysis.

17.5 Connected Components

As an illustration of ϵ-neighbor method, we characterize the abnormal brain network in autism using the connected components of a graph. For the sub-

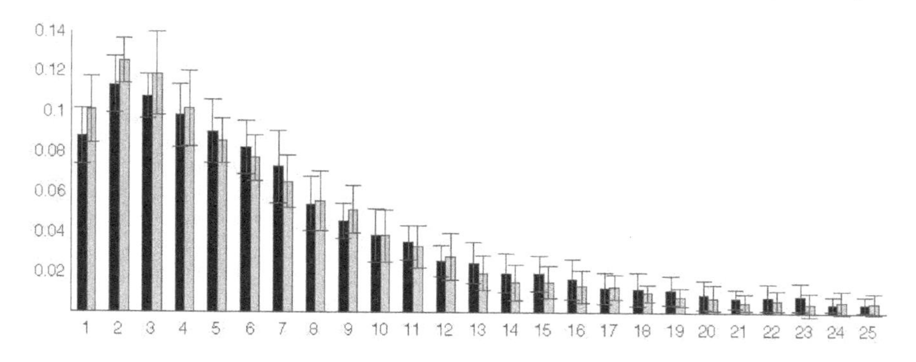

FIGURE 17.5
Degree distribution for autistic (gray) and control (black) subjects. Autistic subjects show under-connectivity which is demonstrated by higher concentration of low degree nodes.

sequent analysis on connected components, 6 mm-neighbor graph was used. This particular resolution is chosen since it is the largest integer resolution that produces the node numbers below 1000. The number of connected components of a graph is a topological invariant that measures the number of structurally independent or disjoint subnetworks. It can be interpreted as the zeroth Betti number β_0 in algebraic topology [116]. The connected components can be identified using the Dulmage-Mendelsohn decomposition [289], which has been widely used for decomposing sparse matrices into block triangular forms in speeding up matrix operations. Figure 17.7 shows connected components in 4 brain networks. All nodes in the same connected component are colored identically. Most of nodes belong to the largest connected component indicating the brain network is highly connected. There are only 4% of nodes that are not connected to the largest connected component while the remaining 96% are all connected on average. We have plotted those 4% of nodes that are not part of the largest connected component for all subjects (Figure 17.9). Figure 17.9 shows the clustering pattern between the groups differs. Although these nodes are scattered in most parts of brain, a high concentration of clustering occurs on the right parietal lobe for the control subjects. We tested if the size of the largest connected components differs between the groups. At 6 mm resolution, control subjects have 642.86 ± 68.60 nodes in the largest component while autistic subjects have 607.12 ± 39.39 nodes. Note that we do not need to account for brain size difference since networks are constructed in the normalized space. The cluster size difference is significant (p-value $= 0.079$). We expect a larger sample size would increase the statistical significance.

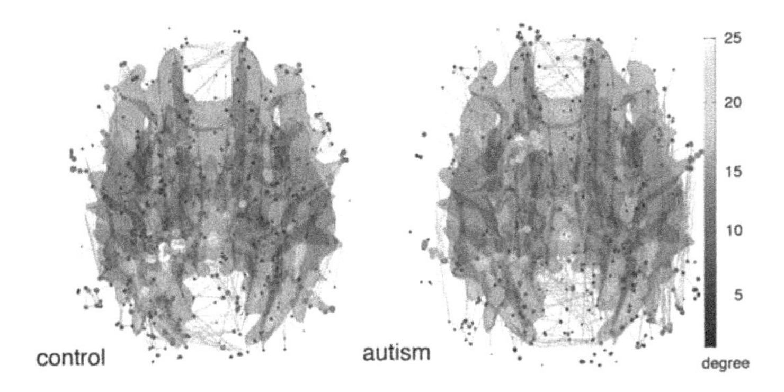

FIGURE 17.6
Node degrees are shown as colored spheres superimposed on top of the white matter surface. The size of a sphere at a node is proportional to the degree. (See color insert.)

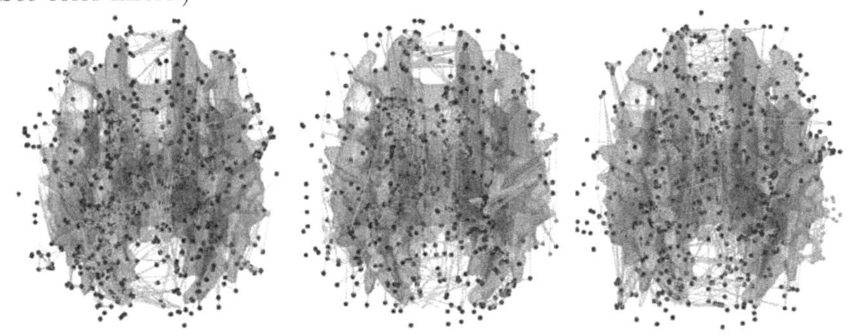

FIGURE 17.7
All nodes in the same connected component are colored identically. The brain network is characterized by a giant connected dominant component. (See color insert.)

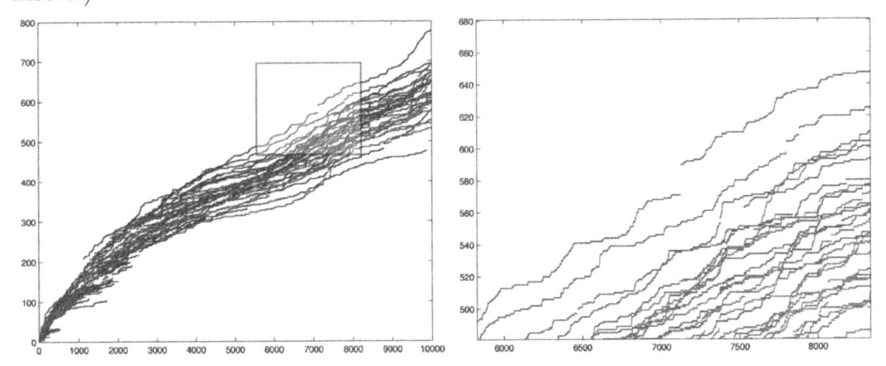

FIGURE 17.8
The size of the largest connected component (vertical) over the ϵ-filtration showing group difference (control = blue, autism= red). (See color insert.)

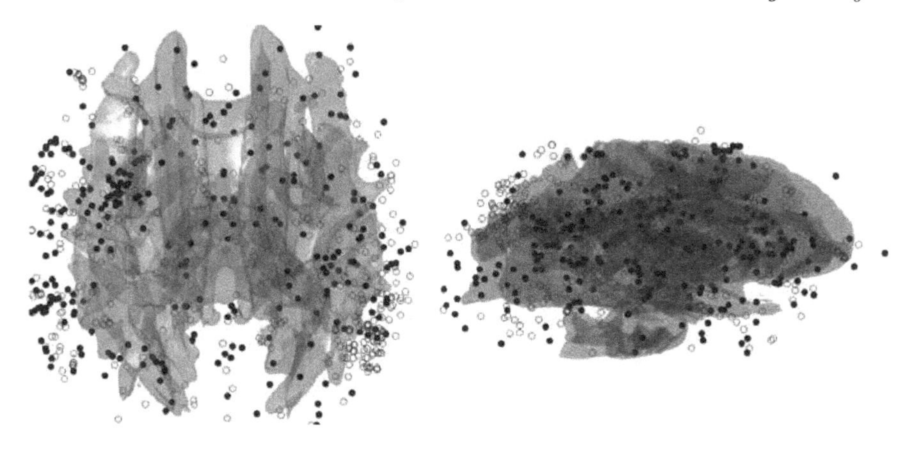

FIGURE 17.9
Superimposition of all 4% of nodes that do not belong to the largest cluster for all 30 subjects. Nodes from the control are colored as white circles while nodes from the autistic subjects are colored as solid black circles. The control subjects show high clustering on the right parietal lobe.

17.6 ϵ-Filtration

The analysis in the previous section characterizes networks at the last iteration of the ϵ-neighbor construction. In this section, we present the idea of quantifying the network over all iterations *via* ϵ-filtration, which is motivated by the Rips filtration in persistent homology.

For each given ϵ, we have Rips complex $\mathcal{G}_\epsilon$. By increasing the ϵ value, we have the Rips filtration, a sequence of larger Rips complexes:

$$\mathcal{G}_{\epsilon_1} \subset \mathcal{G}_{\epsilon_2} \subset \mathcal{G}_{\epsilon_3} \subset \cdots$$

for $\epsilon_1 \leq \epsilon_2 \leq \epsilon_3 \leq \cdots$ [114, 138, 176, 416]. During the Rips filtration, the topological features such as the Betti numbers change. The topological change over the filtration can be visualized by using the barcode. In the barcode, we plot the zeroth Betti number β_0 over the changing ϵ value. The resulting barcode is a decreasing function of ϵ and its decreasing pattern can be used to discriminate groups [220]. Although the Rips filtration completely characterizes the topological change of a network, it is difficult to biologically interpret what it really means to have a changing network over scale ϵ.

Similar to the Rips filtration, the ϵ-filtration is a sequence of networks obtained from the ϵ-neighbor method. At the k-th iteration, we have a network $\mathcal{G}_k$. As the number of iteration increases, we are generating a sequence of larger networks

$$\mathcal{G}_1 \subset \mathcal{G}_2 \subset \mathcal{G}_3 \subset \cdots,$$

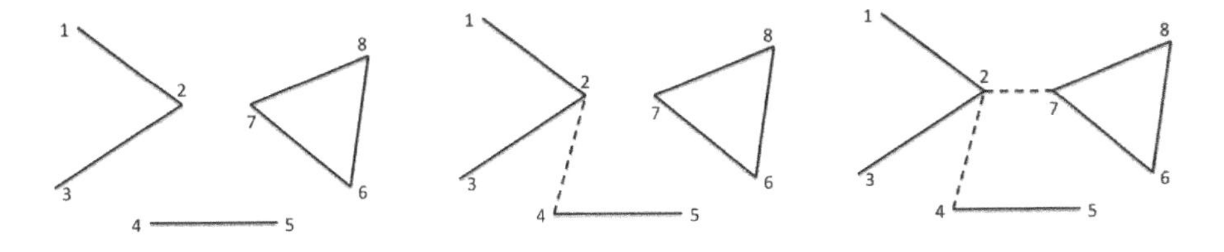

FIGURE 17.10
An example of how the size of connected component changes in the ϵ-filtration. Originally a network consists of three disjoint components $\{1,2,3\},\{4,5\},\{6,7,8\}$. The size of the largest components is 3. In the next iteration, we connect the nodes 2 and 4 and obtain two components $\{1,2,3,4,5\},\{6,7,8\}$. The size of the largest component is 5. In the 2nd iteration, we connect the nodes 2 and 7 and eventually obtain a giant component of size 8. The node 2 is more than a hub and connects all disjoint components together. By removing the node 2, the giant components disintegrate into three disjoint components. The sudden jumps in Figure 17.8 are showing the introduction of such nodes in the network.

which we will call the ϵ-filtration. On the other hand, the ϵ-filtration is much easier to interpret since it shows the actual process of network construction.

We computed the size of the largest component in each iteration in the ϵ-filtration (Figure 17.8). It is always an increasing function and at about 6000 iterations, we begin to see a significant group difference in the increasing pattern. For instance, at 6000, 7000 and 8000 iterations, we have the p-values of 0.058, 0.038, 0.04 respectively. The control network (blue) is integrating into the largest component much faster than the autistic network (red). The growth of the size of the largest component is higher in controls. A schematic understanding of the underlying process is given in Figure 17.10. The control subjects are expected to have hub nodes that speed up the integration of disjoint components into the largest component.

17.7 Numerical Implementation

Based on the ϵ-neighbor algorithm [66], the brain connectivity of DTI fibers can be obtained as an adjacency matrix. Here we present `MATLAB` implementation of the ϵ-neighbor algorithm and the corresponding visualization tools. The `MATLAB` script below is given in `http://brainimaging.waisman.wisc.edu/~chung/BIA`.

17.7.1 Fiber Bundle Visualization

Let's load the fractional anisotropy (FA) map of the template used in [66].

```
nii=load_nii('mean_diffeomorphic_initial6_fa.nii');
d=nii.img;
surf = isosurface(d);
coord=surf.vertices;
temp=coord;
temp(:,2)=coord(:,1);
temp(:,1)=coord(:,2);
surf.vertices=temp;
figure_patch(surf,[0.74 0.71 0.61],0.2);
```

Since there are too many tracts, we subsample for every 30 tracts and plot them.

```
load SL.mat
for i=1:30:10000
    tract=SL{i}';
    if size(tract,2)> 10
        hold on
        plot3(tract(1,:),tract(2,:),tract(3,:),'b')
    end;
end;
```

To obtain the two end points of tracts and color them as read, we run

```
endp=get_endp_tracts(SL,30);
hold on; plot3(endp(:,1,1),endp(:,1,2), endp(:,1,3),'.r')
hold on; plot3(endp(:,2,1),endp(:,2,2), endp(:,2,3),'.r')
```

The resulting image is shown in Figure 1.11. Figure 1.11 shows the white matter fiber bundles obtained from a streamline-based algorithm. The tracts are sparsely subsampled for better visualization. The end points are colored as red. The surface is the isosurface of the template FA map so some tracts are expected to be outside of the surface. The ϵ-neighbor method will use the proximity of the end points in constructing the network graph.

17.7.2 ϵ-Neighbor Network Construction

The ϵ-neighbor algorithm is run directly on the list of fiber bundles `SL.mat` with the radius 10 mm using

```
[adj, prob, vertices]=tract2graph(SL,10);
figure; imagesc(-adj); colormap('bone');
```

adj is the adjacency matrix, prob is the fiber concentration density, which is computed by counting the number of tracts that is connecting to the given node within the ϵ-neighbor. vertices is the coordinates of the constructed nodes. Figure 17.4 shows the scalable 3D connectivity graphs constructed from the proposed algorithm and the corresponding adjacency matrices. The nodes are indexed by numbers. From the left to right, graphs are at 20, 10 and 6 mm resolution.

The ϵ-neighbor algorithm constructs adj matrix as follows. The algorithm starts with the sorted list of tracts. The sorting is based on the length of tracts. So we expect the end point list endp is sorted in decreasing order. Then we start with the empty adj. Then simply add the end points of the longest tract endp(1,1,:) and endp(1,2,:) as new vertices and connect them.

```
adj=[];
nodelist=[squeeze(endp(1,1,:))'; squeeze(endp(1,2,:))'];

adj(1,2)=1;
adj(2,1)=1;

prob(1)=1;
prob(2)=1;
```

The network with the two vertices and an edge serves as the initial graph. Then we add the next longest tract to the network by changing adj. nodelist stores all the end points that added into the graph. We need to perform this operation in an iterative fashion. To speed up the computation, one may opt to use the k-d tree algorithm which will drastically speed up the computation.

Consider adding the second tract to the existing graph, i.e. i = 2. For the two end points node1 = endp(2,1,:) and node2 = endp(2,2,:), we determine if there is any node in nodelist that is close to the end points. The closed nodes are identified as ind1 and ind2.

```
n_node=size(nodelist,1);
node1=squeeze(endp(i,1,:))';
repnode=repmat(node1,n_node,1);
diff=sqrt(sum((nodelist-repnode).^2,2));

ind1=find(diff<res);

node2=squeeze(endp(i,2,:))';
repnode=repmat(node2,n_node,1);
diff=sqrt(sum((nodelist-repnode).^2,2));
ind2=find(diff<res);
```

There are four possibilities associated with ind1 and ind2 as shown in Figure 17.3.

Case 1. `ind1` and `ind2` are not empty. It only happens if `node1` and `node2` are close to existing nodes in the graph. In this case we do nothing except change the edge weight `prob`.

Case 2. Only `ind2` is empty so `node1` is close to existing nodes in the graph while `node2` is not. Then we add `node2` to `nodelist` and change `adj`.

Case 3. Only `ind1` is empty so `node2` is close to existing nodes in the graph while `node1` is not. We add `node1` to `nodelist` and change `adj`.

Case 4. `ind1` and `ind2` are empty so we add them to `nodelist` and change `adj`.

These four cases are implemented as follows.

```
if ~isempty(ind1)
     if isempty(ind2)
         %Case 2
         adj(ind1,n_node+1)=1;
         adj(n_node+1,ind1)=1;
         nodelist=[nodelist; node2];
         prob(ind1)=prob(ind1)+1;
         prob(n_node+1)=1;
     else
         %Case 1
         prob(ind1)=prob(ind1)+1;
         prob(ind2)=prob(ind2)+1;
     end;
 else
     if isempty(ind2)
     %Case 4
         adj(n_node+1,n_node+2)=1;
         adj(n_node+2,n_node+1)=1;
         nodelist=[nodelist; node1; node2];
         prob(n_node+1)=1;
         prob(n_node+2)=1;
     else
         %Case 3
         adj(ind2,n_node+1)=1;
         adj(n_node+1,ind2)=1;
         nodelist=[nodelist; node1];
         prob(ind2)=prob(ind2)+1;
         prob(n_node+1)=1;
     end;
 end;
```

17.7.3 Network Computation

With the constructed network, we will briefly show how to compute various network characteristics. To compute the degree of nodes, we sum the adjacency matrix either along the row or the column direction. The node degree can be displayed in 3D:

```
degree=sum(adj);
figure_graph_color(adj,vertices,degree);
colormap('hot');
```

The plotting function uses the MATLAB command scatter3 that draws a 3D sphere as a node. This gives a network visualization similar to Figure 17.6.

For identifying all the disjoint components in the network, we can use the Dulmage-Mendelson permutation implemented as dmperm.m. The same component is colored identically. For large radius in tract2graph.m, all the nodes are merged into a single component so all the nodes are colored the same. The visualization is given in Figure 17.7.

```
[adj, prob, vertices]=tract2graph(SL,6);
adj = adj + eye(size(adj,1));
[p,q,r,s] = dmperm(adj);

ck=length(r);
for i=1:(ck-1)
    cp=p(r(i):r(i+1)-1);
    clr=10*i*ones(length(cp),1);
    figure_graph_color(adj(cp,cp),vertices(cp,:),clr);
    hold on;
end;
```

17.8 Discussion

We have presented a novel data-driven connectivity graph construction method for DTI. The proposed ϵ-neighbor network construction can avoid the problem of parcellation and arbitrary connectivity matrix thresholding. The method is applied in showing abnormal connectivity patterns in autistic subjects in the ϵ-filtration. The autistic network shows a slower integration rate than the control network in the filtration demonstrating reduced network integration capability and efficiency.

It is possible to modify our ϵ-neighbor construction in such a way that the nodes do not have to coincide with the end points of tracts. Then we can define nodes in a template so that every subject has an identical number of nodes. This simplifies the problem of performing local network inference at each node but requires sufficiently accurate tensor image registration to work and requires additional validation.

18

Mixed Effects Models

The main problem with all the previous statistical models studied in this book assumes the parameters of the model to be fixed and not vary across subjects and scans. Even if we have repeated scans of the same subject, the fixed effect models treat the repeated scans as independent scans ignoring the dependency between scans. In this chapter, we introduce the mixed effects models that can explicitly model such dependency.

18.1 Introduction

Nonlinear mixed effects models came to be used when the classical approach of fitting a fixed effects model to individual subjects and then summarizing the parameter estimates for the population proved to be inadequate [281]. The inadequacy of the fixed effects model is caused by the fact that the within-subject dependency is ignored in longitudinally collected data. However, mixed effects models can take into account the within-subject dependency in the parameters.

Extensive literature on mixed effects models is available [125, 251, 253, 281]. There are three advantages of the mixed effects model over the usual fixed effects model. It explicitly models individual growth patterns. It accommodates an unequal number of follow-up image scans per subject and unequal intervals between scans.

As an illustration for this chapter, apply the proposed method in determining the effect of family income on the growth of hippocampus in children. The data set is published in [72] and saved in `hippocampus.mat`. There are a total of 124 children and 82 of them have repeat scans two years later. The data set contains the left hippocampus surface mesh template `hippoleft`. Since the mesh template is fairly noisy, we will perform the Laplace-Beltrami operator based heat kernel smoothing with the bandwidth $\sigma = 0.5$ and $k = 500$ number of eigenfunctions. The first four eigenfunctions are shown on smoothed template in Figure 18.1.

```
load hippocampus.mat

[A, C] =FEM(hippoleft);
```

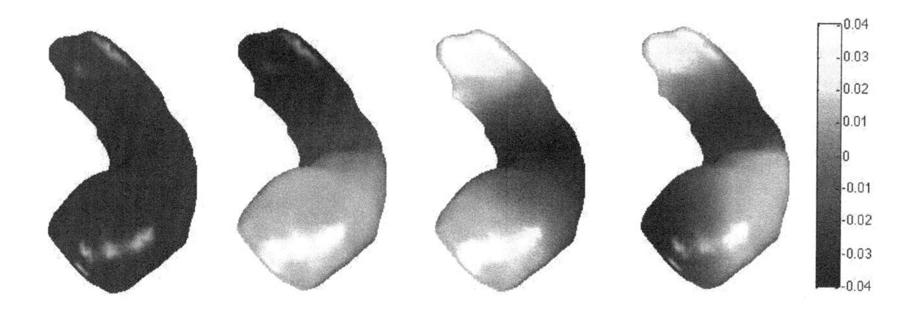

FIGURE 18.1
First four eigenfunctions of the Laplace-Beltrami operator on the left hippocampus surface. Up to 500 eigenfunctions are used in performing heat kernel smoothing along the surface to filter out noise.

```
[V1,D1] = eigs(C,A, 500, 'sa');

hippolefts = lb_smooth([],hippoleft,0.5, 500, V1, D1);

for i=1:4
    figure;
    figure_trimesh(hippolefts,V1(:,i),'rwb')
    colormap('hot')
    view([0 90]); camlight
    caxis([-0.04 0.04]);
end;
```

18.2 Mixed Effects Models

Let us follow notations given in [251]. Longitudinal data on the i-th subject will be modeled as

$$Y_i = X_i\beta + Z_i\gamma_i + e_i, \tag{18.1}$$

where Y_i is the longitudinal outcome from an image, β are fixed effects shared by all subjects, γ_i are subject specific random effects and $e_i \sim N(0, \sigma^2)$ are independent and identically distributed noise. X_i and Z_i are design matrix corresponding to the fixed and random effects. For example, we can have

$$Z_i = (1, age) \text{ and } \gamma_i = (\gamma_{i1}, \gamma_{i2})'.$$

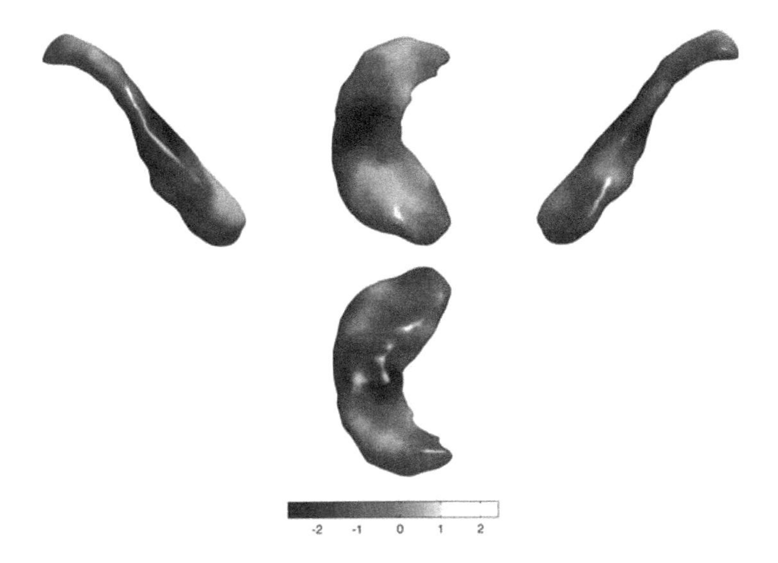

FIGURE 18.2
T-statistic results on the group difference while accounting for age and gender for all 206 scans in the fixed effects model. We are treating even the repeat scans to be independent.

The linear mixed effects model assumes the individual growth trend with an individualized linear model

$$Z_i\gamma_i = \gamma_{i1} + \texttt{age} \cdot \gamma_{i2}.$$

In the data set `hippocampus.mat`, 82 subjects have the second scans so it is necessary to explicitly model the within-subject variability for them. We assume the i-th subject has repeated scans $y_{i1}, \cdots, y_{in_i}$ at a given fixed mesh vertex. Since each subject has only one or two scans, n_i is either 1 or 2. If the data is balanced (identical number of scans per subject), the mixed effects model is straightforward but the complication arises when the data is not balanced.

18.2.1 Fixed Effects Model

In the fixed effects model, we simply treat the within-subject repeated scans as statistically independent. So we have a linear model containing fixed effect term $\texttt{age}_{ij}$:

$$y_{ij} = \beta_0 + \beta_1 \texttt{age}_{ij} + \epsilon_{ij}, \tag{18.2}$$

where ϵ_{ij} is assumed to follow Gaussian. The population average is modeled modeled linearly linearly with intercept β_0 and slope β_1.

In this example, we treat the within-subject repeated scans as statistically independent. Then using `SurfStat` package, we can perform the T-test on the significance of `group` variable as follows.

```
Age=term(age);
Gender=term(gender);
Group=term(group);

leftsurf.coord= hippolefts.vertices';
leftsurf.tri=hippolefts.faces';

left_disp2 = sqrt(sum(left_disp.^2,3));

slm0 = SurfStatLinMod( left_disp2, 1+ ...
Gender + Age + Group, leftsurf);
slm = SurfStatT(slm0, group);
figure_origami(hippolefts,slm.t)
colormap('hot')
```

The result is shown in Figure 18.2. To determine the multiple comparisons corrected thresholding, we use the random field theory.

```
pvalue= [0.01 0.05 0.1];

>>threshold=randomfield_threshold(slm, pvalue)

threshold =
    3.9354    3.4267    3.1815
```

The T-statistic value corresponding the p-value of 0.1 is 3.18 indicating there is no region of hippocampus that even has a significance of 0.1. So the fixed effect model does not show any group effect.

18.2.2 Random Effects Model

In the model (18.2), every subject has identical growth trajectory $\beta_0 + \beta_1 \texttt{age}$, which is unrealistic. Biologically every subject is expected to have its own unique growth trajectory. So we assume each subject to have its own intercept $\beta_0 + \gamma_{i0}$ and slope $\beta_1 + \gamma_{i1}$:

$$y_{ij} = \beta_0 + \gamma_{i0} + (\beta_1 + \gamma_{i1})\texttt{age}_{ij} + \epsilon_{ij}. \tag{18.3}$$

It is reasonable to assume $\gamma = (\gamma_{i0}, \gamma_{i1})'$ to be multivariate normal. The model (18.3) can be decomposed into fixed and random effect terms:

$$y_{ij} = (\beta_0 + \beta_1 \texttt{age}_{ij}) + (\gamma_{i0} + \gamma_{i1}\texttt{age}_{ij}) + \epsilon_{ij}. \tag{18.4}$$

The mixed effects model can be written in a more matrix form. Consider the general model

$$Y_i = X_i\beta + Z_i\gamma_i + \epsilon_i, \tag{18.5}$$

where $Y_i = (y_{i1}, \cdots, y_{in_i})'$, $\beta = (\beta_0, \beta_1)'$ and $\gamma_i = (\gamma_{i0}, \gamma_{i1})'$. X_i and Z_i are $n_i \times 2$ design matrices corresponding to the fixed and the random effect respectively for the i-th subject. (18.5) can be also written in a single matrix form:

$$Y = X\beta + Z\gamma + \epsilon,$$

We assume $\gamma_i \sim N(0, \Gamma)$ and $\epsilon_i \sim N(0, \Sigma_i)$. Hierarchically we can also model (18.5) as

$$Y_i|\gamma_i \sim N(X_i\beta + Z_i\gamma_i, \Sigma_i), \ \gamma_i \sim N(0, \Gamma).$$

The size of the covariance matrix is $n_i \times n_i$. The within-subject variability is expected to be smaller than between-subject variability and explicitly modeled by Σ_i. The covariance of γ_i and ϵ are expected to have block diagonal structure such that there is no correlation among the scans of different subjects while there is high correlation between the scans of the same subject:

$$\mathbb{V}\begin{pmatrix} \gamma_i \\ \epsilon_i \end{pmatrix} = \begin{pmatrix} \Gamma & 0 \\ 0 & \Sigma_i \end{pmatrix}.$$

The covariance of Y_i is given by

$$\mathbb{V}Y_i = Z_i\Gamma Z_i' + \Sigma_i.$$

The random-effect contribution is $Z_i\Gamma Z_i'$ while the within-subject contribution is Σ_i.

In `SurfStat`, in order to explicitly model the subject dependency, it is necessary to use `var2fac.m` and `random.m` functions as follows. `subject` is an identification number keeping track of which scans belong to which subject.

```
Age=term(age);
Gender=term(gender);
Group=term(group);
Subject = term(var2fac(subject));


M1 = 1+ Gender + Age + Group + random(Subject) + I;
slm = SurfStatLinMod(left_disp2, M1, leftsurf);
slm = SurfStatT( slm, group);
figure_origami(hippolefts,slm.t)
```

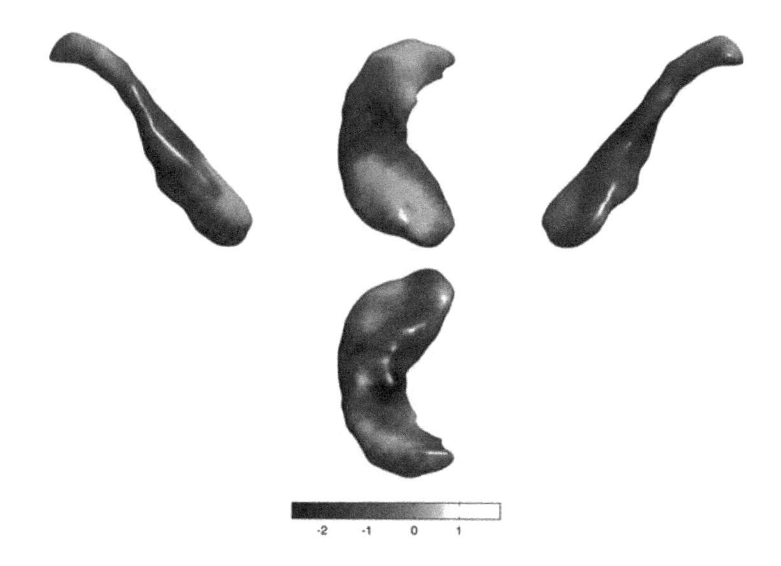

FIGURE 18.3
T-statistic results on the group difference while accounting for age and gender for all 206 scans in the random effects model. We are explicitly modeling the dependency between the repeat scans. There is some difference with the fixed effects results in Figure 18.2 but the overall patterns are similar.

```
colormap('hot')

pvalue= [0.0001 0.001 0.01 0.05 0.1];
>>threshold=randomfield_threshold(slm, pvalue)

threshold =

    4.9455    4.3482    3.6572    3.0775    2.7849
```

Again we are not detecting any regions of group difference in the mixed effects model. So we tested for the interaction difference.

18.2.3 Restricted Maximum Likelihood Estimation

The parameters and the covariance matrices are estimated by maximizing the likelihood function. Let $f(y_i|\gamma_i)$ and $f(\gamma_i)$ are density functions for $Y_i|\gamma_i$ and γ_i. The marginal density of Y_i is then given by

$$f(y_i) = \int f(y_i|\gamma_i) f(\gamma_i)\, d\gamma_i.$$

It can be shown that Y_i is again a multivariate normal

$$Y_i \sim N(X_i\beta, V_i),$$

where

$$V_i(\alpha) = Z_i\Gamma Z_i' + \Sigma_i.$$

The covariance V_i is given by some parameters α. The likelihood function is given by

$$L(\alpha,\beta) \propto \prod_{i=1}^{n} |V_i|^{-1/2} exp\Big[-\frac{1}{2}(y_i - X_i\beta)'V_i^{-1}(y_i - X_i\beta)\Big]. \tag{18.6}$$

For any parameters α, the estimate

$$\widehat{\beta}(\alpha) = \Big(\sum_{i=1}^{n} X_i V_i^{-1} X_i'\Big)^{-1} \sum_{i=1}^{n} X_i V_i^{-1} Y_i,$$

maximizes the likelihood (18.6). It can be shown that

$$\mathbb{E}\,\widehat{\beta}(\alpha) = \beta$$

and

$$\mathbb{V}\,\widehat{\beta}(\alpha) = \Big(\sum_{i=1}^{n} X_i V_i^{-1} X_i'\Big)^{-1}.$$

Maximizing the restricted likelihood without β produces the *restricted maximum likelihood* (REML) estimates for covariance parameters α [125, 281].

The REML estimate of α is as follows. Consider any full rank matrix K satisfying $KX = 0$. Then the marginal distribution of $Z = KY$ does not depend on β:

$$Z = K\epsilon \sim N(0, KVK).$$

Once V is estimated by REML, β is estimated by plugging $\widehat{V}$.

The most widely used tools for fitting the mixed effects model is the `nlme` library in `R` statistical package, which has been extensively covered in [281]. However, there is no need to use `R` to fit the mixed effects model. Keith Worsley has implemented REML procedure in `SurfStat` package that runs on top of `MATLAB` [395, 80].

18.2.4 Case Study: Longitudinal Image Analysis

There are few studies that relate the longitudinal reduction in right hippocampus volume to stress and affective disorder. The severity of stress level in children negatively correlates with the change of right hippocampus volume in longitudinally collected images [59]. Bipolar patients have significant smaller

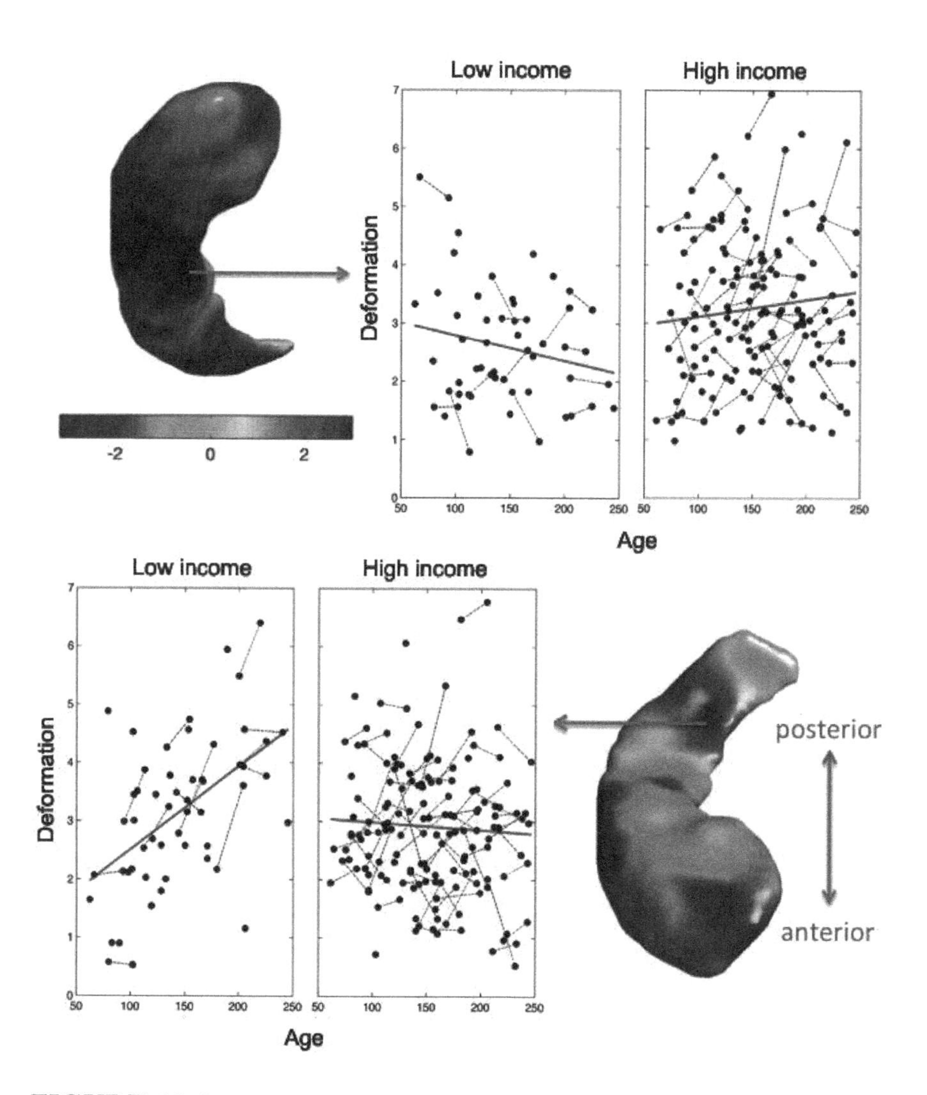

FIGURE 18.4
T-statistic map showing significant growth rate difference between children from low-income and high-income families. The significance of the growth rate difference is determined by the interaction term in a linear model. Highly focalized regions of group difference were detected in the right hippocampus (corrected p-value =0.03). The posterior region is enlarging while the midbody and the anterior parts are shrinking in children from low-income families. On the other hand the pattern is the opposite for children from high-income families. (See color insert.)

right hippocampus volume compared to normal control subjects [348]. However, the limitation of these traditional volumetric studies is that even if we can determine the volume difference, there is no way of determining if the volume difference is diffuse over the whole hippocampus or localized within small regions of the hippocampus. In this section, we introduce the mixed effects modeling framework that enables localized longitudinal hippocampus shape characterization and is able to overcome the limitations of the traditional volumetric studies.

Figure 18.4 scatter plots show the age distribution of the study. The first scans are taken at 11.6 ± 3.7 years while the second scans are taken at 14 ± 3.9 years. The symmetric diffeomorphic image normalization and template construction were performed on MRI [19]. The left and right hippocampi were manually segmented in the template using the protocol outlined in [308]. The marching cubes algorithm was then used to obtain the surface mesh model. On the template surface, we have a displacement vector field of mapping from the template to individual subject. The length of the displacement vector measures the amount of growth from the template. Since the length measurement is noisy, surface-based smoothing is necessary. We have used heat kernel smoothing to smooth out noise.

Other than deformation fields and `age` and `gender`, we also have the variable `income` which is a factor indicating whether the subjects are from high- or low-income families. T1-weighted MRIs were collected using a 3T GE SIGNA scanner on 124 children and adolescents from high- ($> 75000\$$; $n = 86$) and low-income ($< 35000\$$, $n = 38$) parents respectively. In addition to this cross-sectional data, longitudinal data was available for 82 of these subjects ($n = 66$, $> 75000\$$; $n = 16$, $< 35000\$$). This second MRI scan was acquired about 2 years later. The total number of MRI scans is 206.

We will take the length of the surface displacement with respect to the template as the response variable. If we have age information in a cross-sectional data, we can still set up a linear model involving the age term `age` like

$$\texttt{deformation} = \beta_0 + \beta_1 \texttt{age} + \beta_2 \texttt{group} + \epsilon.$$

However, this is not truly a longitudinal model but a cross-sectional model. The fixed effects model is shown in Figure 18.2, where we did not detect any signal. Having the age term in the model does not make the model longitudinal. The main difference between the longitudinal and cross-sectional models is whether if we can explicitly incorporate the dependence of repeated measurements (multiple scans of the same subject). This can be done by introducing a random effect term in the above fixed effect model.

Since 82 subjects have the second scans after about 2 years later, it is necessary to explicitly model the within-subject variability that is expected to be smaller than between-subject variability. For each scan at a given mesh vertex p, we have the following mixed effects model containing fixed effect terms (`age`, `group`) and a random effect term (`subject`):

$$\texttt{deformation} = \beta_0 + \beta_1\texttt{age} + \beta_2\texttt{group} + \beta_3\texttt{age}\cdot\texttt{group} + \gamma\texttt{subject} + \epsilon,$$

where ϵ and γ are Gaussian noise. The covariance of γ and ϵ are expected to have block structures such that there is no correlation among the scans of different subjects while there is high correlation between the scans of the same subject. The parameters are estimated using the restricted maximum likelihood (REML) method [281, 125]. The REML method is implemented in `SurfStat` package [80].

We did not detect any statistically significant group difference (β_2) at 0.01 level (corrected) in both the left and right hippocampi (Figure 18.3). However, we obtained highly focalized regions of group difference in the growth rate (interaction term β_3) in the right hippocampus (corrected pvalue =0.03). The posterior region is enlarging while the midbody and the anterior parts are shrinking in children from low-income families (Figure 18.4). On the other hand the pattern is opposite for children from high-income families.

This is the first study localizing the regions of hippocampus growth difference in children from high- and low- income families. Note that the right hippocampus is involved in the active maintenance of associations with spatial information [279]. Future studies investigating the relationship between family socioeconomic status and spatial information processing measures are warranted.

18.2.5 Functional Mixed Effects Models

It is also possible to extend a linear mixed effects model to incorporate more complex nonlinear growth patterns by taking functional covariates into the model. [143] modeled the clinical outcome Y_i of the i-th subject as

$$Y_i = X\beta + \int_0^1 W_i(p)f(p)\,dp + \epsilon_i, \tag{18.7}$$

where $W_i(p)$ is the functional observation at position p and f is the smooth functional parameter that has to be estimated. β is the fixed effect shared by all subjects. (18.7) is related to the standard impulse-response model, which has been often used in modeling fMRI responses. In the standard impulse-response model, the impulse function W_i is usually discrete and the response Y_i is continuous. The fixed effects model (18.7) is further generalized by incorporating the subject specific random effect terms in [142].

Let the j-th measurement of the i-th subject be Y_{ij}. We also have the corresponding functional observations $W_{ij}(p)$. Then Y_{ij} is modeled as

$$Y_{ij} = X_{ij}\beta + Z_{ij}\gamma_i + \int_0^1 W_{ij}(p)f(p)\,dp + \epsilon_i, \tag{18.8}$$

where X_i is the fixed effects and Z_i is the subject specific random effects.

$f(p)$ is the vector of functional effect that has to be estimated. The functional effect $f(p)$ are population level parameters and do not vary across different subjects.

Using the Karhunen-Loeve (KL) expansion, $W_{ij}(p)$ is decomposed as

$$W_{ij}(p) = \sum_k c_{ijk} \psi_k(p), \tag{18.9}$$

where c_{ijk} are uncorrelated random variables and ψ_k are KL-basis. The main limitation of this model is that the model is based on a single scalar outcome per subject. So this is not a true functional mixed effects model.

On the other hand, [413] and [414] are proposing a more general functional mixed effects model:

$$Y_{ij}(p) = \mu(p) + W_{ij}(p), \tag{18.10}$$

where $\mu(p)$ is the population level fixed functional effect and $W_i(p)$ is the subject specific random functional effect. W_{ij} is then decomposed using the similar KL-decomposition (18.9).

Bibliography

[1] F. Abell, M. Krams, J. Ashburner, R. Passingham, K. Friston, R. Frackowiak, F. Happé, C. Frith, and U. Frith. The neuroanatomy of autism: a voxel-based whole brain analysis of structural scans. *NeuroReport*, 10:1647–1651, 1999.

[2] R.J. Adler. *The Geometry of Random Fields*. John Wiley & Sons, 1981.

[3] R.J. Adler. *An Introduction to Continuity, Extrema, and Related Topics for General Gaussian Processes*. IMS, Hayward, CA, 1990.

[4] R.J. Adler. On excursion sets, tube formulas and maxima of random fields. *Annals of Applied Probability*, 10:1–74, 2000.

[5] R.J. Adler, G. Samorodnitsky, and T. Gadrich. The expected number of level crossings for stationary, harmonisable, symmetric, stable processes. *The Annals of Applied Probability*, 3:553–575, 1993.

[6] R.J. Adler and J.E. Taylor. *Random fields and geometry*. Springer Verlag, 2007.

[7] D. Aldous. *Probability Approximations via the Poisson Clumping Heuristic*. Springer-Verlag, New York, 1989.

[8] D.C. Alexander and G.J. Barker. Optimal imaging parameters for fibre-orientation estimation in diffusion MRI. *NeuroImage*, 27:357–367, 2005.

[9] T.W. Anderson. *An Introduction to Multivariate Statistical Analysis*. Wiley, 2nd edition, 1984.

[10] A. Andrade, F. Kherif, J. Mangin, K.J. Worsley, A. Paradis, O. Simon, S. Dehaene, D. Le Bihan, and J-B. Poline. Detection of fmri activation using cortical surface mapping. *Human Brain Mapping*, 12:79–93, 2001.

[11] P. R. Andresen, F. L. Bookstein, K. Conradsen, B. K. Ersbøll, J. L. Marsh, and S. Kreiborg. Surface-bounded growth modeling applied to human mandibles. *IEEE Transactions on Medical Imaging*, 19:1053–1063, 2000.

[12] S. Angenent, S. Hacker, A. Tannenbaum, and R. Kikinis. On the laplace-beltrami operator and brain surface flattening. *IEEE Transactions on Medical Imaging*, 18:700–711, 1999.

[13] G.B. Arfken. *Mathematical Methods for Physicists.* Academic Press, 5th edition, 2000.

[14] J. Ashburner and K. Friston. Voxel-based morphometry - the methods. *NeuroImage*, 11:805–821, 2000.

[15] J. Ashburner and K. Friston. Why voxel-based morphometry should be used. *NeuroImage*, 14:1238–1243, 2001.

[16] J Ashburner, C. Good, and K.J. Friston. Tensor based morphometry. *NeuroImage*, 11S:465, 2000.

[17] J. Ashburner, C. Hutton, R. S. J. Frackowiak, I. Johnsrude, C. Price, and K. J. Friston. Identifying global anatomical differences: deformation-based morphometry. *Human Brain Mapping*, 6:348–357, 1998.

[18] J. Ashburner, P. Neelin, D.L. Collins, A.C. Evans, and K.J. Friston. Incorporating prior knowledge into image registration. *NeuroImage*, 6:344–352, 1997.

[19] B.B. Avants, C.L. Epstein, M. Grossman, and J.C. Gee. Symmetric diffeomorphic image registration with cross-correlation: Evaluating automated labeling of elderly and neurodegenerative brain. *Medical Image Analysis*, 12:26–41, 2008.

[20] E.H. Aylward, N.J Minshew, G. Goldstein, N.A. Honeycutt, A.M. Augustine, K.O. Yates, P.E. Bartra, and G.D. Pearlson. Mri volumes of amygdala and hippocampus in nonmentally retarded autistic adolescents and adults. *Neurology*, 53:2145–2150, 1999.

[21] I. Babuška, R. Tempone, and G.E. Zouraris. Galerkin finite element approximations of stochastic elliptic partial differential equations. *Siam Journal of Numerical Analysis*, 42:800–825, 2004.

[22] O. Banerjee, L.E. Ghaoui, A. d'Aspremont, and G. Natsoulis. Convex optimization techniques for fitting sparse Gaussian graphical models. In *Proceedings of the 23rd international conference on Machine learning*, page 96, 2006.

[23] N. Barnea-Goraly, H. Kwon, V. Menon, S. Eliez, L. Lotspeich, and A.L. Reiss. White matter structure in autism: preliminary evidence from diffusion tensor imaging. *Biological Psychiatry*, 55:323–326, 2004.

[24] S. Baron-Cohen, H.A. Ring, S. Wheelwright, E.T. Bullmore, M.J. Brammer, A. Simmons, and S.C. Williams. Social intelligence in the normal and autistic brain: An fMRI study. *European Journal of Neuroscience*, 11:1891–1898, 1999.

[25] T.R. Barrick, C.E. Mackay, S. Prima, F. Maes, D. Vandermeulen, T.J. Crow, and N. Roberts. Automatic analysis of cerebral asymmetry: an exploratory study of the relationship between brain torgue and planum temporale asymmetry. *NeuroImage*, 24:678–691, 2005.

[26] P. Barta, M.I. Miller, and A. Qiu. A stochastic model for studying the laminar structure of cortex from mri. *IEEE Transactions on Medical Imaging*, 24:728–742, 2005.

[27] A. Bartesaghi and G. Sapiro. A system for the generation of curves on 3d brain images. *Human brain mapping*, 14(1):1–15, 2001.

[28] C.N. Barton and E.C. Cramer. Hypothesis testing in multivariate linear models with randomly missing data. *Communications in Statistics-Simulation and Computation*, 18:875–895, 1989.

[29] P.J. Basser, J. Mattiello, and D. LeBihan. MR diffusion tensor spectroscopy and imaging. *Biophys J.*, 66:259–267, 1994.

[30] P.J. Basser, S. Pajevic, C. Pierpaoli, J. Duda, and A. Aldroubi. In vivo tractography using dt-mri data. *Magnetic Resonance in Medicine*, 44:625–632, 2000.

[31] P.J. Basser and C. Pierpaoli. Microstructural and physiological features of tissues elucidated by quantitative-diffusion-tensor MRI. *Journal of Magnetic Resonance, Series B*, 111:209–219, 1996.

[32] P.G. Batchelor, F. Calamante, J.D. Tournier, D. Atkinson, D.L. Hill, and A. Connelly. Quantification of the shape of fiber tracts. *Magnetic Resonance in Medicine*, 55:894–903, 2006.

[33] P.G. Batchelor, D.L.G. Hill, F. Calamante, and D. Atkinson. Study of connectivity in the brain using the full diffusion tensor from MRI. *Lecture Notes in Computer Science*, pages 121–133, 2001.

[34] T.E.J. Behrens, H.J. Berg, S. Jbabdi, M.F.S. Rushworth, and M.W. Woolrich. Probabilistic diffusion tractography with multiple fibre orientations: what can we gain? *NeuroImage*, 34:144–155, 2007.

[35] Y. Benjamini and Y. Hochberg. Controlling the false discovery rate: a practical and powerful approach to multiple testing. *J. R. Stat. Soc, Ser. B*, 57:289–300, 1995.

[36] Y. Benjamini and D. Yekutieli. The control of the false discovery rate in multiple testing under dependency. *Annals of statistics*, pages 1165–1188, 2001.

[37] N. Berline, E. Getzler, and M. Vergne. *Heat kernels and Dirac operators.* Springer-Verlag, 1991.

[38] J. Bernal-Rusiel, M. Atienza, and J. Cantero. Detection of focal changes in human cortical thickness: spherical wavelets versus gaussian smoothing. *NeuroImage*, 41:1278–1292, 2008.

[39] M.A. Bernstein, K.E. King, X.J. Zhou, and W. Fong. Handbook of MRI pulse sequences. *Medical Physics*, 32:1452, 2005.

[40] D. Betounes. *Partial differential equations for computational science: with Maple and vector analysis.* Springer, 1998.

[41] E.D. Bigler, D.D. Blatter, C.V. Anderson, S.C. Johnson, S.D. Gale, R.O. Hopkins, and B. Burnett. Hippocampal volume in normal aging and traumatic brain injury. *American Journal of Neuroradiology*, 18:11, 1997.

[42] M.A. Blanco, M. Florez, and M. Bermejo. Evaluation of the rotation matrices in the basis of real spherical harmonics. *Journal of Molecular Structure: THEOCHEM*, 419:19–27, 1997.

[43] M. Bocher. Introduction to the theory of Fourier's series. *Ann. Math*, 7:81–152, 1906.

[44] F.L. Bookstein. Principal warps: thin-plate splines and the decomposition of deformations. *IEEE Transactions on Pattern Analysis and Machine Intelligence*, 11:567–585, 1989.

[45] F.L. Bookstein. *Morphometric Tools for Landmark Data: Geometry and Biology.* Cambridge University Press, Cambridge, 1991.

[46] W.M. Boothby. *An Introduction to Differential Manifolds and Riemannian Geometry.* Academic Press, London, 2nd edition, 1986.

[47] R.A. Boyles. On the convergence of the EM algorithm. *Journal of the Royal Statistical Society. Series B (Methodological)*, 45:47–50, 1983.

[48] P. Brambilla, M.A. Nicoletti, R.B. Sassi, A.G. Mallinger, E. Frank, D.J. Kupfer, M.S. Keshavan, and J.C. Soares. Magnetic resonance imaging study of corpus callosum abnormalities in patients with bipolar disorder. *Biological psychiatry*, 54:1294–1297, 2003.

[49] C. Brechbuhler, G. Gerig, and O. Kubler. Parametrization of closed surfaces for 3d shape description. *Computer Vision and Image Understanding*, 61:154–170, 1995.

[50] C. Brezinski. Extrapolation algorithms for filtering series of functions, and treating the Gibbs phenomenon. *Numerical Algorithms*, 36:309–329, 2004.

[51] M. M. Bronstein and I. Kokkinos. Scale-invariant heat kernel signatures for non-rigid shape recognition. In *IEEE Conference on Computer Vision and Pattern Recognition (CVPR)*, pages 1704–1711, 2010.

[52] T. Bulow. Spherical diffusion for 3D surface smoothing. *IEEE Transactions on Pattern Analysis and Machine Intelligence*, 26:1650–1654, 2004.

[53] A. Cachia, J.-F. Mangin, D. Riviére, F. Kherif, N. Boddaert, A. Andrade, D. Papadopoulos-Orfanos, J.-B. Poline, I. Bloch, M. Zilbovicius, P. Sonigo, F. Brunelle, and J. Régis. A primal sketch of the cortex mean curvature: a morphogenesis based approach to study the variability of the folding patterns. *IEEE Transactions on Medical Imaging*, 22:754–765, 2003.

[54] A. Cachia, J.-F. Mangin, D. Riviére, D. Papadopoulos-Orfanos, F. Kherif, I. Bloch, and J. Régis. A generic framework for parcellation of the cortical surface into gyri using geodesic Voronoï diagrams. *Image Analysis*, 7:403–416, 2003.

[55] J. Cao. The size of the connected components of excursion sets of χ^2, t and F fields. *Advances in Applied Probability*, 31:579–595, 1999.

[56] J. Cao and K.J. Worsley. The detection of local shape changes via the geometry of Hotelling's T2 fields. *Annals of Statistics*, 27:925–942, 1999.

[57] J. Cao and KJ Worsley. Applications of random fields in human brain mapping. *Spatial Statistics: Methodological Aspects and Applications*, 159:170–182, 2001.

[58] G. Carlsson and F. Memoli. Persistent clustering and a theorem of J. Kleinberg. *arXiv preprint arXiv:0808.2241*, 2008.

[59] V.G. Carrion, C.F. Weems, and A.L. Reiss. Stress predicts brain changes in children: A pilot longitudinal study on youth stress, posttraumatic stress disorder, and the hippocampus. *Pediatrics*, 119:509, 2007.

[60] M. Catani, R.J. Howard, S. Pajevic, and D.K. Jones. Virtual in vivo interactive dissection of white matter fasciculi in the human brain. *NeuroImage*, 17:77–94, 2002.

[61] J. Cates, P.T. Fletcher, M. Styner, H.C. Hazlett, and R. Whitaker. Particle-Based Shape Analysis of Multi-Object Complexes. In *International Conference on Medical Image Computing and Computer-Assisted Intervention (MICCAI)*, volume 11, pages 477–485, 2008.

[62] D.S. Chandrasekharaiah and L. Debnath. *Continuum Mechanics*. Academic Press, San Diego, 1994.

[63] P. Chaudhuri and J. S. Marron. Scale space view of curve estimation. *The Annals of Statistics*, 28:408–428, 2000.

[64] M.K. Chung. *Statistical Morphometry in Neuroanatomy*. Ph.D. Thesis, McGill University, 2001.

[65] M.K. Chung. *Computational Neuroanatomy: The Methods.* World Scientific, Singapore, 2012.

[66] M.K. Chung, N. Adluru, K.M. Dalton, A.L. Alexander, and R.J. Davidson. Scalable brain network construction on white matter fibers. In *Proc. of SPIE*, volume 7962, page 79624G, 2011.

[67] M.K. Chung, N. Adluru, J.E. Lee, M. Lazar, J.E. Lainhart, and A.L. Alexander. Cosine series representation of 3d curves and its application to white matter fiber bundles in diffusion tensor imaging. *Statistics and Its Interface*, 3:69–80, 2010.

[68] M.K. Chung, P. Bubenik, and P.T. Kim. Persistence diagrams of cortical surface data. *Proceedings of the 21st International Conference on Information Processing in Medical Imaging (IPMI), Lecture Notes in Computer Science (LNCS)*, 5636:386–397, 2009.

[69] M.K. Chung, K.M. Dalton, A.L. Alexander, and R.J. Davidson. Less white matter concentration in autism: 2D voxel-based morphometry. *NeuroImage*, 23:242–251, 2004.

[70] M.K. Chung, K.M. Dalton, and R.J. Davidson. Tensor-based cortical surface morphometry via weighted spherical harmonic representation. *IEEE Transactions on Medical Imaging*, 27:1143–1151, 2008.

[71] M.K. Chung, K.M. Dalton, L. Shen, A.C. Evans, and R.J. Davidson. Weighted Fourier representation and its application to quantifying the amount of gray matter. *IEEE Transactions on Medical Imaging*, 26:566–581, 2007.

[72] M.K. Chung, J.L. Hanson, R.J. Davidson, and S.D. Pollak. Effect of family income on hippocampus growth: longitudinal study. *17th Annual Meeting of the Organization for Human Brain Mapping*, (2697), 2011.

[73] M.K. Chung, R. Hartley, K.M. Dalton, and R.J. Davidson. Encoding cortical surface by spherical harmonics. *Statistica Sinica*, 18:1269–1291, 2008.

[74] M.K. Chung, M. Lazar, A.L. Alexander, Y. Lu, and R.J. Davidson. Probabilistic connectivity measure in diffusion tensor imaging via anisotropic kernel smoothing. *University of Wisconsin, Department of Statistics, Technical Report*, 1081, 2003.

[75] M.K. Chung, H. Lee, P.T. Kim, and J.C. Ye. Sparse topological data recovery in medical images. In *Biomedical Imaging: From Nano to Macro, 2011 IEEE International Symposium on*, pages 1125–1129, 2011.

[76] M.K. Chung, S. Robbins, K.M. Dalton, R.J. Davidson, A.L. Alexander, and A.C. Evans. Cortical thickness analysis in autism with heat kernel smoothing. *NeuroImage*, 25:1256–1265, 2005.

[77] M.K. Chung, S. Robbins, and A.C. Evans. Unified statistical approach to cortical thickness analysis. *Information Processing in Medical Imaging (IPMI), Lecture Notes in Computer Science*, 3565:627–638, 2005.

[78] M.K. Chung, V. Singh, P.T. Kim, K.M. Dalton, and R.J. Davidson. Topological Characterization of Signal in Brain Images Using Min-Max Diagrams. *MICCAI, Lecture Notes in Computer Science (LNCS)*, 5762:158–166, 2009.

[79] M.K. Chung and J. Taylor. Diffusion smoothing on brain surface via finite element method. In *Proceedings of IEEE International Symposium on Biomedical Imaging (ISBI)*, volume 1, pages 432–435, 2004.

[80] M.K. Chung, K.J. Worsley, M.N. Brendon, K.M. Dalton, and R.J. Davidson. General Multivariate Linear Modeling of Surface Shapes Using SurfStat. *NeuroImage*, 53:491–505, 2010.

[81] M.K. Chung, K.J. Worsley, T. Paus, D.L. Cherif, C. Collins, J. Giedd, J.L. Rapoport, and A.C. Evans. A unified statistical approach to deformation-based morphometry. *NeuroImage*, 14:595–606, 2001.

[82] M.K. Chung, K.J. Worsley, S. Robbins, and A.C. Evans. Tensor-based brain surface modeling and analysis. In *IEEE Conference on Computer Vision and Pattern Recognition (CVPR)*, volume I, pages 467–473, 2003.

[83] M.K. Chung, K.J. Worsley, S. Robbins, T. Paus, J. Taylor, J.N. Giedd, J.L. Rapoport, and A.C. Evans. Deformation-based surface morphometry applied to gray matter deformation. *NeuroImage*, 18:198–213, 2003.

[84] J.D. Clayden, A.J. Storkey, and M.E. Bastin. A probabilistic model-based approach to consistent white matter tract segmentation. *IEEE Transactions on Medical Imaging*, 11:1555–1561, 2007.

[85] D.L. Collins, P. Neelin, T.M. Peters, and A.C. Evans. Automatic 3d intersubject registration of mr volumetric data in standardized talairach space. *J. Comput. Assisted Tomogr.*, 18:192–205, 1994.

[86] D.L. Collins, T. Paus, A. Zijdenbos, K.J. Worsley, J. Blumenthal, J.N. Giedd, J.L. Rapoport, and A.C. Evans. Age related changes in the shape of temporal and frontal lobes: An mri study of children and adolescents. *Soc. Neurosci. Abstr.*, 24:304, 1998.

[87] W.J. Conover. *Practical Nonparametric Statistics.* Wiley, New York, 1980.

[88] T.E. Conturo, N.F. Lori, T.S. Cull, E. Akbudak, A.Z. Snyder, J.S. Shimony, R.C. McKinstry, H. Burton, and M.E. Raichle. Tracking neuronal fiber pathways in the living human brain. In *Natl Acad Sci USA*, volume 96, pages 10422–10427, 1999.

[89] A. Convit, P McHugh, O.T. Wolf, M.J. de Leon, M. Bobinkski, S. De Santi, A. Roche, and W. Tsui. Mri volume of the amygdala: a reliable method allowing separation from the hippocampal formation. *Psychiatry Res.*, 90:113–123, 1999.

[90] P.A. Cook, Y. Bai, S. Nedjati-Gilani, K.K. Seunarine, M.G. Hall, G.J. Parker, and D.C. Alexander. Camino: Open-source diffusion-MRI reconstruction and processing. In *14th Scientific Meeting of the International Society for Magnetic Resonance in Medicine*, page p. 2759, 2006.

[91] I. Corouge, S. Gouttard, and G. Gerig. Towards a shape model of white matter fiber bundles using diffusion tensor MRI. In *IEEE International Symposium on Biomedical Imaging: Nano to Macro*, pages 344–347, 2004.

[92] R. Courant and D. Hilbert. *Methods of Mathematical Physics.* Interscience, New York, english edition, 1953.

[93] E. Courchesne. Brainstem, cerebellar and limbic neuroanatomical abnormalities in autism. *Current Opinion in Neurobiology*, 7:269–278, 1997.

[94] E. Courchesne, H.J. Chisum, J. Townsend, A. Cowles, J. Covington, B. Egaas, M. Harwood, S. Hinds, and G.A. Press. Normal brain development and aging: quantitative analysis at in vivo MR imaging in healthy volunteers. *Radiology*, 216:672–682, 2000.

[95] H.D. Critchley, E.M. Daly, E.T. Bullmore, S.C. Williams, Van Amelsvoort T., and D.M. Robertson. The functional neuroanatomy of social behaviour: Changes in cerebral blood flow when people with autistic disorder process facial expressions. *Brain*, 123:2203–2212, 2000.

[96] J.G. Csernansky, L. Wang, S.C. Joshi, J. Tilak Ratnanather, and M.I. Miller. Computational anatomy and neuropsychiatric disease: probabilistic assessment of variation and statistical inference of group difference, hemispheric asymmetry, and time-dependent change. *NeuroImage*, 23:56–68, 2004.

[97] A.M. Dale and B. Fischl. Cortical surface-based analysis I. segmentation and surface reconstruction. *NeuroImage*, 9:179–194, 1999.

[98] K.M. Dalton, B.M. Nacewicz, T. Johnstone, H.S. Schaefer, M.A. Gernsbacher, H.H. Goldsmith, A.L. Alexander, and R.J. Davidson. Gaze fixation and the neural circuitry of face processing in autism. *Nature Neuroscience*, 8:519–526, 2005.

[99] C. Davatzikos. Spatial transformation and registration of brain images using elastically deformable models. *Comput. Vis. Image Understanding*, 66:207–222, 1997.

[100] C. Davatzikos and R.N. Bryan. Using a deformable surface model to obtain a shape representation of the cortex. *Proceedings of the IEEE International Conference on Computer Vision*, 9:2122–2127, 1995.

[101] F. de Goes, S. Goldenstein, and L. Velho. A hierarchical segmentation of articulated bodies. *Computer Graphics Forum*, 27:1349–1356, 2008.

[102] M.C. de Lara. Geometric and symmetry properties of a nondegenerate diffusion process. *The Annals of Probability*, pages 1557–1604, 1995.

[103] V. de Silva and R. Ghrist. Homological sensor networks. *Notic Amer Math Soc*, 54:10–17, 2007.

[104] Y.A. Dementieva, D.D. Vance, S.L. Donnelly, L.A. Elston, C.M. Wolpert, S.A. Ravan, G.R. DeLong, R.K. Abramson, H.H. Wright, and M.L. Cuccaro. Accelerated head growth in early development of individuals with autism. *Pediatric neurology*, 32:102–108, 2005.

[105] A.P. Dempster, N.M. Laird, D.B. Rubin, et al. Maximum likelihood from incomplete data via the EM algorithm. *Journal of the Royal Statistical Society. Series B (Methodological)*, 39:1–38, 1977.

[106] P.A.M. Dirac. *The principles of quantum mechanics.* Oxford University Press, USA, 1981.

[107] M.P. do Carmo. *Riemannian Geometry.* Prentice-Hall, Inc., 1992.

[108] D.L. Donoho and Y. Tsaig. *Fast solution of l_1-norm minimization problems when the solution may be sparse.* Citeseer, 2006.

[109] E.R. Dougherty. *Random Processes for Image and Signal Processing.* IEEE Press, 1999.

[110] D.A. Drew. *Theory of multicomponent fluids.* Springer-Verlag, New York, 1991.

[111] I.L. Dryden and K.V. Mardia. *Statistical Shape Analysis.* John Wiley & Sons, 1998.

[112] A.T. Du, N. Schuff, L.L. Chao, J. Kornak, W.J. Jagust, J.H. Kramer, B.R. Reed, B.L. Miller, D. Norman, H.C. Chui, et al. Age effects on atrophy rates of entorhinal cortex and hippocampus. *Neurobiology of Aging*, 27:733–740, 2006.

[113] H. Edelsbrunner, M-L. Dequent, Y. Mileyko, and O. Pourquie. Assessing periodicity in gene expression as measured by microarray data. *Preprint*, 2008.

[114] H. Edelsbrunner and J. Harer. Persistent homology - a survey. *Contemporary Mathematics*, 453:257–282, 2008.

[115] H. Edelsbrunner and John Harer. *Computational Topology: An Introduction.* American Mathematical Society Press, 2009.

[116] H. Edelsbrunner, D. Letscher, and A. Zomorodian. Topological persistence and simplification. *Discrete and Computational Geometry*, 28:511–533, 2002.

[117] B. Egaas, E. Courchesne, and O. Saitoh. Reduced size of corpus callosum in autism. *Archives of Neurology*, 52:794, 1995.

[118] L.C. Evans. *Partial differential equations.* American Mathematical Society, 1998.

[119] W. Feller. *Introduction to probability theory and its applications, Vol. 1.* Wiley, New York, 1968.

[120] B. Fischl and A.M. Dale. Measuring the thickness of the human cerebral cortex from magnetic resonance images. *Proceedings of the National Academy of Sciences (PNAS)*, 97:11050–11055, 2000.

[121] B. Fischl, M.I. Sereno, R. Tootell, and A.M. Dale. High-resolution intersubject averaging and a coordinate system for the cortical surface. *Hum. Brain Mapping*, 8:272–284, 1999.

[122] B. Flury. *A First Course in Multivariate Statistics.* Springer, 1997.

[123] A. Fornito, A. Zalesky, and E.T. Bullmore. Network scaling effects in graph analytic studies of human resting-state fMRI data. *Frontiers in Systems Neuroscience*, 4:1–16, 2010.

[124] J. Foster and F.B. Richards. The Gibbs phenomenon for piecewise-linear approximation. *American Mathematical Monthly*, 98:47–49, 1991.

[125] J. Fox. *An R and S-Plus companion to applied regression.* Sage Publications, Inc., 2002.

[126] J. Fox, M. Friendly, and G. Monette. Visualizing hypothesis tests in multivariate linear models: the heplots package for R. *Computational Statistics*, 24:233–246, 2009.

[127] J. Friedman, T. Hastie, and R. Tibshirani. Sparse inverse covariance estimation with the graphical lasso. *Biostatistics*, 9:432, 2008.

[128] M. Frigge, D.C. Hoaglin, and B. Iglewicz. Some implementations of the boxplot. *American Statistician*, pages 50–54, 1989.

[129] K.J. Friston, A.P. Holmes, K.J. Worsley, J.-P. Poline, C.D. Frith, and R.S.J. Frackowiak. Statistical parametric maps in functional imaging: a general linear approach. *Human Brain Mapping*, 2:189–210, 1995.

[130] C. Gaser, E. Luders, P.M. Thompson, A.D. Lee, R.A. Dutton, J.A. Geaga, K.M. Hayashi, U. Bellugi, A.M. Galaburda, J.R. Korenberg, D.L. Mills, A.W. Toga, and A.L. Reiss. Increased local gyrification mapped in williams syndrome. *NeuroImage*, 33:46–54, 2006.

[131] C. Gaser, H.-P. Volz, S. Kiebel, S. Riehemann, and H. Sauer. Detecting structural changes in whole brain based on nonlinear deformations - Application to schizophrenia research. *NeuroImage*, 10:107–113, 1999.

[132] A. Gelb. The resolution of the gibbs phenomenon for spherical harmonics. *Mathematics of Computation*, 66:699–717, 1997.

[133] I.M. Gelfand, G.E. Shilov, and E. Saletan. *Generalized Functions*. Academic Press, New York, 1964.

[134] C.R. Genovese, N.A. Lazar, and T. Nichols. Thresholding of statistical maps in functional neuroimaging using the false discovery rate. *NeuroImage*, 15:870–878, 2002.

[135] G. Gerig, M. Styner, D. Jones, D. Weinberger, and J. Lieberman. Shape analysis of brain ventricles using spharm. In *MMBIA*, pages 171–178, 2001.

[136] G. Gerig, M. Styner, and G. Szekely. Statistical shape models for segmentation and structural analysis. In *Proceedings of IEEE International Symposium on Biomedical Imaging (ISBI)*, volume I, pages 467–473, 2004.

[137] J.-M. Geusebroek, A. W. M. Smeulders, and J. van de Weijer. Fast anisotropic gauss filtering. *Lecture Notes in Computer Science*, pages 99–112, 2000.

[138] R. Ghrist. Barcodes: The persistent topology of data. *Bulletin of the American Mathematical Society*, 45:61–75, 2008.

[139] J.W. Gibbs. Fourier's Series. *Nature*, 59:200, 1898.

[140] J.N. Giedd, J. Blumenthal, N.O. Jeffries, J.C. Rajapakse, A.C. Vaituzis, H. Liu, Y.C. Berry, M. Tobin, J. Nelson, and F.X. Castellanos. Development of the human corpus callosum during childhood and adolescence: a longitudinal MRI study. *Progress in Neuro-Psychopharmacology & Biological Psychiatry*, 23:571–588, 1999.

[141] J.N. Giedd, J.W. Snell, N. Lange, J.C. Rajapakse, D. Kaysen, A.C. Vaituzis, Y.C. Vauss, S.D. Hamburger, P.L. Kozuch, and J.L. Rapoport. Quantitative magnetic resonance imaging of human brain development: Ages 4-18. *Cerebral Cortex*, 6:551–160, 1996.

[142] J. Goldsmith, C.M. Crainiceanu, B. Caffo, and D. Reich. Longitudinal penalized functional regression for cognitive outcomes on neuronal tract measurements. *Journal of the Royal Statistical Society: Series C (Applied Statistics)*, 61:in press, 2012.

[143] J. Goldsmith, C.M. Crainiceanu, B.S. Caffo, and D.S. Reich. Penalized functional regression analysis of white-matter tract profiles in multiple sclerosis. *NeuroImage*, 57:431–439, 2011.

[144] P. Golland, W.E.L. Grimson, M.E. Shenton, and R. Kikinis. Deformation analysis for shape based classification. *Information Processing in Medical Imaging (IPMI), Lecture Notes in Computer Science*, 2082:517–530, 2001.

[145] G. Gong, Y. He, L. Concha, C. Lebel, D.W. Gross, A.C. Evans, and C. Beaulieu. Mapping anatomical connectivity patterns of human cerebral cortex using in vivo diffusion tensor imaging tractography. *Cerebral Cortex*, 19:524–536, 2009.

[146] C.D. Good, I. Johnsrude, J. Ashburner, R.N.A. Henson, K.J. Friston, and R.S.J. Frackowiak. Cerebral asymmetry and the effects of sex and handedness on brain structure: a voxel-based morphometric analysis of 465 normal adult human brains. *NeuroImage*, 14:685–700, 2001.

[147] C.D. Good, I.S. Johnsrude, J. Ashburner, R.N.A. Henson, K.J. Friston, and R.S.J. Frackowiak. A voxel-based morphometric study of ageing in 465 normal adult human brains. *NeuroImage*, 14:21–36, 2001.

[148] D. Gottlieb and C.-W. Shu. On the gibbs phenomenon and its resolution. *SIAM Review*, 39:644–668, 1997.

[149] H. Gray. *Henry Gray's Anatomy of the Human Body*. http://en.wikipedia.org/wiki, 1858.

[150] L.D. Griffin. The intrinsic geometry of the cerebral cortex. *Journal of Theoretical Biology*, 166:261–273, 1994.

[151] A. Gruen and D. Akca. Least squares 3D surface and curve matching. *ISPRS Journal of Photogrammetry and Remote Sensing*, 59:151–174, 2005.

[152] X. Gu, Y.L. Wang, T.F. Chan, T.M. Thompson, and S.T. Yau. Genus zero surface conformal mapping and its application to brain surface mapping. *IEEE Transactions on Medical Imaging*, 23:1–10, 2004.

[153] A. Gueziec, X. Pennec, and N. Ayache. Medical image registration using geometeric hashing. *IEEE Computational Science and Engineering*, 4:29–41, 1997.

[154] R.C. Gur, F. Gunning-Dixon, W.B. Bilker, and R.E. Gur. Sex differences in temporo-limbic and frontal brain volumes of healthy adults. *Cerebral Cortex*, 12:998–1003, 2002.

[155] M.E. Gurtin and G.B. McFadden. *On the Evolutation of Phase Boundaries*. Springer-Verlag, New York, 1991.

[156] I. Guskov and Z.J. Wood. Topological noise removal. In *Graphics Interface*, pages 19–26, 2001.

[157] D.J. Hagler Jr., A.P. Saygin, and M.I. Sereno. Smoothing and cluster thresholding for cortical surface-based group analysis of fMRI data. *NeuroImage*, 33:1093–1103, 2006.

[158] P. Hagmann, M. Kurant, X. Gigandet, P. Thiran, V.J. Wedeen, R. Meuli, and J.P. Thiran. Mapping human whole-brain structural networks with diffusion MRI. *PLoS One*, 2(7):e597, 2007.

[159] P. Hagmann, J.P. Thiran, P. Vandergheynst, S. Clarke, R. Meuli, and S. Lausanne. Statistical fiber tracking on DT-MRI data as a potential tool for morphological brain studies. In *ISMRM Workshop on Diffusion MRI: Biophysical Issues*, 2000.

[160] L.C. Hamilton. *Regression with graphics: A second course in applied statistics*. Duxbury Press, Belmont, 1992.

[161] J. Han, J.S. Kim, C.K. Chung, and K.S. Park. Evaluation of smoothing in an iterative lp-norm minimization algorithm for surface-based source localization of meg. *Physics in Medicine and Biology*, 52:4791–4803, 2007.

[162] X. Han, J. Jovicich, D. Salat, A. van der Kouwe, B. Quinn, S. Czanner, E. Busa, J. Pacheco, M. Albert, R. Killiany, et al. Reliability of MRI-derived measurements of human cerebral cortical thickness: The effects of field strength, scanner upgrade and manufacturer. *NeuroImage*, 32:180–194, 2006.

[163] J.L. Hanson, M.K. Chung, B.B. Avants, K.D. Rudolph, E.A. Shirtcliff, J.C. Gee, R.J. Davidson, and S.D. Pollak. Structural variations in prefrontal cortex mediate the relationship between early childhood stress and spatial working memory. *The Journal of Neuroscience*, 32:7917–7925, 2012.

[164] J.L. Hanson, M.K. Chung, B.B. Avants, E.A. Shirtcliff, J.C. Gee, R.J. Davidson, and S.D. Pollak. Early stress is associated with alterations in the orbitofrontal cortex: A tensor-based morphometry investigation of brain structure and behavioral risk. *Journal of Neuroscience*, 30:7466–7477, 2010.

[165] A.Y. Hardan, N.J. Minshew, and M.S. Keshavan. Corpus callosum size in autism. *Neurology*, 55:1033, 2000.

[166] J.C. Hart. Computational topology for shape modeling. In *Proceedings of the International Conference on Shape Modeling and Applications*, pages 36–43, 1999.

[167] D.A. Harville. *Matrix Algebra from a Statistician's Perspective.* Springer-Verlag, New York, 1997.

[168] S. Hayasaka, A.M. Peiffer, C.E. Hugenschmidt, and P.J. Laurienti. Power and sample size calculation for neuroimaging studies by non-central random field theory. *NeuroImage*, 37:721–730, 2007.

[169] M.M. Haznedar, M.S. Buchsbaum, T.C. Wei, P.R. Hof, C Cartwright, C.A. Bienstock, and E. Hollander. Limbic circuitry in patients with autism spectrum disorders studied with positron emission tomography and magnetic resonance imaging. *American Journal of Psychiatry*, 157:1994–2001, 2000.

[170] Y. He, Z.J. Chen, and A.C. Evans. Small-world anatomical networks in the human brain revealed by cortical thickness from MRI. *Cerebral Cortex*, 17:2407–2419, 2007.

[171] C.C. Henery and T.M. Mayhew. The cerebrum and cerebellum of the fixed human brain: efficient and unbiased estimates of volumes and cortical surface areas. *Journal of Anatomy*, 167:167–180, 1989.

[172] V. Hernandez, J. E. Roman, A. Tomas, and V. Vidal. A survey of software for sparse eigenvalue problems. *Universidad Politécnica de Valencia*, Technical Report STR-6, 2006. http://www.grycap.upv.es/slepc.

[173] T.J. Hoffmann, M.K. Chung, K.D. Dalton, A.L. Alexander, G. Wahba, and R.J. Davidson. Subpixel curvature estimation of the corpus callosum via splines and its application to autism. In *10th Annual Meeting of the Organization for Human Brain Mapping*, 2004.

[174] M. Hoffmann-Ostenhof, T. Hoffmann-Ostenhof, and N. Nadirashvili. On the multiplicity of eigenvalues of the laplacian on surfaces. *Annals of Global Analysis and Geometry*, 17:43–48, 1999.

[175] H.H.H. Homeier and E.O. Steinborn. Some properties of the coupling coefficients of real spherical harmonics and their relation to gaunt coefficients. *Journal of Molecular Structure: THEOCHEM*, 368:31–37, 1996.

[176] D. Horak, S. Maletić, and M. Rajković. Persistent homology of complex networks. *Journal of Statistical Mechanics: Theory and Experiment*, 2009:P03034, 2009.

[177] M.A. Howard, P.E. Cowell, J. Boucher, P. Broks, A. Mayes, A. Farrant, and N. Roberts. Convergent neuroanatomical and behavioral evidence of an amygdala hypothesis of autism. *NeuroReport*, 11:2931–2935, 2000.

[178] S. Huang, J. Li, L. Sun, J. Liu, T. Wu, K. Chen, A. Fleisher, E. Reiman, and J. Ye. Learning brain connectivity of Alzheimer's disease from neuroimaging data. In *Proceedings of Neural Information Processing Systems Conference (NIPS)*, volume 22, pages 808–816, 2009.

[179] S. Huang, J. Li, L. Sun, J. Ye, A. Fleisher, T. Wu, K. Chen, E. Reiman, and Alzheimer's Disease NeuroImaging Initative. Learning brain connectivity of Alzheimer's disease by sparse inverse covariance estimation. *NeuroImage*, 50:935–949, 2010.

[180] M. K. Hurdal and K. Stephenson. Cortical cartography using the discrete conformal approach of circle packings. *NeuroImage*, 23:S119–S128, 2004.

[181] A.P. Jackowski, C.M. de Araújo, A.L.T. de Lacerda, M. de Jesus, and J. Kaufman. Neurostructural imaging findings in children with post-traumatic stress disorder: Brief review. *Psychiatry and Clinical Neurosciences*, 63:1–8, 2009.

[182] S. Jbabdi, P. Bellec, R. Toro, J. Daunizeau, M. Pélégrini-Issac, and H. Benali. Accurate anisotropic fast marching for diffusion-based geodesic tractography. *International Journal of Biomedical Imaging*, 2008:1–12, 2008.

[183] M. Jenkinson, P. Bannister, M. Brady, and S. Smith. Improved optimization for the robust and accurate linear registration and motion correction of brain images. *NeuroImage*, 17:825–841, 2002.

[184] A.J. Jerri. *The Gibbs phenomenon in Fourier analysis, splines and wavelet approximations.* Springer, 1998.

[185] P. Jezzard and R.S. Balaban. Correction for geometric distortion in echo planar images from b0 field variations. *Magn. Reson. Med.*, 34:65–73, 2007.

[186] H.J. Jo, J.-M. Lee, J.-H. Kim, Y.-W. Shin, I.-Y. Kim, J.S. Kwon, and S.I. Kim. Spatial accuracy of fmri activation influenced by volume- and surface-based spatial smoothing techniques. *NeuroImage*, 34:550–564, 2007.

[187] S.C. Johnson, L.C. Baxter, L. Susskind-Wilder, D.J. Connor, M.N. Sabbagh, and R.J. Caselli. Hippocampal adaptation to face repetition in healthy elderly and mild cognitive impairment. *Neuropsychologia*, 42:980–989, 2004.

[188] D.K. Jones and P.J. Basser. MR diffusion tensor spectroscopy and imaging. *Magnetic Resonance in Medicine*, 52:979–993, 2004.

[189] D.K. Jones, M. Catani, C. Pierpaoli, S.J. Reeves, S.S. Shergill, M. O'Sullivan, P. Golesworthy, P. McGuire, M.A. Horsfield, A. Simmons, S.C. Williams, and R.J. Howard. Age effects on diffusion tensor magnetic resonance imaging tractography measures of frontal cortex connections in schizophrenia. *Human Brain Mapping*, 27:230–238, 2006.

[190] S.E. Jones, B.R. Buchbinder, and I. Aharon. Three-dimensional mapping of cortical thickness using Laplace's equation. *Human Brain Mapping*, 11:12–32, 2000.

[191] A.A. Joshi, D.W. Shattuck, P.M. Thompson, and R.M. Leahy. A parameterization-based numerical method for isotropic and anisotropic diffusion smoothing on non-flat surfaces. *IEEE Transactions on Image Processing*, 18:1358–1365, 2009.

[192] S. Joshi, B. Davis, M. Jomier, and G. Gerig. Unbiased diffeomorphic atlas construction for computational anatomy. *NeuroImage*, 23:151–160, 2004.

[193] S. Joshi, S. Pizer, P.T. Fletcher, P. Yushkevich, A. Thall, and J.S. Marron. Multiscale deformable model segmentation and statistical shape analysis using medial descriptions. *IEEE Transactions on Medical Imaging*, 21:538–550, 2002.

[194] S.C. Joshi. *Large Deformation Diffeomorphisms and Gaussian Random Fields for Statistical Characterization of Brain Sub-Manifolds*. Ph.D. thesis. Washington University, St. Louis, 1998.

[195] S.C. Joshi, U. Grenander, and M.I. Miller. The geometry and shape of brain sub-manifolds. *International Journal of Pattern Recognition and Artificial Intelligence: Special Issue on Processing of MR Images of the Human*, 11:1317–1343, 1997.

[196] S.C. Joshi, J. Wang, M.I. Miller, D.C. Van Essen, and U. Grenander. On the differential geometry of the cortical surface. *Vision Geometry IV, Vol. 2573, Proceedings of the SPIE's 1995 International Symposium on Optical Science, Engineering and Instrumentation*, pages 304–311, 1995.

[197] N. Kabani, D. Le Goualher, G. MacDonald, and A.C. Evans. Measurement of cortical thickness using an automated 3-D algorithm: a validation study. *NeuroImage*, 13:375–380, 2000.

[198] A. Kelemen, G. Szekely, and G. Gerig. Elastic model-based segmentation of 3-d neuroradiological data sets. *IEEE Transactions on Medical Imaging*, 18:828–839, 1999.

[199] D.G. Kendall. A survey of the statistical theory of shape. *Statistical Science*, 4:87–120, 1989.

[200] D.N. Kennedy, K.M. O'Craven, B.S. Ticho, A.M. Goldstein, N. Makris, and J.W. Henson. Structural and functional brain asymmetries in human situs inversus totalis. *Neurology*, 53:1260–1265, 1999.

[201] A.R. Khan, M.K. Chung, and M.F. Beg. Robust atlas-based brain segmentation using multi-structure confidence-weighted registration. *Lecture Notes on Computer Science*, 5762:549–557, 1999.

[202] N. Khaneja, M.I. Miller, and U. Grenander. Dynamic programming generation of curves on brain surfaces. *IEEE Transactions on Pattern Analysis and Machine Intelligence*, 20:1260–1265, 1998.

[203] S.J. Kiebel, J.P. Poline, K.J. Friston, A.P. Holmes, and K.J. Worsley. Robust smoothness estimation in statistical parametric maps using standardized residuals from the general linear model. *NeuroImage*, 10:756–766, 1999.

[204] D.-E. Kim, K.-J. Park, D. Schellingerhout, S.-W. Jeong, M.-G. Ji, W.J. Choi, Y.-O. Tak, G.-H. Kwan, E.A. Koh, S.-M. Noh, H.Y. Jang, T.-Y. Kim, J.-W. Jeong, J.S. Lee, and H.-K. Choi. A new image-based stroke registry containing quantitative magnetic resonance imaging data. *Cerebrovascular Diseases*, 32:567–576, 2011.

[205] S.-G. Kim, M.K. Chung, S.M. Schaefer, C. van Reekum, and R.J. Davidson. Sparse shape representation using the laplace-beltrami eigenfunctions and its application to modeling subcortical structures. In *The proceedings of IEEE Computer Society Workshop on Mathematical Methods in Biomedical Image Analysis (MMBIA)*, pages 25–32, 2012.

[206] S.-G. Kim, H.K. Lee, M.K. Chung, J.L. Hanson, B.B. Avants, J.C. Gee, R.J. Davidson, and S.D. Pollak. Agreement between the white matter connectivity based on the tensor-based morphometry and the volumetric white matter parcellations based on diffusion tensor imaging. In *The proceedings of IEEE International Symposium on Biomedical Imaging (ISBI)*, pages 42–45, 2012.

[207] S.J. Kim, K. Koh, M. Lustig, S. Boyd, and D. Gorinevsky. An interior-point method for large-scale l1-regularized least squares. *IEEE Journal of Selected Topics in Signal Processing*, 1:606–617, 2007.

[208] S.J. Kim, K. Koh, M. Lustig, S. Boyd, and D. Gorinevsky. An interior-point method for large-scale l_1-regularized least squares. *IEEE Journal of Selected Topics in Signal Processing*, 1:606–617, 2008.

[209] E. Kishon, T. Hastie, and H. Wolfson. 3D curve matching using splines. In *Proceedings of the European Conference on Computer Vision*, pages 589–591, 1990.

[210] M.A. Koch, D.G. Norris, and M. Hund-Georgiadis. An investigation of functional and anatomical connectivity using magnetic resonance imaging. *NeuroImage*, 16:241–250, 2002.

[211] K. Kollakian. Performance analysis of automatic techniques for tissue classification in magnetic resonance images of the human brain. Technical Report Master's thesis, Concordia University, Montreal, Quebec, Canada, 1996.

[212] S. Kovačič and R. Bajcsy. Multiscale/multiresolution representations. *Brain Warping*, pages 45–65, 1999.

[213] E. Kreyszig. *Differential Geometry*. University of Toronto Press, 1959.

[214] S. Kwapien and W.A. Woyczynski. *Random Series and Stochastic Integrals: Single and Multiple*. Birkhauser, 1992.

[215] Z. Lai, J. Hu, C. Liu, V. Taimouri, D. Pai, J. Zhu, J. Xu, and J. Hua. Intra-patient supine-prone colon registration in CT colonography using shape spectrum. In *Medical Image Computing and Computer-Assisted Intervention – MICCAI 2010*, volume 6361 of *Lecture Notes in Computer Science*, pages 332–339, 2010.

[216] C.L. Lawson and R.J. Hanson. *Solving Least Squares Problems*. Prentice-Hall, 1974.

[217] M. Lazar and A.L. Alexander. White matter tractography algorithms error analysis. *NeuroImage*, 20:1140–1153, 2003.

[218] M. Lazar, D.M. Weinstein, J.S. Tsuruda, K.M. Hasan, K. Arfanakis, M.E. Meyerand, B. Badie, H. Rowley, V. Haughton, A. Field, B. Witwer, and A.L. Alexander. White matter tractography using tensor deflection. *Human Brain Mapping*, 18:306–321, 2003.

[219] H. Lee, M.K. Chung, H. Kang, B.-N. Kim, and D.S. Lee. Computing the shape of brain networks using graph filtration and Gromov-Hausdorff metric. *MICCAI, Lecture Notes in Computer Science*, 6892:302–309, 2011.

[220] H. Lee, M.K. Chung, H. Kang, B.-N. Kim, and D.S. Lee. Discriminative persistent homology of brain networks. In *IEEE International Symposium on Biomedical Imaging (ISBI)*, pages 841–844, 2011.

[221] H. Lee, D.S. Lee, H. Kang, B.-N. Kim, and M.K. Chung. Sparse brain network recovery under compressed sensing. *IEEE Transactions on Medical Imaging*, 30:1154–1165, 2011.

[222] J. Lee, D. Hsu, A.L. Alexander, M. Lazar, D. Bigler, and J.E. Lainhart. A study of underconnectivity in autism using DTI: W-matrix tractography. In *Proceedings of ISMRM*, 2008.

[223] A. Leemans, J. Sijbers, S. De Backer, E. Vandervliet, and P. Parizel. Multiscale white matter fiber tract coregistration: A new feature-based approach to align diffusion tensor data. *Magnetic Resonance in Medicine*, 55:1414–1423, 2006.

[224] R.B. Lehoucq, D.C. Sorensen, and C. Yang. *ARPACK Users' Guide: Solution of Large-Scale Eigenvalue Problems with Implicitly Restarted Arnoldi Methods.* SIAM Publications, Philadelphia, 1998.

[225] N. Lepore, C.A. Brun, M.C. Chiang, Y.Y. Chou, R.A. Dutton, K.M. Hayashi, O.L. Lopez, H.J. Aizenstein, A.W. Toga, J.T. Becker, and P.M. Thompson. Multivariate statistics of the jacobian matrices in tensor based morphometry and their application to HIV/AIDS. *Lecture Notes in Computer Science*, pages 191–198, 2006.

[226] J. P. Lerch and A.C. Evans. Cortical thickness analysis examined through power analysis and a population simulation. *NeuroImage*, 24:163–173, 2005.

[227] J.P. Lerch, K. Worsley, W.P. Shaw, D.K. Greenstein, R.K. Lenroot, J. Giedd, and A.C. Evans. Mapping anatomical correlations across cerebral cortex (MACACC) using cortical thickness from MRI. *NeuroImage*, 31:993–1003, 2006.

[228] M.E. Leventon, W.E.L. Grimson, and O. Faugeras. Statistical shape influence in geodesic active contours. In *IEEE Conference on Computer Vision and Pattern Recognition (CVPR)*, volume 1, pages 316–323, 2000.

[229] B. Lévy and F. Inria-Alice. Laplace-Beltrami eigenfunctions towards an algorithm that "understands" geometry. In *IEEE International Conference on Shape Modeling and Applications, 2006*, page 13, 2006.

[230] T. Lindeberg. *Scale-Space Theory in Computer Vision.* Kluwer Academic Publisher, 1994.

[231] C. Lord, M. Rutter, and A.L. Couteur. Autism diagnostic interview-revised: a revised version of a diagnostic interview for caregivers of individuals with possible pervasive developmental disorders. *Journal of Autism and Developmental Disorders*, pages 659–685, 1994.

[232] W.E. Lorensen and H.E. Cline. Marching cubes: A high resolution 3D surface construction algorithm. In *Proceedings of the 14th Annual Conference on Computer Graphics and Interactive Techniques*, pages 163–169, 1987.

[233] M. Loubele, F. Maes, F. Schutyser, G. Marchal, R. Jacobs, and P. Suetens. Assessment of bone segmentation quality of cone-beam CT versus multislice spiral CT: a pilot study. *Oral Surgery, Oral Medicine, Oral Pathology, Oral Radiology, and Endodontology*, 102:225–234, 2006.

[234] E. Luders, K.L. Narr, P.M. Thompson, D.E. Rex, R.P. Woods, H. DeLuca, L. Jancke, and A.W. Toga. Gender effects on cortical thickness and the influence of scaling. *Human Brain Mapping*, 27:314–324, 2006.

[235] E. Luders, P.M. Thompson, K.L. Narr, A.W. Toga, L. Jancke, and C. Gaser. A curvature-based approach to estimate local gyrification on the cortical surface. *NeuroImage*, 29:1224–1230, 2006.

[236] D. Lundqvist, A. Flykt, and A. Ohman. *Karolinska Directed Emotional Faces*. Department of Neurosciences, Karolinska Hospital, Stockholm, Sweden, 1998.

[237] J.D. MacDonald, N. Kabani, D. Avis, and A.C. Evans. Automated 3-D extraction of inner and outer surfaces of cerebral cortex from MRI. *NeuroImage*, 12:340–356, 2000.

[238] J.K. Mai, J. Assheuer, and G. Paxinos. *Atlas of the Human Brain*. Academic Press, San Diego, 1997.

[239] R Malladi and I. Ravve. Fast difference schemes for edge enhancing Beltrami flow. In *Proceedings of Computer Vision-ECCV, Lecture Notes in Computer Science (LNCS)*, volume 2350, pages 343–357, 2002.

[240] S. Mallat and Z. Zhang. Matching pursuits with time-frequency dictionaries. *IEEE Transactions on Signal Processing*, 41:3397–3415, 1993.

[241] F. Manes, J. Piven, D. Vrancic, V. Nanclares, C. Plebst, and S.E. Starkstein. An MRI study of the corpus callosum and cerebellum in mentally retarded autistic individuals. *Journal of Neuropsychiatry and Clinical Neurosciences*, 11:470, 1999.

[242] G. Marrelec, A. Krainik, H. Duffau, M. Pélégrini-Issac, S. Lehéricy, J. Doyon, and H. Benali. Partial correlation for functional brain interactivity investigation in functional MRI. *NeuroImage*, 32:228–237, 2006.

[243] J.E. Marsden and T.J.R. Hughes. *Mathematical Foundations of Elasticity*. Dover Publications, Inc., 1983.

[244] M. Mather, T. Canli, T. English, S. Whitfield, P. Wais, K. Ochsner, J.D.E. Gabrieli, and L.L. Carstensen. Amygdala responses to emotionally valenced stimuli in older and younger adults. *Psychological Science*, 15:259–263, 2004.

[245] R. Mazumder and T. Hastie. Exact covariance thresholding into connected components for large-scale graphical lasso. *The Journal of Machine Learning Research*, 13:781–794, 2012.

[246] A.B. McMillan, B.P. Hermann, S.C. Johnson, R.R. Hansen, M. Seidenberg, and M.E. Meyerand. Voxel-based morphometry of unilateral temporal lobe epilepsy reveals abnormalities in cerebral white matter. *NeuroImage*, 23:167–174, 2004.

[247] F. Mémoli, G. Sapiro, and T. Thompson. Implicit brain imaging. *NeuroImage*, 23:S179–S188, 2004.

[248] M.I. Miller, A. Banerjee, G.E. Christensen, S.C. Joshi, N. Khaneja, U. Grenander, and L. Matejic. Statistical methods in computational anatomy. *Statistical Methods in Medical Research*, 6:267–299, 1997.

[249] M.I. Miller, A.B. Massie, J.T. Ratnanather, K.N. Botteron, and J.G. Csernansky. Bayesian construction of geometrically based cortical thickness metrics. *NeuroImage*, 12:676–687, 2000.

[250] M.I. Miller and A. Qiu. The emerging discipline of computational functional anatomy. *Neuroimage*, 45(1S):S16–S39, 2009.

[251] J.K. Milliken and S.D. Edland. Mixed effect models of longitudinal alzheimer's disease data: a cautionary note. *Statist. Med.*, 19:1617–1629, 2000.

[252] J. Milnor. *Morse Theory.* Princeton University Press, 1973.

[253] G. Molenberghs and G. Verbeke. *Models for Discrete Longitudinal Data.* Springer, 2005.

[254] S. Mori, B.J. Crain, V.P. Chacko, and P.C. van Zijl. Three-dimensional tracking of axonal projections in the brain by magnetic resonance imaging. *Annals of Neurology*, 45:256–269, 1999.

[255] S. Mori, W.E. Kaufmann, C. Davatzikos, B. Stieltjes, L. Amodei, K. Fredericksen, G.D. Pearlson, E.R. Melhem, M. Solaiyappan, G.V. Raymond, H.W. Moser, and P.C. van Zijl. Imaging cortical association tracts in the human brain using diffusion-tensor-based axonal tracking. *Magnetic Resonance in Medicine*, 47:215–223, 2002.

[256] S. Mori and P.C.M. van Zijl. Fiber tracking: principles and strategies-a technical review. *NMR in Biomedicine*, 15:468–480, 2002.

[257] B.M. Nacewicz, K.M. Dalton, T. Johnstone, M.T. Long, E.M. McAuliff, T.R. Oakes, A.L Alexander, and R.J. Davidson. Amygdala volume and nonverbal social impairment in adolescent and adult males with autism. *Arch. Gen. Psychiatry*, 63:1417–1428, 2006.

[258] E.A. Nadaraya. On estimating regression. *Theory Probab. Appl.*, 9:157–159, 1964.

[259] D. Nain, M. Styner, M. Niethammer, J.J. Levitt, M.E. Shenton, G. Gerig, A. Bobick, and A. Tannenbaum. Statistical shape analysis of brain structures using spherical wavelets. In *IEEE Symposium on Biomedical Imaging ISBI*, 2007.

[260] T. Neumann-Haefelin, H.J. Wittsack, F. Wenserski, M. Siebler, R.J. Seitz, U. Mödder, and H.J. Freund. Diffusion- and perfusion-weighted MRI: the DWI/PWI mismatch region in acute stroke. *Stroke*, 30:1591–1597, 1999.

[261] T.E. Nichols and A.P. Holmes. Nonparametric permutation tests for functional neuroimaging: a primer with examples. *Human Brain Mapping*, 15:1–25, 2002.

[262] M. Niethammer, M. Reuter, F. Wolter, S. Bouix, N. Peinecke, M. Koo, and M.E. Shenton. Global Medical Shape Analysis Using the Laplace-Beltrami Spectrum. *Lecture Notes in Computer Science*, 4791:850, 2007.

[263] T.R. Oakes, J. Koger, and R.J. Davidson. Automated whole-brain segmentation. *NeuroImage*, 9:237, 1999.

[264] R.G. O'Brien and K.E. Muller. Unified power analysis for t-tests through multivariate hypotheses. *Applied analysis of variance in behavioral science*, pages 297–344, 1993.

[265] L. O'Donnell, S. Haker, and C.F. Westin. New approaches to estimation of white matter connectivity in diffusion tensor MRI: Elliptic PDEs and geodesics in a tensor-warped space. *Medical Image Computing and Computer-Assisted Intervention (MICCAI)*, pages 459–466, 2002.

[266] L.J. O'Donnell, M. Kubicki, M.E. Shenton, M.H. Dreusicke, W.E. Grimson, and C.F. Westin. A method for clustering white matter fiber tracts. *American Journal of Neuroradiology*, 27:1032–1036, 2006.

[267] L.J. O'Donnell and C.F. Westin. Automatic tractography segmentation using a high-dimensional white matter atlas. *IEEE Transactions on Medical Imaging*, 26:1562–1575, 2007.

[268] B. Øksendal. *Stochastic differential equations: an introduction with applications*. Springer, 2010.

[269] M.R. Osborne, B. Presnell, and B.A. Turlach. A new approach to variable selection in least squares problems. *IMA Journal of Numerical Analysis*, 20:389–404, 2000.

[270] M. Ozkan, B.M. Dawant, and R.J. Maciunas. Neural-network-based segmentation of multi-modal medical images: a comparative and prospective study. *IEEE Transactions on Medical Imaging*, 12:534–544, 1993.

[271] G.J.M. Parker, C.A.M. Wheeler-Kingshott, and G.J. Barker. Estimating distributed anatomical connectivity using fast marching methods and diffusion tensor imaging. *IEEE Transactions on Medical Imaging*, 21:505–512, 2002.

[272] W. Paul and J. Baschnagel. *Stochastic Processes from Physics to Finance.* Springer-Verlag, Berlin, 1999.

[273] T. Paus, A. Zijdenbos, K.J. Worsley, D.L. Collins, J. Blumenthal, J.N. Giedd, J.L. Rapoport, and A.C. Evans. Structural maturation of neural pathways in children and adolescents: In vivo study. *Science*, 283:1908–1911, 1999.

[274] J. Peng, P. Wang, N. Zhou, and J. Zhu. Partial correlation estimation by joint sparse regression models. *Journal of the American Statistical Association*, 104:735–746, 2009.

[275] P. Perona and J. Malik. Scale-space and edge detection using anisotropic diffusion. *IEEE Trans. Pattern Analysis and Machine Intelligence*, 12:629–639, 1990.

[276] E. Persoon and K.S. Fu. Shape discrimination using Fourier descriptors. *IEEE Transactions on Systems, Man and Cybernetics*, 7:170–179, 1977.

[277] B.S. Peterson, P.A. Feineigle, L.H. Staib, and J.C. Gore. Automated measurement of latent morphological features in the human corpus callosum. *Human Brain Mapping*, 12:232–245, 2001.

[278] A. Pfefferbaum, D.H. Mathalon, E.V. Sullivan, J.M. Rawles, R.B. Zipursky, K.O. Lim, et al. A quantitative magnetic resonance imaging study of changes in brain morphology from infancy to late adulthood. *Archives of Neurology*, 51:874, 1994.

[279] C. Piekema, R.P.C. Kessels, R.B. Mars, K.M. Petersson, and G. Fernández. The right hippocampus participates in short-term memory maintenance of object-location associations. *NeuroImage*, 33:374–382, 2006.

[280] K. Pierce, R.A. Muller, J. Ambrose, G. Allen, and E. Courchesne. Face processing occurs outside the fusiform "face area" in autism: evidence from functional mri. *Brain*, 124:2059–2073, 2001.

[281] J.C. Pinehiro and D.M. Bates. *Mixed Effects Models in S and S-Plus.* Springer, 3rd edition, 2002.

[282] J. Piven, S. Arndt, J. Bailey, and N. Andreasen. Regional brain enlargement in autism: a magnetic resonance imaging study. *Journal of the American Academy of Child & Adolescent Psychiatry*, 35:530–536, 1996.

[283] J. Piven, J. Bailey, B.J. Ranson, and S. Arndt. An MRI study of the corpus callosum in autism. *American Journal of Psychiatry*, 154:1051, 1997.

[284] S.M. Pizer, D.S. Fritsch, P.A. Yushkevich, V.E. Johnson, and E.L. Chaney. Segmentation, registration, and measurement of shape variation via image object shape. *IEEE Transactions on Medical Imaging*, 18:851–865, 1999.

[285] D.A. Pizzagalli, T.R. Oakes, A.S. Fox, M.K. Chung, C.L. Larson, H.C. Abercrombie, S.M. Schaefer, R.M. Benca, and R.J. Davidson. Functional but not structural subgenual prefrontal cortex abnormalities in melancholia. *Molecular Psychiatry*, 9:393–405, 2004.

[286] M.D. Plumbley. Geometry and homotopy for l_1 sparse representations. *Proceedings of SPARS*, 5:206–213, 2005.

[287] J-B Poline and B.M. Mazoyer. Analysis of individual brain activation maps using hierarchical description and multiscale detection. *IEEE Transactions on Medical Imaging*, 13:702–710, 1994.

[288] J.-B. Poline, K.J. Worsley, A.P. Holmes, R.S.J. Frackowiak, and K.J. Friston. Estimating smoothness in statistical parametric maps: Variability of P values. *Journal of Computer Assisted Tomography*, 19:788–796, 1995.

[289] A. Pothen and C.J. Fan. Computing the block triangular form of a sparse matrix. *ACM Transactions on Mathematical Software (TOMS)*, 16:324, 1990.

[290] A. Qiu, D. Bitouk, and M.I. Miller. Smooth functional and structural maps on the neocortex via orthonormal bases of the laplace-beltrami operator. *IEEE Transactions on Medical Imaging*, 25:1296–1396, 2006.

[291] A. Qiu, T. Brown, B. Fischl, J. Ma, and M.I. Miller. Atlas generation for subcortical and ventricular structures with its applications in shape analysis. *IEEE Transactions on Image Processing*, 19:1539 –1547, 2010.

[292] A. Qiu and M.I. Miller. Multi-structure network shape analysis via normal surface momentum maps. *NeuroImage*, 42:1430–1438, 2008.

[293] J.C. Rajapakse, J.N. Giedd, C. DeCarli, J.W. Snell, A. McLaughlin, Y.C. Vauss, A.L. Krain, S. Hamburger, and J.L. Rapoport. A technique for single-channel MR brain tissue segmentation: Application to a pediatric sample. *Magnetic Resonance Imaging*, 14:1053–1065, 1996.

[294] J.O. Ramsay and B.W. Silverman. *Functional Data Analysis*. Springer-Verlag, 1997.

[295] N. Raz, U. Lindenberger, K.M. Rodrigue, K.M. Kennedy, D. Head, A. Williamson, C. Dahle, D. Gerstorf, and J.D. Acker. Regional brain changes in aging healthy adults: general trends, individual differences and modifiers. *Cerebral Cortex*, 15:1676, 2005.

[296] M. Reuter. Hierarchical shape segmentation and registration via topological features of Laplace-Beltrami eigenfunctions. *International Journal of Computer Vision*, 89:287–308, 2010.

[297] M. Reuter, F.-E. Wolter, and N. Peinecke. Laplace-Beltrami spectra as Ôshape-dnaÕ of surfaces and solids. *Computer-Aided Design*, 38:342–366, 2006.

[298] M. Reuter, F.-E. Wolter, M. Shenton, and M. Niethammer. Laplace-Beltrami eigenvalues and topological features of eigenfunctions for statistical shape analysis. *Computer-Aided Design*, 41:739–755, 2009.

[299] S.O. Rice. Mathematical analysis of random noise. *Bell System Tech. J*, 23:282–332, 1944.

[300] A.L. Riess, M.T. Abrams, H.S. Singer, J.L. Ross, and M.B. Denckla. Brain development, gender and iq in children: A volumetric imaging study. *Brain*, 119:1763–1774, 1996.

[301] S.M. Robbins. Anatomical standardization of the human brain in euclidean 3-space and on the cortical 2-manifold. Technical Report PhD thesis, School of Computer Science, McGill University, Montreal, Quebec, Canada, 2003.

[302] C.P. Robert and G. Casella. *Monte Carlo statistical methods*. Springer Verlag, 2004.

[303] D.C. Rojas, J.A. Smith, T.L. Benkers, S.L. Camou, M.L. Reite, and S.J. Rogers. Hippocampus and amygdala volumes in parents of children with autistic disorder. *The Canadian Journal of Statistics*, 28:225–240, 2000.

[304] S. Rosenberg. *The Laplacian on a Riemannian Manifold*. Cambridge University Press, 1997.

[305] A. Rosenfeld and A.C. Kak. Digital picture processing. *Academic Press*, 1982.

[306] S.N. Roy. On a heuristic method of test construction and its use in multivariate analysis. *Ann. Math. Statist.*, 24:220–238, 1953.

[307] W. Rudin. *Functional Analysis*. McGraw-Hill, 1991.

[308] B.D. Rusch, H.C. Abercrombie, T.R. Oakes, S.M. Schaefer, and R.J. Davidson. Hippocampal morphometry in depressed patients and control subjects: relations to anxiety symptoms. *Biological Psychiatry*, 50:960–964, 2001.

[309] R.M. Rustamov. Laplace-Beltrami eigenfunctions for deformation invariant shape representation. In *Proceedings of the fifth Eurographics symposium on Geometry processing*, page 233, 2007.

[310] Z.S. Saad, R.C. Reynolds, B. Argall, S. Japee, and R.W. Cox. Suma: an interface for surface-based intra-and inter-subject analysis with afni. In *IEEE International Symposium on Biomedical Imaging (ISBI)*, pages 1510–1513, 2004.

[311] A. Sacan, O. Ozturk, H. Ferhatosmanoglu, and Y. Wang. Lfm-pro: A tool for detecting significant local structural sites in proteins. *Bioinformatics*, 6:709–716, 2007.

[312] M.N.O. Sadiku. A simple introduction to finite element analysis of electromagnetic problems. *IEEE Transactions on Education*, 32:85–93, 1989.

[313] M.N.O. Sadiku. *Numerical Techniques in Electromagnetics.* CRC Press, 1992.

[314] J. Schäfer and K. Strimmer. A shrinkage approach to large-scale covariance matrix estimation and implications for functional genomics. *Statistical Applications in Genetics and Molecular Biology*, 4:32, 2005.

[315] V. Schmidt and E. Spodarev. Joint estimators for the specific intrinsic volumes of stationary random sets. *Stochastic Processes and their Applications*, 115:959–981, 2005.

[316] C.M. Schumann, J. Hamstra, B.L. Goodlin-Jones, L.J. Lotspeich, H. Kwon, M.H. Buonocore, C.R. Lammers, A.L. Reiss, and D.G. Amaral. The amygdala is enlarged in children but not adolescents with autism; the hippocampus is enlarged at all ages. *Journal of Neuroscience*, 24:6392–6401, 2004.

[317] F. Ségonne, J. Pacheco, and B. Fischl. Geometrically accurate topology-correction of cortical surfaces using nonseparating loops. *IEEE Transactions on Medical Imaging*, 26:518–529, 2007.

[318] S. Seo, M.K. Chung, K.M. Dalton, and R.J. Davidson. Multivariate cortical shape modeling based on sparse representation. In *17th Annual Meeting of the Organization for Human Brain Mapping (HBM)*, page 648, 2011.

[319] S. Seo, M.K. Chung, and H.K. Vorperian. Heat kernel smothing using laplace-beltrami eigenfunctions. In *Medical Image Computing and Computer-Assisted Intervention – MICCAI 2010*, volume 6363 of *Lecture Notes in Computer Science*, pages 505–512, 2010.

[320] J.A. Sethian. *Level Set Methods and Fast Marching Methods: Evolving Interfaces in Computational Geometry, Fluid Mechanics, Computer Vision and Material Science*. Cambridge University Press, 2002.

[321] D.W. Shattuck and R.M. Leahy. Automated graph-based analysis and correction of cortical volume topology. *IEEE Transactions on Medical Imaging*, 20:1167–1177, 2001.

[322] L. Shen and M.K. Chung. Large-scale modeling of parametric surfaces using spherical harmonics. In *Third International Symposium on 3D Data Processing, Visualization and Transmission (3DPVT)*, 2006.

[323] L. Shen, J. Ford, F. Makedon, and A. Saykin. Surface-based approach for classification of 3d neuroanatomical structures. *Intelligent Data Analysis*, 8:519–542, 2004.

[324] L. Shen, A. Saykin, M.K. Chung, H. Huang, J. Ford, F. Makedon, T.L. McHugh, and C.H. Rhodes. Morphometric analysis of genetic variation in hippocampal shape in mild cognitive impairment: Role of an il-6 promoter polymorphism. In *Life Science Society Computational Systems Bioinformatics Conference*, 2006.

[325] Y. Shi, I. Dinov, and A. W. Toga. Cortical shape analysis in the Laplace-Beltrami feature space. In *12th International Conference on Medical Image Computing and Computer Assisted Intervention (MICCAI 2009)*, volume 5762 of *Lecture Notes in Computer Science (LNCS)*, pages 208–215, 2009.

[326] Y. Shi, R. Lai, K. Kern, N. Sicotte, I. Dinov, and A. W. Toga. Harmonic surface mapping with Laplace-Beltrami eigenmaps. In *11th International Conference on Medical Image Computing and Computer Assisted Intervention (MICCAI 2008)*, volume 5242 of *Lecture Notes in Computer Science (LNCS)*, pages 147–154, 2008.

[327] D.O. Siegmund and K.J. Worsley. Testing for a signal with unknown location and scale in a stationary gaussian random field. *Annals of Statistics*, 23:608–639, 1996.

[328] P.P. Silvester and R.L. Ferrari. *Finite Elements for Electrical Engineers*. Cambridge University Press, 1983.

[329] G. Singh, F. Memoli, T. Ishkhanov, G. Sapiro, G. Carlsson, and D.L. Ringach. Topological analysis of population activity in visual cortex. *Journal of Vision*, 8:1–18, 2008.

[330] J.G. Sled, A.P. Zijdenbos, and A.C. Evans. A nonparametric method for automatic correction of intensity nonuniformity in mri data. *IEEE Transactions on Medical Imaging*, 17:87–97, 1988.

[331] C.G. Small. *The Statistical Theory of Shape*. Springer, New York, 1996.

[332] S.M. Smith. Fast robust automated brain extraction. *Human Brain Mapping*, 17:143–155, 2002.

[333] N. Sochen, R. Kimmel, and R. Malladi. A general framework for low level vision. *IEEE Transactions on Image Processing*, 7:310–318, 1998.

[334] E.R. Sowell, P.M. Thompson, K.D. Tessner, and A.W. Toga. Mapping continued brain growth and gray matter density reduction in dorsal frontal cortex: inverse relationships during postadolescent brain maturation. *The Journal of Neuroscience*, 21:8819–8829, 2001.

[335] B.F. Sparks, S.D. Friedman, D.W. Shaw, E.H. Aylward, D. Echelard, A.A. Artru, K.R. Maravilla, J.N. Giedd, J. Munson, G. Dawson, and S.R. Dager. Brain structural abnormalities in young children with autism spectrum disorder. *Neurology*, 59:184–192, 2002.

[336] H. Späth. Fitting affine and orthogonal transformations between two sets of points. *Mathematical Communications*, 9:27–34, 2004.

[337] P. Staempfli, T. Jaermann, G.R. Crelier, S. Kollias, A. Valavanis, and P. Boesiger. Resolving fiber crossing using advanced fast marching tractography based on diffusion tensor imaging. *NeuroImage*, 30:110–120, 2006.

[338] L.H. Staib and J.S. Duncan. Boundary finding with parametrically deformable models. *IEEE Transactions on Pattern Analysis and Machine Intelligence*, 14:1061–1075, 1992.

[339] I. Stakgold. *Boundary Value Problems of Mathematical Physics*. Society for Industrial Mathematics, 2000.

[340] R.G. Steen, R.J. Ogg, W.E. Reddick, and P.B. Kingsley. Age-related changes in the pediatric brain: quantitative MR evidence of maturational changes during adolescence. *American Journal of Neuroradiology*, 18:819–828, 1997.

[341] C.F. Stevens. *The Six Core Theories of Modern Physics*. The MIT Press, 1995.

[342] S. A. Stratemann, J. C. Huang, K. Maki, D. C. Hatcher, and A. J. Miller. Evaluating the mandible with cone-beam computed tomography. *American Journal of Orthodontics and Dentofacial Orthopedics*, 137:S58–S70, 2010.

[343] M. Styner, G. Gerig, S. Joshi, and S. Pizer. Automatic and robust computation of 3D medial models incorporating object variability. *International Journal of Computer Vision*, 55:107–122, 2003.

[344] M. Styner, I. Oguz, S. Xu, C. Brechbuhler, D. Pantazis, J. Levitt, M. Shenton, and G. Gerig. Framework for the statistical shape analysis of brain structures using spharm-pdm. In *Insight Journal, Special Edition on the Open Science Workshop at MICCAI*, pages 1–20, http:hdl.handle.net/1926/215, 2006.

[345] E.V. Sullivan, L. Marsh, D.H. Mathalon, K.O. Lim, and A. Pfefferbaum. Age-related decline in MRI volumes of temporal lobe gray matter but not hippocampus. *Neurobiology of Aging*, 16(4):591–606, 1995.

[346] E.V. Sullivan, L. Marsh, and A. Pfefferbaum. Preservation of hippocampal volume throughout adulthood in healthy men and women. *Neurobiology of Aging*, 26:1093, 2005.

[347] J. Sun, M. Ovsjanikov, and L. J. Guibas. A concise and provably informative multi-scale signature based on heat diffusion. *Comput. Graph. Forum*, 28:1383–1392, 2009.

[348] V.W. Swayze 2nd, N.C. Andreasen, R.J. Alliger, W.T. Yuh, and J.C. Ehrhardt. Subcortical and temporal structures in affective disorder and schizophrenia: a magnetic resonance imaging study. *Biological psychiatry*, 31:221–240, 1992.

[349] H. Tager-Flusberg and R.M. Joseph. Identifying neurocognitive phenotypes in autism. *Philosophical Transactions: Biological Sciences*, 358:303–314, 2003.

[350] J. Talairach and P. Tournoux. Co-planar stereotactic atlas of the human brain: 3-dimensional proportional system. an approach to cerebral imaging. *Thieme, Stuttgart*, 1988.

[351] B. Tang, G. Sapiro, and V. Caselles. Direction diffusion. In *The Proceedings of the Seventh IEEE International Conference on Computer Vision*, pages 2:1245–1252, 1999.

[352] T. Tasdizen, R. Whitaker, P. Burchard, and S. Osher. Geometric surface smoothing via anisotropic diffusion of normals. In *Geometric Modeling and Processing*, pages 687–693, 2006.

[353] G. Taubin. Geometric Signal Processing on Polygonal Meshes. In *EUROGRAPHICS*, 2000.

[354] J.E. Taylor and K.J. Worsley. Detecting sparse signals in random fields, with an application to brain mapping. *Journal of the American Statistical Association*, 102:913–928, 2007.

[355] J.E. Taylor and K.J. Worsley. Random fields of multivariate test statistics, with applications to shape analysis. *Annals of Statistics*, 36:1–27, 2008.

[356] C.R. Tench, P.S. Morgan, L.D. Blumhardt, and C. Constantinescu. Improved white matter fiber tracking using stochastic labeling. *Magn. Res. Med.*, 48:677–683, 2002.

[357] J.-P. Thirion and G. Calmon. Deformation analysis to detect and quantify active lesions in 3d medical image sequences. *IEEE Transactions on Medical Imaging*, 18:429–441, 1999.

[358] D.W. Thompson. *On Growth and Form.* Cambridge University Press, New York, 1961.

[359] P. M. Thompson, D. MacDonald, M. S. Mega, C.J. Holmes, A.C. Evans, and A.W Toga. Detection and mapping of abnormal brain structure with a probabilistic atlas of cortical surfaces. *Journal of Computer Assisted Tomography*, 21:567–581, 1997.

[360] P.M. Thompson, T.D. Cannon, K.L. Narr, T. van Erp, V.P. Poutanen, M. Huttunen, J. Lonnqvist, C.G. Standertskjold-Nordenstam, J. Kaprio, M. Khaledy, et al. Genetic influences on brain structure. *Nature Neuroscience*, 4:1253–1258, 2001.

[361] P.M. Thompson, J.N. Giedd, R.P. Woods, D. MacDonald, A.C. Evans, and A.W Toga. Growth patterns in the developing human brain detected using continuum-mechanical tensor mapping. *Nature*, 404:190–193, 2000.

[362] P.M. Thompson, K.M. Hayashi, G. de Zubicaray, A.L. Janke, S.E. Rose, J. Semple, M.S. Hong, D.H. Herman, D. Gravano, D.M. Doddrell, and A.W. Toga. Mapping hippocampal and ventricular change in alzheimer disease. *NeuroImage*, 22:1754–1766, 2004.

[363] P.M. Thompson and A.W. Toga. A surface-based technique for warping 3-dimensional images of the brain. *IEEE Transactions on Medical Imaging*, 15:1–16, 1996.

[364] P. Thottakara, M. Lazar, S.C. Johnson, and A.L. Alexander. Probabilistic connectivity and segmentation of white matter using tractography and cortical templates. *NeuroImage*, 29:868–878, 2006.

[365] R. Tibshirani. Regression shrinkage and selection via the LASSO. *Journal of the Royal Statistical Society. Series B (Methodological)*, 58:267–288, 1996.

[366] N.H. Timm and T.A. Mieczkowski. *Univariate and Multivariate General Linear Models: Theory and Applications using SAS Software.* SAS Publishing, 1997.

[367] B. Timsari and R. Leahy. An optimization method for creating semi-isometric flat maps of the cerebral cortex. In *The Proceedings of SPIE, Medical Imaging*, 2000.

[368] L.A. Tupler and M.D. De Bellis. Segmented hippocampal volume in children and adolescents with posttraumatic stress disorder. *Biological psychiatry*, 59:523–529, 2006.

[369] P.A. Valdés-Sosa, J.M. Sánchez-Bornot, A. Lage-Castellanos, M. Vega-Hernández, J. Bosch-Bayard, L. Melie-García, and E. Canales-Rodríguez. Estimating brain functional connectivity with sparse multivariate autoregression. *Philosophical Transactions of the Royal Society B: Biological Sciences*, 360:969–981, 2005.

[370] B. Vallet and B. Lévy. Spectral geometry processing with manifold harmonics. *Computer Graphics Forum*, 27:251–260, 2008.

[371] C.N. Vidal, T.J. DeVito, K.M. Hayashi, D.J. Drost, P.C. Williamson, B. Craven-Thuss, D. Herman, Y. Sui, A.W. Toga, R. Nicolson, and P.M. Thompson. Detection and visualization of corpus callosum deficits in autistic children using novel anatomical mapping algorithms. In *Proc. International Society for Magnetic Resonance in Medicine*, 2003.

[372] G. Wahba. *Spline Models for Observational Data.* SIAM, New York, 1990.

[373] K.B. Walhovd, L.T. Westlye, I. Amlien, T. Espeseth, I. Reinvang, N. Raz, I. Agartz, D.H. Salat, D.N. Greve, B. Fischl, et al. Consistent neuroanatomical age-related volume differences across multiple samples. *Neurobiology of Aging*, 2009.

[374] F.-Y. Wang. Sharp explict lower bounds of heat kernels. *Annals of Probability*, 24:1995–2006, 1997.

[375] L. Wang, J.S. Swank, I.E. Glick, M.H. Gado, M.I. Miller, J.C. Morris, and J.G. Csernansky. Changes in hippocampal volume and shape across time distinguish dementia of the Alzheimer type from healthy aging. *NeuroImage*, 20:667–682, 2003.

[376] Y. Wang, Y. Song, P. Rajagopalan, T. An, K. Liu, Y.Y. Chou, B. Gutman, A.W. Toga, P.M. Thompson, and The Alzheimer's Disease Neuroimaging Initiative. Surface-based tbm boosts power to detect disease effects on the brain: An n= 804 adni study. *NeuroImage*, 56:1993–2010, 2011.

[377] Y. Wang, J. Zhang, B. Gutman, T.F. Chan, J.T. Becker, H.J. Aizenstein, O.L. Lopez, R.J. Tamburo, A.W. Toga, and P.M. Thompson. Multivariate tensor-based morphometry on surfaces: Application to mapping ventricular abnormalities in HIV/AIDS. *Neuroimage*, 49:2141–2157, 2010.

[378] M. Wardetzky. Convergence of the cotangent formula: An overview. In A.I. Bobenko, J.M. Sullivan, P. Schröder, and G.M. Ziegler, editors, *Discrete Differential Geometry*, volume 38 of *Oberwolfach Seminars*, pages 275–286. Birkhäuser Basel, 2008.

[379] G.S. Watson. Smooth regression analysis. *Sankhyā: The Indian Journal of Statistics, Series A*, 26:359–372, 1964.

[380] H. Wilbraham. On a certain periodic function. *Cambridge and Dublin Mathematical Journal*, 3:198–201, 1848.

[381] M.B. Wilk and R. Gnanadesikan. Probability plotting methods for the analysis of data. *Biometrika*, 55:1, 1968.

[382] S.F. Witelson. The brain connection: the corpus callosum is larger in left-handers. *Science*, 229:665, 1985.

[383] S.F. Witelson. Hand and sex differences in the isthmus and genu of the human corpus callosum: a postmortem morphological study. *Brain*, 112:799, 1989.

[384] A. Witkin. Scale-space filtering. In *Int. Joint Conference on Artificial Intelligence*, pages 1019–1021, 1983.

[385] Z. Wood, H. Hoppe, M. Desbrun, and P. Schröder. Removing excess topology from isosurfaces. *ACM Transactions on Graphics (TOG)*, 23:190–208, 2004.

[386] R.P. Woods, S.T. Grafton, C.J. Holmes, S.R. Cherry, and J.C. Mazziotta. Automated image registration: I. General methods and intrasubject, intramodality validation. *Journal of Computer Assisted Tomography*, 22:139–152, 1998.

[387] K.J. Worlsey, J-B. Poline, A.C. Vandal, and K.J. Friston. Test for distributed, non-focal brain activations. *NeuroImage*, 2:173–181, 1995.

[388] K.J. Worsley. Local maxima and the expected euler characteristic of excursion sets of χ^2, f and t fields. *Advances in Applied Probability*, 26:13–42, 1994.

[389] K.J. Worsley. Detecting activation in fMRI data. *Statistical Methods in Medical Research.*, 12:401–418, 2003.

[390] K.J. Worsley, M. Andermann, T. Koulis, D. MacDonald, and A.C. Evans. Detecting changes in nonisotropic images. *Human Brain Mapping*, 8:98–101, 1999.

[391] K.J. Worsley, J. Cao, T. Paus, M. Petrides, and A.C. Evans. Applications of random field theory to functional connectivity. *Human Brain Mapping*, 6:364–7, 1998.

[392] K.J. Worsley, AC Evans, S. Marrett, and P. Neelin. A three-dimensional statistical analysis for CBF activation studies in human brain. *Journal of Cerebral Blood Flow and Metabolism*, 12:900–900, 1992.

[393] K.J. Worsley, S. Marrett, P. Neelin, and A.C. Evans. Searching scale space for activation in pet images. *Human Brain Mapping*, 4:74–90, 1996.

[394] K.J. Worsley, S. Marrett, P. Neelin, A.C. Vandal, K.J. Friston, and A.C. Evans. A unified statistical approach for determining significant signals in images of cerebral activation. *Human Brain Mapping*, 4:58–73, 1996.

[395] K.J. Worsley, J.E. Taylor, F. Carbonell, M.K. Chung, E. Duerden, B. Bernhardt, O. Lyttelton, M. Boucher, and A.C. Evans. SurfStat: A Matlab toolbox for the statistical analysis of univariate and multivariate surface and volumetric data using linear mixed effects models and random field theory. *NeuroImage*, 47:S102, 2009.

[396] K.J. Worsley, J.E. Taylor, F. Tomaiuolo, and J. Lerch. Unified univariate and multivariate random field theory. *NeuroImage*, 23:S189–195, 2004.

[397] C.F.J. Wu. On the convergence properties of the EM algorithm. *The Annals of Statistics*, 11:95–103, 1983.

[398] Y. Xu, D.J. Valentino, A.I. Scher, I. Dinov, L.R. White, P.M. Thompson, L.J. Launer, and A.W. Toga. Age effects on hippocampal structural changes in old men: the haas. *NeuroImage*, 40:1003–1015, 2008.

[399] A.M. Yaglom. *Correlation Theory of Stationary and Related Random Functions Vol. I: Basic Results*. Springer-Verlag, 1987.

[400] X. Yang, A. Goh, and A. Qiu. Locally linear diffeomorphic metric embedding (LLDME) for surface-based anatomical shape modeling. *NeuroImage*, 56:149–161, 2011.

[401] A. Yezzi and J.L. Prince. A PDE approach for measuring tissue thickness. In *IEEE Computer Society Conference on Computer Vision and Pattern Recognition (CVPR)*, volume 1, pages I–87, 2001.

[402] A. Yezzi and J.L. Prince. An Eulerian PDE approach for computing tissue thickness. *IEEE Transactions on Medical Imaging*, 22:1332–1339, 2003.

[403] E. Yoruk, B. Acar, and R. Bammer. A physical model for DT-MRI based connectivity map computation. *Lecture Notes in Computer Science*, 3749:213, 2005.

[404] R. A. Yotter, R. Dahnke, and C. Gaser. Topological correction of brain surface meshes using spherical harmonics. In *Medical Image Computing and Computer-Assisted Intervention (MICCAI)*, volume 5762 of *Lecture Notes in Computer Science*, pages 125–132. Springer, 2009.

[405] P. Yu, P.E. Grant, Y. Qi, X. Han, F. Segonne, R. Pienaar, E. Busa, J. Pacheco, N. Makris, R.L. Buckner, et al. Cortical Surface Shape Analysis Based on Spherical Wavelets. *IEEE Transactions on Medical Imaging*, 26:582, 2007.

[406] P.A. Yushkevich, H. Zhang, T.J. Simon, and J.C. Gee. Structure-specific statistical mapping of white matter tracts using the continuous medial representation. In *IEEE 11th International Conference on Computer Vision (ICCV)*, pages 1–8, 2007.

[407] A. Zalesky, A. Fornito, I.H. Harding, L. Cocchi, M. Yücel, C. Pantelis, and E.T. Bullmore. Whole-brain anatomical networks: Does the choice of nodes matter? *NeuroImage*, 50:970–983, 2010.

[408] H. Zhang, O. van Kaick, and R. Dyer. Spectral methods for mesh processing and analysis. In *EUROGRAPHICS*, pages 1–22, 2007.

[409] H. Zhang, O. van Kaick, and R. Dyer. Spectral mesh processing. *Computer Graphics Forum*, 29:1865–1894, 2010.

[410] H. Zhang, P.A. Yushkevich, D.C. Alexander, and J.C. Gee. Deformable registration of diffusion tensor MR images with explicit orientation optimization. *Medical Image Analysis*, 10:764–785, 2006.

[411] H. Zhu, J.G. Ibrahim, N. Tang, D.B. Rowe, X. Hao, R. Bansal, and B.S. Peterson. A statistical analysis of brain morphology using wild bootstrapping. *IEEE Transactions on Medical Imaging*, 26:954–966, 2007.

[412] A.P. Zijdenbos, A. Jimenez, and A.C Evans. Pipelines: Large scale automatic analysis of 3D brain data sets. *NeuroImage*, 7S:783, 1998.

[413] V. Zipunnikov, B. Caffo, D.M. Yousem, C. Davatzikos, B.S. Schwartz, and C. Crainiceanu. Functional principal components model for high-dimensional brain imaging. *NeuroImage*, 58:772–784, 2011.

[414] V. Zipunnikov, S. Greven, B. Caffo, D.S. Reich, and C. Crainiceanu. Longitudinal high-dimensional data analysis. *Johns Hopkins University, Dept. of Biostatistics Working Papers*, 2011.

[415] A.J. Zomorodian. *Topology for computing*, volume 16 of *Cambridge Monographs on Applied and Computational Mathematics*. Cambridge University Press, Cambridge, 2009.

[416] A.J. Zomorodian and G. Carlsson. Computing persistent homology. *Discrete and Computational Geometry*, 33:249–274, 2005.

Index